AN INTRODUCTION TO THERMOMECHANICS

NORTH-HOLLAND SERIES IN

APPLIED MATHEMATICS AND MECHANICS

VOLUME 21

1977

NORTH-HOLLAND PUBLISHING COMPANY
AMSTERDAM·NEW YORK·OXFORD

AN INTRODUCTION TO THERMOMECHANICS

Hans ZIEGLER

Swiss Federal Institute of Technology, Zürich

1977

NORTH-HOLLAND PUBLISHING COMPANY
AMSTERDAM · NEW YORK · OXFORD

PUBLISHERS:

NORTH-HOLLAND PUBLISHING COMPANY
AMSTERDAM · OXFORD · NEW YORK

SOLE DISTRIBUTORS FOR THE U.S.A. AND CANADA:
ELSEVIER/NORTH-HOLLAND INC.
52 VANDERBILT AVENUE
NEW YORK, N.Y. 10017

Library of Congress Catalog Card Number : 76-973
North-Holland ISBN 07204 0432 0

Library of Congress Cataloging in Publication Data

Ziegler, Hans, 1910-An introduction to thermomechanics.
(North-Holland series in applied mathematics and mechanics; 21)
Bibliography: 2 pp.
Inducles index.
1. Thermodynamics. 2. Continuum mechanics.
I. Title.
QC311.~~ZE~~ Z5 531 76-973
ISBN 0-444-11080-1 *(American Elsevier)*

PRINTED IN HUNGARY

PREFACE

Continuum mechanics deals with deformable bodies. In its early stages it was confined to a few special materials and to particular situations, namely to ideal liquids or to elastic solids under isothermal or adiabatic conditions. In these special cases it is possible to solve the basic problem, i.e. to determine the flow and pressure distribution or the deformation and stress field in purely mechanical terms. This is due to the fact that the solution can be developed from a set of differential equations which does not contain the energy theorem. Misuse of term 'theorem'?

From the viewpoint of general continuum mechanics, however, problems of this type are singular. Anyone working in this field knows that sooner or later he gets involved in thermodynamics. The reason for this is that in general a complete set of differential equations contains the energy theorem. Since part of the energy exchange takes place as heat flow, the appropriate form of the energy theorem is the first fundamental law of thermodynamics, and it becomes clear therefore that it is generally impossible to separate the mechanical aspect of a problem from the thermodynamic processes accompanying the motion. To obtain a solution, the fundamental laws of both mechanics and thermodynamics must be applied. In gas dynamics and in thermoelasticity this has long been recognized.

This situation has its counterpart in thermodynamics. Until recently the interest in this field was almost exclusively focused on particularly simple bodies, mainly on gases, and on reversible processes. Besides, the object of investigation has always been a finite volume, e.g. a mole, and the state within the body has been tacitly assumed to be the same throughout the whole volume. This attitude has not been modified by the creation of statistical and quantum mechanics.

It is clear, though, that the thermodynamic state within a finite volume is a function of time and position, and that thermodynamics should therefore be conceived as a field theory. Moreover, the investigation of other materials requires not only the extension of thermodynamics to irreversible processes, but also the introduction of strain, strain rate, and stress tensors in place of volume and pressure. This implies that thermodynamics cannot be separated from continuum mechanics. Even a correct formulation of the first fundamental law is impossible as long as the mechanical motion is not taken into account.

In view of these statements it becomes clear that continuum mechanics and thermodynamics are inseparable: a general theory of continuum mechanics always includes thermodynamics and vice versa. The entire field is truly interdisciplinary and requires a unified treatment, which may properly be denoted as *thermomechanics*. Such a unified treatment is the topic of this book.

The strong interdependence of continuum mechanics and thermodynamics was generally recognized about two decades ago. Various schools have since contributed to thermomechanics, each from its point of view and in its own language or formalism. It is not the aim of this book to report on the various approaches nor to compare them. The book is intended as an introduction to this fascinating field, based on the simplest possible approach.

Part of the book (Chapters 4 and 14 through 18) is a synopsis of the author's contributions to thermomechanics, published from 1957 onwards, occasionally with the assistance of Dr. Jürg Nänni and Professor Christoph Wehrli. It is clear that in a synopsis of this type many points which once seemed essential but have lost their importance can be dropped, and it is equally obvious that many thoughts which originally appeared vague have since assumed a more concise form. Incidentally, in a field which is still in a state of development a certain amount of controversy cannot be avoided; in this respect I assume full responsibility for the contents of the book.

Besides an introduction to the theory of cartesian tensors the first three chapters are concerned with the mechanical laws governing the motion of a continuum. They are based on considerations of mass geometry, on the principle of virtual power and on a general form of the reaction principle. It is well known that the most general approach to continuum mechanics makes use of the displacement field and of material, and hence curvilinear, coordinates. I have stated, however, that I was looking for the simplest representation, free of mathematical difficulties, which so often obscure the physical contents. Since physics deserves priority in an introduction of this type, a treatment based on the velocity field has many advantages and has therefore been preferred. This kind of approach has been presented in a masterly fashion by Prager in his "Introduction to Mechanics of Continua", and since there is not much point in making changes just for the sake of riginality, the first three chapters and certain portions of the subsequent applications are rather similar to the corresponding parts of Prager's book.

Chapter 4 deals with thermodynamics. Its principal topic is the formulation of the classical results in terms of a field theory. In this general form there is no need any more for the usual restriction to the immediate vicinity of equilibrium states; it becomes possible to discuss actual processes (barring extremely fast ones), and the inclusion of irreversible processes presents no difficulties. Here, then are the characteristic points of the present treatment: The stress appears as the sum of a quasiconservative and a dissipative stress. The first is a state function, dependent on the free energy; the second is connected with the dissipation function. In view of later developments (Chapter 14) the role of the two functions is emphasized. The deformation history is represented in the simplest possible manner, namely by internal parameters. The presentation, although reflecting my own views and preferences, remains within the well-established limits of classical thermodynamics.

Chapter 5 deals with the characteristic properties of various materials. A rough classification of bodies is presented, and the constitutive equations of some continua are discussed. The general theorems established in the preceding chapters, supplemented by the proper constitutive relations, determine the thermomechanical behavior of a given body. This is illustrated in Chapters 6 through 11, which deal with the application of the theory to various types of continua.

Chapters 12 and 13 contain a short outline of general tensors and their application in the study of large displacements. The representation follows the lines of Green and Zerna in their excellent book on "Theoretical Elasticity". The inclusion of this material makes it possible, in particular, to point out (a) the importance of a proper choice of the strain measure and of the corresponding stress, and (b) the difference between covariant and contravariant components of a tensor, essential for the proof of the orthogonality condition in Chapter 14.

Up to and including Chapter 13 the subject matter, in spite of a personal tinge in the presentation, remains within confines that appear to be generally accepted by now. The remainder of the book transgresses these traditional limits. Chapter 14 returns to the basis of thermodynamics. The classical theory, restricted to reversible processes, tacitly excludes gyroscopic forces. With exactly the same right they may be excluded in the irreverible case. This implies that the dissipative stresses are determined by the dissipation function alone much in the same way as the quasiconservative forces depend on the free energy. For certain systems, to be called elementary, the connection between dissipative stresses and dissipation function turns out to have the form of an *orthogonality condition*, and it follows that two scalar functions, the free energy and the dissipation function (or the rate of entropy production), completely govern any kind of process.

Chapter 15 shows that the orthogonality condition is equivalent to a number of extremum principles, among them a principle of maximal rate of entropy production. This last principle suggests a generalization of the orthogonality condition for systems of the so-called complex type. It will be referred to as the *orthogonality principle*, and it is easy to see that it reduces to Onsager's symmetry relations in the linear case. Chapters 16 through 18 finally are concerned with applications of the orthogonality condition and the orthogonality principle to various types of continua.

As already mentioned, I have tried to keep the mathematical formalism as simple as possible. I assume, however, that the reader is familiar with vector algebra and analysis, with the basic laws of mechanics and thermodynamics, with the elements of geometry in *n*-dimensional space and of the theory of functions, and with the notion of convexity. To provide the reader with a means of testing his grasp of the matter, problems have been added at the end of each section wherever this was possible.

I am greatly indebted to Professors William Prager and Warner T. Koiter who have both critically read the manuscript and provided numerous suggestions for improvement. A special word of thanks is due to my son, lic. phil. Hansheinrich Ziegler, for his valuable linguistic assistance in the preparation the text. I finally express my gratitude to Dipl. Ing. Carlo Spinedi for his help in proofreading, and to the Daniel Jenny Foundation for support in the preparation of the drawings.

Zürich, November 1975 Hans Ziegler

TABLE OF CONTENTS

CHAPTER 1

MATHEMATICAL PRELIMINARIES

In order to describe the *configuration* of an arbitrary body, we need a *reference system*, e.g. a rigid body or frame serving as a basis for the observer. Any quantitative treatment requires a *coordinate system* fixed to this reference frame. Our first task is to develop the mathematical tools needed for the description of the motion or, more generally, of any process in which the body in consideration takes part. The mathematical framework must be consistent with the fact that the choice of the coordinate system is arbitrary. In consequence, our starting point must be the study of coordinate transformations. Restricting ourselves in this chapter to cartesian coordinate systems, we will develop the concept of the cartesian tensor.

1.1. Cartesian tensors

Let us refer (Fig. 1.1) the three-dimensional physical space to a given reference frame and here to a *cartesian*, i.e. rectangular and rectilinear

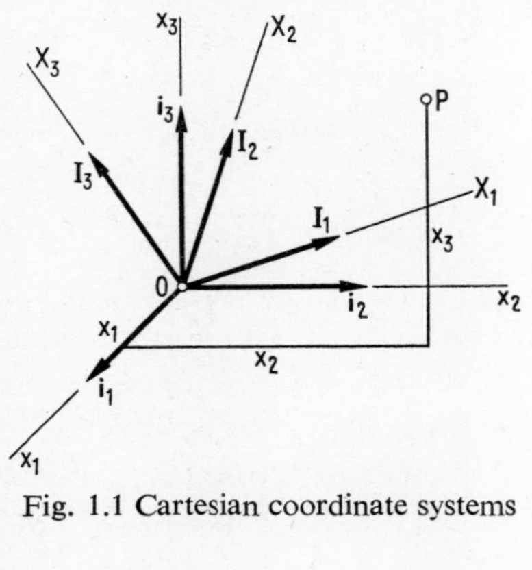

Fig. 1.1 Cartesian coordinate systems

coordinate system x_1, x_2, x_3 with unit vectors $\boldsymbol{i}_1$, $\boldsymbol{i}_2$, $\boldsymbol{i}_3$ along the coordinate axes. The axes X_p ($p = 1, 2, 3$) with unit vectors $\boldsymbol{I}_p$ define another cartesian coordinate system with the same origin O. Denoting the cosines between the axes X_p and x_i by c_{pi}, we have, for arbitrary indices p and i between 1 and 3,

$$c_{pi} = \cos(X_p, x_i) = \boldsymbol{I}_p \cdot \boldsymbol{i}_i \,. \tag{1.1}$$

Let P be a point with coordinates x_i in the first system. Its coordinates in the second system are the projections of the radius vector (or, equivalently, of the sequence of straight segments representing the x_i) onto the axes X_p. Making use of (1.1), we obtain

$$\begin{aligned} X_1 &= c_{11}x_1 + c_{12}x_2 + c_{13}x_3 \,, \\ X_2 &= c_{21}x_1 + c_{22}x_2 + c_{23}x_3 \,, \\ X_3 &= c_{31}x_1 + c_{32}x_2 + c_{33}x_3 \end{aligned} \tag{1.2}$$

as coordinate transformations between the two coordinate systems. It is easy to see that the inversions are

$$\begin{aligned} x_1 &= c_{11}X_1 + c_{21}X_2 + c_{31}X_3 \,, \\ x_2 &= c_{12}X_1 + c_{22}X_2 + c_{32}X_3 \,, \\ x_3 &= c_{13}X_1 + c_{23}X_2 + c_{33}X_3 \,. \end{aligned} \tag{1.3}$$

A more compact way to write (1.2) and (1.3) is

$$X_p = \sum_{i=1}^{3} c_{pi} x_i \,, \qquad x_i = \sum_{p=1}^{3} c_{pi} X_p \,, \tag{1.4}$$

where p is free in the first equation, and i in the second one. We may even dispense of the summation symbol by adopting, once and for all, the so-called *summation convention* stipulating that whenever a letter index appears twice in a product the sum is to be taken over this index. We thus write, in place of (1.4),

$$X_p = c_{pi} x_i, \qquad x_i = c_{pi} X_p \,. \tag{1.5}$$

It is clear that an index appearing once in a term of an equation like (1.5) must appear in every single term. On the other hand, the summation index is sometimes called a *dummy index* since it may be replaced by any other letter. Such a replacement may become necessary to avoid indices appearing more than twice. To insert $(1.5)_2$ into $(1.5)_1$, e.g., it is necessary to write

$(1.5)_2$ in the form

$$x_i = c_{qi} X_q . \tag{1.6}$$

Thus,

$$X_p = c_{pi} c_{qi} X_q \quad \text{and similarly} \quad x_i = c_{pi} c_{pj} x_j , \tag{1.7}$$

where the right-hand sides are double sums.

It is obvious that the coefficient of X_q in $(1.7)_1$ must be 1 for $q = p$ and 0 for $q \neq p$. A similar statement holds for $(1.7)_2$. Introducing the so-called *Kronecker symbol*

$$\delta_{pq} = \begin{cases} 1 & (p = q) \\ 0 & (p \neq q) \end{cases}, \tag{1.8}$$

we thus have

$$c_{pi} c_{qi} = \delta_{pq}, \qquad c_{pi} c_{pj} = \delta_{ij} . \tag{1.9}$$

These equations might be interpreted as orthonormality conditions; they are valid only in orthogonal coordinate systems.

The c_{pi} may be written as a matrix

$$c_{pi} = \begin{pmatrix} c_{11} & c_{12} & c_{13} \\ c_{21} & c_{22} & c_{23} \\ c_{31} & c_{32} & c_{33} \end{pmatrix} . \tag{1.10}$$

Here the first index indicates the line, the second the column in which a given element is situated. For any fixed value of p the c_{pi}, appearing in the p-th line of the matrix (1.10), are, according to (1.1), the components of the unit vector $\boldsymbol{I}_p$ in the coordinate system x_i. Thus, the determinant of the matrix is the double product

$$\det c_{pi} = \boldsymbol{I}_1 \cdot (\boldsymbol{I}_2 \times \boldsymbol{I}_3). \tag{1.11}$$

It follows that

$$\det c_{pi} = \pm 1, \tag{1.12}$$

where the positive sign corresponds to the case where both coordinate systems are right- or left-handed, the negative sign to the case where one of them is right-handed and the other one left-handed. In the first case the second coordinate system is obtained from the first one by a rotation about O; in the second case, a reflexion on a plane passing through O must be added.

Making once more use of (1.1), we obtain

$$\boldsymbol{I}_p = (\boldsymbol{I}_p \cdot \boldsymbol{i}_i)\boldsymbol{i}_i = c_{pi} \boldsymbol{i}_i, \qquad \boldsymbol{i}_i = (\boldsymbol{i}_i \cdot \boldsymbol{I}_p)\boldsymbol{I}_p = c_{pi} \boldsymbol{I}_p . \tag{1.13}$$

Comparing this to (1.5), we note that the base vectors of the two cartesian coordinate systems transform as the coordinates of a point (or, equivalently, as the components of its radius vector). In non-cartesian coordinate systems, this would not be true.

Our present interpretation of (1.2) is this: P is a point fixed in space, i.e. in our reference frame, and (1.2) connects its coordinates in different cartesian systems. Another interpretation, to be used later, considers (1.2) as representing a displacement with respect to the reference frame: the coordinate system is fixed, and the X_p are the instantaneous positions of the points with original positions x_i. The displacement is obviously a rotation about O, possibly combined with a reflection on a plane passing through O.

A *scalar* λ is a quantity which is independent of the coordinate system. Denoting the corresponding quantity in the system X_p by Λ, we thus have

$$\Lambda = \lambda. \tag{1.14}$$

[Margin note: Not a def'n! Add: Having no components]

A *vector* v has a direction and hence three components v_i. The vector itself is independent of the coordinate system; its components transform as the coordinates of a point (the end point of v when the coordinate origin is chosen as the starting point), i.e. according to (1.5):

$$V_p = c_{pi}v_i, \qquad v_i = c_{pi}V_p. \tag{1.15}$$

Thus, a vector might be defined as a triplet of components transforming according to (1.15), and this definition might be used to obtain some of the rules of vector algebra, supplying, e.g. the product λv of a scalar and a vector or the scalar product $u \cdot v$ of two vectors.

Generalizing (1.15), let us define a *cartesian tensor* of order n as a set of 3^n components $t_{ij\ldots l}$ transforming according to

$$T_{pq\ldots s} = c_{pi}c_{qj}\ldots c_{sl}t_{ij\ldots l}, \qquad t_{ij\ldots l} = c_{pi}c_{qj}\ldots c_{sl}T_{pq\ldots s}. \tag{1.16}$$

Note that the order of the tensor is given by the number of its indices. In accordance with this definition, a scalar λ may be considered as a tensor of order zero. A vector is a tensor of order one, symbolically denoted by v. We will henceforth prefer the *index notation*, representing a vector by the symbol v_i of its general component and keeping in mind, of course, that the components transform if the coordinate system is changed.

We will be most often concerned with tensors of the second order, denoted symbolically by t and in index notation by t_{ij}. Here the transformations (1.16) are

$$T_{pq} = c_{pi}c_{qj}t_{ij}, \qquad t_{ij} = c_{pi}c_{qj}T_{pq}. \tag{1.17}$$

The 9 components of a second-order tensor may be arranged in matrix form:

$$t_{ij} = \begin{pmatrix} t_{11} & t_{12} & t_{13} \\ t_{21} & t_{22} & t_{23} \\ t_{31} & t_{32} & t_{33} \end{pmatrix}. \tag{1.18}$$

For other tensors this is not true; the corresponding arrangement of the 27 components of a third-order tensor, e.g., is a three-dimensional block.

Writing the Kronecker symbol (1.8) as a matrix, we obtain

$$\delta_{ij} = \begin{pmatrix} 1 & 0 & 0 \\ 0 & 1 & 0 \\ 0 & 0 & 1 \end{pmatrix}. \tag{1.19}$$

If we interpret the elements of this diagonal matrix as components in a coordinate system x_i, (1.19) defines a second-order tensor. On account of $(1.17)_1$ and (1.9), its components in an arbitrary coordinate system X_p are

$$\Delta_{pq} = c_{pi}c_{qj}\delta_{ij} = c_{pi}c_{qi} = \delta_{pq}, \tag{1.20}$$

i.e. they are the same in any coordinate system. An arbitrary tensor with components that are invariant is called *isotropic*. Examples treated so far are the scalar and the Kronecker tensor.

Problem

Show (by means of a few simple coordinate transformations) that any isotropic tensor of order two has the form $\lambda\delta_{ij}$.

1.2. Tensor algebra

In this section we will briefly discuss the principal rules of tensor algebra. In some cases we will restrict ourselves to typical examples which are easily generalized, and we will leave part of the proofs to the problem section.

Let $r_{ijk\ldots m}$ and $s_{ijk\ldots m}$ be two tensors of equal but arbitrary order. Adding corresponding components, we obtain another tensor of the same order, $t_{ijk\ldots m} = r_{ijk\ldots m} + s_{ijk\ldots m}$, called the *sum* of the original tensors (Problem 1).

Given two tensors of arbitrary order, e.g. r_{ijk} and s_{lm}, let us form the products or their components. These products define another tensor $t_{ijklm} = r_{ijk}s_{lm}$, called the *product* of the original tensors. Its order is the sum of the orders of the given tensors (P2). Special cases are the product of a scalar and a tensor ($t_{ij} = \lambda r_{ij}$) and the tensor obtained by multiplying the components of several vectors ($t_{ijk} = u_i v_j w_k$).

Let $r_{ijkl\ldots p}$ be an arbitrary tensor of order n. Picking the components in which two given indices are equal ($r_{ijil\ldots p}$) and applying the summation convention, we obtain another tensor ($t_{jl\ldots p} = r_{ijil\ldots p}$) of order $n-2$. The process is called *contraction* with respect to the two indices in question (P3). A simple example is the trace of a second order tensor, tr $\boldsymbol{t} = t_{ii}$, which is itself a scalar.

In particular, the process of contraction may be applied to a product with respect to indices taken from each of the two factors ($r_{ijk}s_{lj} = t_{ikl}$). An example is the scalar product $u_i v_i$ of two vectors. If one of the two factors is a second-order tensor and the other the Kronecker tensor ($t_{ij}\delta_{jk} = t_{ik}$), the operation yields the original tensor. Thus, δ_{ij} is also called the *unit tensor* of order two. Other examples are the *powers* of a second-order tensor $\boldsymbol{t}$, symbolically denoted by $\boldsymbol{t}^2, \boldsymbol{t}^3, \ldots$ and defined as the second order tensors $t_{ip}t_{pj}, t_{ip}t_{pq}t_{qj}, \ldots$.

It sometimes happens that, given a set of 3^n quantities $t(i, j, \ldots, l)$, the question arises whether they define a tensor. It is clear that this question can be answered by checking whether the $t(i, j, \ldots, l)$ transform according to (1.16). An easier means is to use the so-called *quotient law*. A typical form of this law states that, e.g. $t(i, j, k)$ are the components of a tensor t_{ijk} if $t(i, j, k)u_i v_j w_k$ is a scalar for any choice of vectors u_i, v_j, w_k. In fact, if this is the case, $(1.15)_2$ yields

$$T(p, q, r)U_p V_q W_r = t(i, j, k)u_i v_j w_k = t(i, j, k)c_{pi}U_p c_{qj}V_j c_{rk}W_k. \quad (1.21)$$

Since U_p, V_q, W_r are arbitrary, it follows from (1.21) that

$$T(p, q, r) = c_{pi}c_{qj}c_{rk}t(i, j, k). \quad (1.22)$$

This is in fact the transformation $(1.16)_1$ for $n = 3$. Another form of the quotient law states that the set $t(i, j, k)$ defines a tensor t_{ijk} if $t(i, j, k)r_{ij}$ is a vector for any choice of the tensor r_{ij} (P4). Other versions of the quotient law are easily inferred from these examples.

A tensor is called *symmetric* with respect to two indices if the exchange of these indices does not alter the components. If the exchange inverts the signs of the components, the tensor is called *antimetric.* In the case of a second-order tensor t_{ij}, the only symmetry relation is $t_{ji} = t_{ij}$. The matrix representation (1.18) shows that the symmetric tensor t_{ij} has only six independent components. On the other hand, the only antimetry condition for t_{ij} is $t_{ji} = -t_{ij}$. Since this implies $t_{11} = \ldots = 0$ (three dots in general indicating cyclic permutation), there remain only three independent components. It is easy to see that these properties are independent of the coordinate system (P6).

By means of the identity

$$t_{ij} = \tfrac{1}{2}(t_{ij}+t_{ji})+\tfrac{1}{2}(t_{ij}-t_{ji}) \tag{1.23}$$

the second-order tensor t_{ij} appears decomposed into its symmetric and antimetric parts

$$t_{(ij)} = \tfrac{1}{2}(t_{ij}+t_{ji}), \qquad t_{[ij]} = \tfrac{1}{2}(t_{ij}-t_{ji}) \tag{1.24}$$

respectively. In the case of two tensors, r_{ij} and s_{ij}, it is easy to see that

$$r_{(ij)}s_{[ij]} = 0. \tag{1.25}$$

It follows immediately that

$$r_{ij}s_{ij} = (r_{(ij)}+r_{[ij]})\,(s_{(ij)}+s_{[ij]}) = r_{(ij)}s_{(ij)}+r_{[ij]}s_{[ij]}\,. \tag{1.26}$$

Three arbitrary non-complanar vectors $\boldsymbol{u}, \boldsymbol{v}, \boldsymbol{w}$ form a right- or left-handed vector system. Since the determinant

$$D = \begin{vmatrix} u_1 & u_2 & u_3 \\ v_1 & v_2 & v_3 \\ w_1 & w_2 & w_3 \end{vmatrix} \tag{1.27}$$

is equal to the triple product $\boldsymbol{u}\cdot(\boldsymbol{v}\times\boldsymbol{w})$, it represents the volume V of the block[1] formed by the three vectors, preceded by the positive sign if the vector system and the coordinate system are both right- or left-handed and by the negative sign if one of them is right-handed, the other one left-handed. For given vectors, V is a scalar, whereas D changes sign in a transformation from a right-handed to a left-handed coordinate system. We therefore call D a *pseudo-scalar* (the simplest version of a *pseudo-tensor*). We will not discuss this concept here, but rather avoid it by restriction to right-handed coordinate systems.

[1] a shorter word for "parallelepiped", suggested by Flügge in [1].

Any permutation of the three digits 1, 2, 3 may be obtained by successive interchanges of two adjacent digits. According as the number of necessary steps is even or odd, the permutation itself is called an even or an odd permutation of 1, 2, 3. Let us define a set of 27 symbols e_{ijk} by stipulating that their values are 1, -1, or 0 according as the sequence i, j, k is either an even permutation of 1, 2, 3, an odd one, or no permutation at all. In other words, $e_{123} = \ldots = 1$, $e_{132} = \ldots = -1$, and $e_{233} = \ldots = e_{223} = \ldots = e_{111} = \ldots = 0$. By means of these symbols, the determinant (1.27) may be written as

$$D = e_{ijk} u_i v_j w_k \tag{1.28}$$

for any set of vectors and any choice of the coordinate system. In fact, D is defined in many texts by (1.28). Since we have restricted ourselves to right-handed coordinate systems, D is a scalar. From (1.28) it follows in connection with the quotient law that e_{ijk} is an isotropic third-order tensor, sometimes called the *permutation tensor* or the *alternating* tensor. It can be shown (P8) that

$$\begin{aligned} e_{pij} e_{pkl} &= \delta_{ik}\delta_{jl} - \delta_{il}\delta_{jk}, \\ e_{pqi} e_{pqj} &= 2\delta_{ij}, \\ e_{pqr} e_{pqr} &= 6. \end{aligned} \tag{1.29}$$

Let s_{jk} be an arbitrary second-order tensor, and let us associate with it a vector

$$t_i = \tfrac{1}{2} e_{ijk} s_{jk}, \tag{1.30}$$

denoted as its *dual vector*. The components of t_i are obviously

$$t_1 = \tfrac{1}{2}(s_{23} - s_{32}) = s_{[23]}, \ \ldots \tag{1.31}$$

and hence are identical with the components of the antimetric part of s_{ij}. On account of (1.30) and $(1.29)_1$,

$$\begin{aligned} e_{jk} t_k &= \tfrac{1}{2} e_{ijk} e_{kpq} s_{pq} = \tfrac{1}{2} e_{kij} e_{kpq} s_{pq} \\ &= \tfrac{1}{2}(\delta_{ip}\delta_{jq} - \delta_{iq}\delta_{jp}) s_{pq} = \tfrac{1}{2}(s_{ij} - s_{ji}) = s_{[ij]}. \end{aligned} \tag{1.32}$$

Thus, the relation

$$s_{ij} = s_{[ij]} = e_{ijk} t_k \tag{1.33}$$

associates an antimetric tensor s_{ij} with a given vector t_k, which is called its *dual tensor*.

The dual vector of $u_j v_k$ is

$$w_i = \tfrac{1}{2} e_{ijk} u_j v_k \,. \tag{1.34}$$

Since its components are $w_1 = (u_2 v_3 - u_3 v_2)/2, \ldots$, we have, in symbolic notation,

$$\boldsymbol{w} = \tfrac{1}{2} \boldsymbol{u} \times \boldsymbol{v}. \tag{1.35}$$

On the other hand, (1.34) is equivalent to

$$w_i = \tfrac{1}{2} e_{kij} u_j v_k = \tfrac{1}{2} U_{ki} v_k = -\tfrac{1}{2} U_{ij} v_j \,, \tag{1.36}$$

where U_{ij}, according to (1.33), is dual to u_k. Thus, the vector product $\boldsymbol{u} \times \boldsymbol{v}$ may be written as $-U_{ij} v_j$.

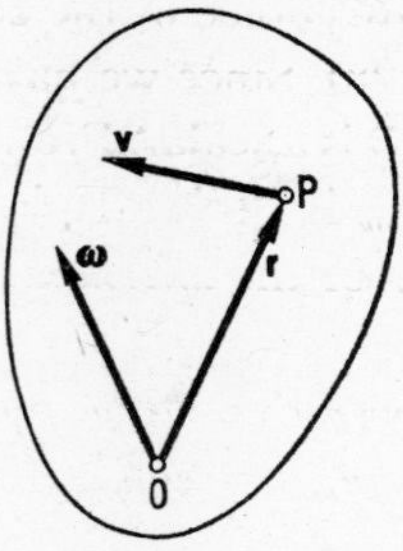

Fig. 1.2 Gyro

If, e.g. ω is the instantaneous angular velocity of a gyro (Fig. 1.2) with fixed point O, the velocity $\boldsymbol{v}$ of the point P with radius vector $\boldsymbol{r}$ from O is $\boldsymbol{v} = \boldsymbol{\omega} \times \boldsymbol{r}$. It may also be expressed by

$$v_i = e_{ijk} \omega_j x_k = -\Omega_{ij} x_j \,, \tag{1.37}$$

where the x_k are the coordinates of P and $\Omega_{ij} = e_{ijk} \omega_k$ is dual to ω_k.

Problems

1. Show that the sum of two tensors, $t_{ijk} = r_{ijk} + s_{ijk}$, is a tensor.
2. Show that the product of two tensors, $t_{ijklm} = r_{ijk} s_{lm}$, is a tensor.
3. Show that contraction of the tensor r_{ijklm} with respect to k and m yields a third-order tensor.
4. Prove that the set $t(i, j, k)$ defines a tensor t_{ijk} if $t(i, j, k) r_{ij}$ is a vector for any choice of the tensor r_{ij}.
5. Prove another form of the quotient law.

6. Show that the properties of symmetry and antimetry of the tensor t_{ijklm} with respect to j and l are independent of the coordinate system.

7. Let D be the determinant of a second-order tensor s_{ij}, written as a matrix. Verify the identity $e_{ijk}D = e_{lmn}s_{il}s_{jm}s_{kn}$.

8. Prove the identities (1.29).

9. Show that the moment of inertia of a body for an axis with direction cosines μ_i, passing through the origin O, has the form $I = I_{ij}\mu_i\mu_j$, where I_{ij} is the symmetric tensor defined by the moments of inertia $I_{11}, \ldots$ and the negative products of inertia $-I_{23}, \ldots$ with respect to the coordinate system x_i. Assume that the body is a gyro with fixed point O and angular velocity ω_i, and find its angular momentum H_i and its kinetic energy T.

1.3. Principal axes

In this section we restrict ourselves to symmetric cartesian second-order tensors, and we will be mainly concerned with finding a coordinate system in which the components of t_{ij} are particularly simple.

Let μ_j denote a unit vector of arbitrary direction. By means of the equation

$$s_i^{(\mu)} = t_{ij}\mu_j \tag{1.38}$$

the tensor t_{ij} associates a vector $s_i^{(\mu)}$ with the direction μ_j. If, in particular, μ_j has the direction of the coordinate axis x_j, the i-th component of the vector (1.38) becomes

$$s_i^{(j)} = t_{ij}\,. \tag{1.39}$$

The component t_{ij} of the given tensor may therefore be interpreted as the i-th component of the vector $\boldsymbol{s}^{(j)}$ associated with the coordinate axis x_j.

Let us ask for a direction μ_j that coincides with that of the corresponding vector $s_i^{(\mu)}$. If it exists, it will be called a *principal direction*, and it will satisfy the relation

$$s_i^{(\mu)} = t_{ij}\mu_j = t\mu_i\,, \tag{1.40}$$

where t is a scalar (positive, negative, or zero). The second equation (1.40) is equivalent to

$$(t_{ij} - t\delta_{ij})\mu_j = 0. \tag{1.41}$$

Since j is a summation index, whereas i is arbitrary, (1.41) represents three homogeneous linear equations, called the *characteristic system*, for the unknowns μ_j. Furthermore, since μ_j is a unit vector,

$$\mu_j \mu_j = 1, \tag{1.42}$$

and the trivial solution $\mu_j = 0$ must be discarded. A nontrivial solution, however, only exists if the determinant of the coefficients in (1.41) vanishes, i.e. if the *characteristic equation*

$$\det (t_{ij} - t\delta_{ij}) = 0 \tag{1.43}$$

is satisfied.

Before proceeding to solve (1.43), let us show that an apparently quite different problem yields the same characteristic system. On account of (1.38), the projection of the vector $s_i^{(\mu)}$ onto the direction μ_i is the scalar

$$p = s_i^{(\mu)} \mu_i = t_{ij} \mu_i \mu_j . \tag{1.44}$$

Obviously p is a function of the direction μ_j, and we may consequently ask for the directions for which p is stationary. This question stipulates an extremum problem, subject to the side condition (1.42) and solved by setting

$$\frac{\partial}{\partial \mu_p} (t_{ij} \mu_i \mu_j - t \mu_i \mu_i) = \frac{\partial}{\partial \mu_p} [(t_{ij} - t\delta_{ij}) \mu_i \mu_j] = 0, \tag{1.45}$$

where t is a Lagrangean multiplier. Carrying out the differentiation and making use of the symmetry of t_{ij}, we obtain the equation

$$2(t_{ip} - t\delta_{ip}) \mu_i = 0, \tag{1.46}$$

which is in fact equivalent to (1.41). Multiplication of both sides of (1.41) by μ_i yields

$$(t_{ij} - t\delta_{ij})\, \mu_i \mu_j = 0 \tag{1.47}$$

or, on account of (1.42) and (1.44), $t = p$. It follows that the Lagrangean multiplier belonging to a solution of (1.41) represents the corresponding stationary value of the projection (1.44).

Proceeding now to the solution of the characteristic equation (1.43), we write it in the form

$$\begin{vmatrix} t_{11} - t & t_{12} & t_{13} \\ t_{21} & t_{22} - t & t_{23} \\ t_{31} & t_{32} & t_{33} - t \end{vmatrix} = 0. \tag{1.48}$$

Developing the left-hand side and ordering with respect to powers of t, we have

$$-t^3+t^2(t_{11}+\ \ldots)-t[(t_{22}t_{33}-t_{23}^2)+\ \ldots]+\det t_{ij}=0 \tag{1.49}$$

or

$$t^3-t_{(1)}t^2-t_{(2)}t-t_{(3)}=0, \tag{1.50}$$

where the coefficients are

$$\begin{aligned} t_{(1)} &= t_{11}+\ \ldots, \\ t_{(2)} &= -t_{22}t_{33}-\ \ldots\ +t_{23}^2+\ \ldots, \\ t_{(3)} &= t_{11}t_{22}t_{33}-t_{11}t_{23}^2-\ \ldots+2t_{23}t_{31}t_{12}=\det t_{ij}. \end{aligned} \tag{1.51}$$

A more concise form (P1) is

$$\begin{aligned} t_{(1)} &= t_{ii}, \\ t_{(2)} &= \tfrac{1}{2}(t_{ij}t_{ji}-t_{ii}t_{jj}), \\ t_{(3)} &= \tfrac{1}{6}(2t_{ij}t_{jk}t_{ki}-3t_{ij}t_{ji}t_{kk}+t_{ii}t_{jj}t_{kk}). \end{aligned} \tag{1.52}$$

The characteristic equation (1.50) is of the third degree in t. It has three roots, called the *principal values* of the tensor t_{ij}. As stationary values of p, they are independent of the coordinate system. According to the lemma of Viéta, the coefficients $t_{(1)}$, $t_{(2)}$, and $t_{(3)}$ in (1.50) may be expressed in terms of the principal values and hence are themselves independent of the coordinate system. They are referred to as the *basic invariants* of t_{ij}, and they can be expressed, according to (1.52), in terms of the traces of $\boldsymbol{t}$, $\boldsymbol{t}^2$, and $\boldsymbol{t}^3$.

One of the roots of (1.50) is always real. Let us denote it as the first principal value t_{I}. For $t = t_{\mathrm{I}}$ the characteristic system (1.41) has at least one real solution μ_j^{I} satisfying (1.42). This solution defines the first principal direction of t_{ij}. Let us introduce a new coordinate system x_i' the first axis of which coincides with the principal direction μ_j^{I}. In this system the second equation (1.40) takes the form $t_{i1}' = t_{\mathrm{I}}\delta_{i1}$. We thus have $t_{11}' = t_{\mathrm{I}}$, $t_{21}' = t_{31}' = 0$, and the characteristic equation (1.48), written in the system x_i', reduces to

$$\begin{vmatrix} t_{\mathrm{I}}-t & 0 & 0 \\ 0 & t_{22}'-t & t_{23}' \\ 0 & t_{32}' & t_{33}'-t \end{vmatrix} = 0 \tag{1.53}$$

or

$$(t_{\mathrm{I}}-t)\,[t^2-(t_{22}'+t_{33}')t+t_{22}'t_{33}'-t_{23}'^2] = 0. \tag{1.54}$$

The remaining principal values t_{II}, t_{III} are obtained by equating the expression in the square bracket to zero. The discriminant of the corresponding quadratic equation is

$$(t'_{22}+t'_{33})^2-4(t'_{22}t'_{33}-t'^2_{23}) = (t'_{22}-t'_{33})^2+4t'^2_{23}. \tag{1.55}$$

Since it is non-negative, t_{II} and t_{III} are real, and it follows that a symmetric tensor of order two only admits real principal values.

Let us denote the principal directions corresponding to t_{II} and t_{III} by μ_j^{II} and μ_j^{III} respectively. They are also real, and since they are solutions of the characteristic system, we have

$$(t_{ij}-t_{\mathrm{II}}\delta_{ij})\mu_j^{\mathrm{II}} = 0, \qquad (t_{ij}-t_{\mathrm{III}}\delta_{ij})\mu_j^{\mathrm{III}} = 0. \tag{1.56}$$

Multiplying the first of these equations by μ_i^{III}, the second one by μ_i^{II}, and subtracting the results, we obtain

$$(t_{\mathrm{II}}-t_{\mathrm{III}})\mu_i^{\mathrm{II}}\mu_i^{\mathrm{III}} = 0. \tag{1.57}$$

It follows that the principal directions corresponding to different principal values are orthogonal. In consequence, the tensor has a unique system of principal axes provided the three principal values are different. If $t_{\mathrm{II}} = t_{\mathrm{III}}$, the discriminant (1.55) must vanish; hence $t'_{23} = 0$ and $t'_{22} = t'_{33} = t_{\mathrm{II}} = t_{\mathrm{III}}$. It follows that the coordinate system x'_i and in consequence any coordinate system containing the axis x'_1 defines a principal system. As long as t_{I} is different from $t_{\mathrm{II}} = t_{\mathrm{III}}$, the principal axis x'_i is unique; otherwise, i.e. if $t_{\mathrm{I}} = t_{\mathrm{II}} = t_{\mathrm{III}}$, any coordinate system defines a system of principal axes.

In principal axes the tensor t_{ij} is represented by a diagonal matrix,

$$\boldsymbol{t} = \begin{pmatrix} t_{\mathrm{I}} & 0 & 0 \\ 0 & t_{\mathrm{II}} & 0 \\ 0 & 0 & t_{\mathrm{III}} \end{pmatrix}, \tag{1.58}$$

and the basic invariants (1.51) become

$$t_{(1)} = t_{\mathrm{I}}+\ldots, \qquad t_{(2)} = -t_{\mathrm{II}}t_{\mathrm{III}}-\ldots, \qquad t_{(3)} = t_{\mathrm{I}}t_{\mathrm{II}}t_{\mathrm{III}}. \tag{1.59}$$

It is obvious that also the powers of $\boldsymbol{t}$, defined in Section 1.2 as $t_{ip}t_{pj}$, $t_{ip}t_{pq}t_{qj}$, ..., are represented by diagonal matrices

$$\boldsymbol{t}^n = \begin{pmatrix} t_{\mathrm{I}}^n & 0 & 0 \\ 0 & t_{\mathrm{II}}^n & 0 \\ 0 & 0 & t_{\mathrm{III}}^n \end{pmatrix}. \tag{1.60}$$

Their principal axes are those of $\boldsymbol{t}$, and their principal values are the powers of t_{I}, ...

Since the principal values $t_{\mathrm{I}}, \ldots$ satisfy the characteristic equation (1.50), we have

$$t_{\mathrm{I}}^3 = t_{(1)}t_{\mathrm{I}}^2 + t_{(2)}t_{\mathrm{I}} + t_{(3)}\,, \quad \ldots \quad . \tag{1.61}$$

The three equations (1.61) may be compressed into a single one, the so-called *Hamilton–Cayley equation*

$$\boldsymbol{t}^3 = t_{(1)}\boldsymbol{t}^2 + t_{(2)}\boldsymbol{t} + t_{(3)}\boldsymbol{\delta}, \tag{1.62}$$

where $\boldsymbol{\delta}$ denotes the unit tensor. In principal axes (1.62) is in fact equivalent to (1.61). As a tensor equation, (1.62) remains valid in the form

$$t_{ip}t_{pq}t_{qj} = t_{(1)}t_{ip}t_{pj} + t_{(2)}t_{ij} + t_{(3)}\delta_{ij} \tag{1.63}$$

in any cartesian coordinate system.

By means of (1.62) it is possible to express any power of $\boldsymbol{t}$ in terms of $\boldsymbol{t}^2$, $\boldsymbol{t}$, and $\boldsymbol{\delta}$. The fourth power, e.g., is given by

$$\begin{aligned} \boldsymbol{t}^4 &= t_{(1)}(t_{(1)}\boldsymbol{t}^2 + t_{(2)}\boldsymbol{t} + t_{(3)}\boldsymbol{\delta}) + t_{(2)}\boldsymbol{t}^2 + t_{(3)}\boldsymbol{t} \\ &= (t_{(1)}^2 + t_{(2)})\boldsymbol{t}^2 + (t_{(1)}t_{(2)} + t_{(3)})\boldsymbol{t} + t_{(1)}t_{(3)}\boldsymbol{\delta}. \end{aligned} \tag{1.64}$$

Here the coefficients of $\boldsymbol{t}^2$, $\boldsymbol{t}$, and $\boldsymbol{\delta}$ are polynomials in the basic invariants, of degree 2 to 4 in t_{ij} since, according to (1.52), $t_{(1)}$, $t_{(2)}$, and $t_{(3)}$ are of the first, second and third degree respectively. In a similar way $\boldsymbol{t}^5, \ldots$ may be reduced. It follows that any power series in $\boldsymbol{t}$,

$$\boldsymbol{s} = A\boldsymbol{\delta} + B\boldsymbol{t} + C\boldsymbol{t}^2 + D\boldsymbol{t}^3 + \ldots, \tag{1.65}$$

can be reduced to three terms

$$\boldsymbol{s} = f\boldsymbol{\delta} + g\boldsymbol{t} + h\boldsymbol{t}^2 \tag{1.66}$$

with coefficients f, g, h that are power series in the basic invariants $t_{(1)}$, $t_{(2)}$, $t_{(3)}$.

In (1.65) and (1.66) the tensor $\boldsymbol{s}$ is a function of $\boldsymbol{t}$. Let us now consider two arbitrary symmetric second-order tensors $\boldsymbol{s}$ and $\boldsymbol{t}$, and let us assume that $\boldsymbol{s}$ depends on $\boldsymbol{t}$ and hence is a *tensor function.* If $\boldsymbol{s}$ can be expanded in powers of $\boldsymbol{t}$, the tensor function has the form (1.66) or, equivalently,

$$s_{ij} = f\delta_{ij} + gt_{ij} + ht_{ik}t_{kj}\,, \tag{1.67}$$

where $f = f(t_{(1)}, t_{(2)}, t_{(3)}), \ldots$.

Let us point out two properties of the tensor function defined by (1.67). In the first place, the components of s_{ij} corresponding to given components t_{ij} are independent of the coordinate system; $\boldsymbol{s}(\boldsymbol{t})$ is therefore called an *isotropic tensor function.* In the second place, (1.67) is not the most general tensor function, even if we restrict ourselves to the case where each component s_{ij} is a power series in the t_{ij}. In fact, the expansion

$$s_{ij} = C_{ij}+C_{ijkl}t_{kl}+C_{ijklmn}t_{kl}t_{mn}+\ \ldots, \tag{1.68}$$

where C_{ij}, C_{ijkl}, ... are constant tensors, is more general than (1.65). It is not generally reduceable to three terms, and the components s_{ij} corresponding to a given set of t_{ij} depend on the orientation of the coordinate system unless the tensors C_{ij}, C_{ijkl}, ... are isotropic. However, comparing (1.65) and (1.68), we suspect that the first of these equations and hence (1.66) is the most general *isotropic* tensor function $\boldsymbol{s}(\boldsymbol{t})$, where the two tensors are symmetric. This can in fact be proved (see for instance [2]).

A *deviator* is a symmetric tensor $\boldsymbol{t}'$ the trace of which is zero. Its basic invariants (1.52) reduce to

$$t'_{(1)} = t'_{ii} = 0, \qquad t'_{(2)} = \tfrac{1}{2}t'_{ij}t'_{ji}, \qquad t'_{(3)} = \tfrac{1}{3}t'_{ij}t'_{jk}t'_{ki}. \tag{1.69}$$

They are expressed, by means of (1.69), in terms of the traces of $\boldsymbol{t}'^2$ and $\boldsymbol{t}'^3$. If t_{ij} is an arbitrary symmetric tensor,

$$t'_{ij} = t_{ij}-\tfrac{1}{3}t_{kk}\delta_{ij} \tag{1.70}$$

is a deviator since $t'_{ii} = 0$. On the other hand, by the inversion of (1.70),

$$t_{ij} = t'_{ij}+\tfrac{1}{3}t_{kk}\delta_{ij}, \tag{1.71}$$

the tensor t_{ij} appears decomposed into a deviator and an isotropic tensor. On account of (1.70), the principal axes of t'_{ij} coincide with the ones of t_{ij}, and the principal values of t'_{ij} are $t_{\mathrm{I}}-t_{kk}/3$, ... The principal axes of the isotropic part of t_{ij} are arbitrary, and the three principal values are $t_{kk}/3$.

Problems

1. Verify the expressions (1.52) for the basic invariants.
2. Show that the decomposition (1.71) of a symmetric tensor into a deviator and an isotropic tensor is unique.

1.4. Tensor analysis

Let us consider a region in space, referred to a cartesian coordinate system x_i, and let us assume that a tensor $t_{j\ldots l}$ of arbitrary order, not necessarily symmetric, is associated with each point x_i. Such a region will be called a *tensor field.* We assume that the functions $t_{j\ldots l}(x_i)$ are single-valued, continuous and differentiable as many times as necessary. In a transformation (1.5) of the coordinate system the tensor components transform according to (1.16):

$$T_{q\ldots s}(X_p) = c_{qj}\ldots c_{sl}t_{j\ldots l}(x_i) = c_{qj}\ldots c_{sl}t_{j\ldots l}(c_{pi}X_p). \tag{1.72}$$

In order to compare the components of the tensor in different points of the field, we need the partial derivatives $\partial t_{j\ldots l}/\partial x_i$ with respect to the coordinates. Since $c_{qj}, \ldots, c_{sl}$ are constant as long as we restrict ourselves to cartesian coordinates, these partial derivatives transform according to

$$\begin{aligned}\frac{\partial}{\partial X_p}T_{q\ldots s} &= c_{qj}\ldots c_{sl}\frac{\partial}{\partial X_p}t_{j\ldots l} = c_{qj}\ldots c_{sl}\frac{\partial}{\partial x_i}t_{j\ldots l}\frac{\partial x_i}{\partial X_p}\\ &= c_{pi}c_{qj}\ldots c_{sl}\frac{\partial}{\partial x_i}t_{j\ldots l},\end{aligned} \tag{1.73}$$

where use has been made of $(1.5)_2$. It follows that the partial derivatives define another tensor, of order $n+1$ if the order of $t_{j\ldots l}$ is n. We are thus justified in extending the index notation by denoting a partial derivative with respect to x_i by an index i, preceded by a comma:

$$\frac{\partial}{\partial x_i}t_{j\ldots l} = t_{j\ldots l,i}. \tag{1.74}$$

The tensor $t_{j\ldots l,i}$ is called the *gradient* of $t_{j\ldots l}$. Its simplest version is the gradient $\varphi_{,i}$ of a scalar φ, symbolically denoted by grad φ. The gradient of a vector $\boldsymbol{v}$, the so-called *vector gradient* $v_{k,j}$, is a second-order tensor, generally asymmetric. Its trace $v_{k,k}$ is a scalar, called the *divergence* of $\boldsymbol{v}$ and symbolically denoted by div $\boldsymbol{v}$. The *curl* of $\boldsymbol{v}$, symbolically denoted by curl $\boldsymbol{v}$, is a vector with components $\partial v_3/\partial x_2 - \partial v_2/\partial x_3, \ldots$ and may be written in the form $e_{ijk}v_{k,j}$. Dividing it by 2, we obtain a vector which, according to (1.30), is dual to the vector gradient $v_{k,j}$, provided we count the differentiation index j as the first one. Here a drawback of the notation (1.74) becomes apparent: to keep the indices in correct sequence, it would be preferable to write, e.g. ${}_{i}t_{j\ldots l}$ instead of $t_{j\ldots l,i}$. However, the notation (1.74) has been generally adopted in tensor analysis. We will use it here too, but make it a rule to treat indices following a comma as if they were the first ones.

The concepts defined above are called *differential operators.* Their index representation makes it particularly simple to calculate operators of products (P1) or combinations of differential operators (P2). If, e.g. the *Laplace operator* $\Delta\varphi$ is written $\varphi_{,ii}$, the identity $\Delta\varphi = \operatorname{div}\operatorname{grad}\varphi$ appears almost trivial. That the curl of a gradient is zero follows immediately if we note that in $e_{ijk}\varphi_{,kj}$ the first factor is antimetric in j and k whereas the second one is symmetric.

In order to derive a few integral theorems, we consider a convex region of volume V with a smooth surface A, situated in a single-valued tensor field $t_{j\ldots l}(x_i)$. To calculate the volume integral of the gradient $t_{j\ldots l,i}$, let us decompose the body into prismatic elements (Fig. 1.3), parallel to the axis x_1,

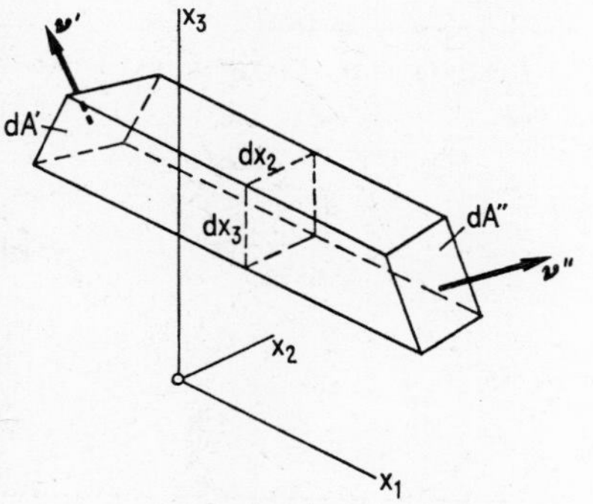

Fig. 1.3 Theorem of Gauss

and of rectangular cross section $dx_2\, dx_3$. Denoting the faces of these prisms by dA', dA'' and their exterior unit normals by $\boldsymbol{\nu}'$, $\boldsymbol{\nu}''$, we have

$$dx_2\, dx_3 = \nu_1''\, dA'' = -\nu_1'\, dA'. \tag{1.75}$$

The contribution of a single prism to the integral

$$\int t_{j\ldots l,1}\, dV \tag{1.76}$$

is obtained by integration of $t_{j\ldots l,1}\, dx_1\, dx_2\, dx_3$ over the length of the prism; it is given by

$$(t_{j\ldots l}'' - t_{j\ldots l}')\, dx_2\, dx_3 = t_{j\ldots l}''\nu_1''\, dA'' + t_{j\ldots l}'\nu_1'\, dA', \tag{1.77}$$

where $t_{j\ldots l}'$ and $t_{j\ldots l}''$ denote the values of $t_{j\ldots l}$ on dA', dA'' respectively. Adding the contributions of all prisms, we easily obtain

$$\int t_{j\ldots l,1}\, dV = \int t_{j\ldots l}\nu_1\, dA \tag{1.78}$$

or in general, since the argument may be repeated for prisms parallel to the other axes,

$$\int t_{j\ldots l,i}\,\mathrm{d}V = \int t_{j\ldots l}\nu_i\,\mathrm{d}A. \tag{1.79}$$

This result, connecting a volume and a surface integral, is called the *theorem of Gauss*. It is easily generalized for regular, i.e. for piecewise smooth surfaces and also for non-convex bodies since any body of this type may be decomposed into convex parts. Applying (1.79) to a scalar, we obtain in symbolic notation

$$\int \operatorname{grad}\varphi\,\mathrm{d}V = \int \boldsymbol{\nu}\varphi\,\mathrm{d}A. \tag{1.80}$$

In a similar manner, application to v_i and to $e_{kij}v_j$ yields

$$\int \operatorname{div}\boldsymbol{v}\,\mathrm{d}V = \int \boldsymbol{\nu}\cdot\boldsymbol{v}\,\mathrm{d}A, \qquad \int \operatorname{curl}\boldsymbol{v}\,\mathrm{d}V = \int \boldsymbol{\nu}\times\boldsymbol{v}\,\mathrm{d}A. \tag{1.81}$$

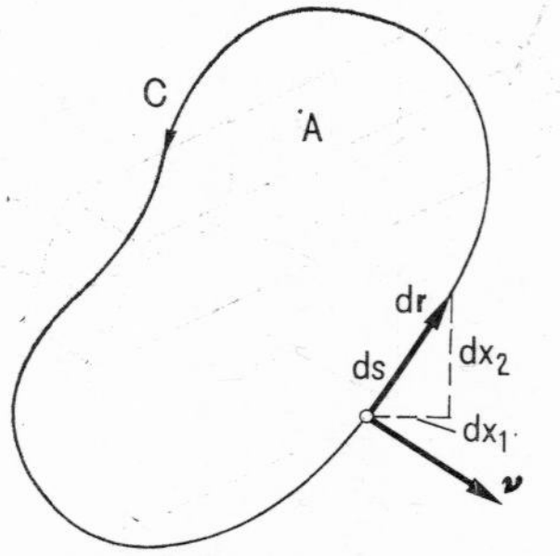

Fig. 1.4 Line element and unit normal

If we specialize (1.79) for a plane area A and its boundary C (Fig. 1.4), we have

$$\int t_{j\ldots l,i}\,\mathrm{d}A = \int t_{j\ldots l}\nu_i\,\mathrm{d}s. \tag{1.82}$$

Applying this to the tensor $e_{3ij}v_j$ and noting that e_{3ij} is constant, we obtain

$$\int (e_{3ij}v_j)_{,i}\,\mathrm{d}A = \int e_{3ij}v_{j,i}\,\mathrm{d}A = \int e_{3ij}v_j\nu_i\,\mathrm{d}s = \int (v_2\nu_1 - v_1\nu_2)\,\mathrm{d}s. \tag{1.83}$$

Since $\nu_1\,\mathrm{d}s = \mathrm{d}x_2$ and $-\nu_2\,\mathrm{d}s = \mathrm{d}x_1$, (1.83) may be written as

$$\int (\operatorname{curl}\boldsymbol{v})_3\,\mathrm{d}A = \int \boldsymbol{v}\cdot\mathrm{d}\boldsymbol{r}, \tag{1.84}$$

where $\mathrm{d}\boldsymbol{r}$ is the vectorial line element of the curve C, taken in the counter-clockwise sense. Figure 1.5 finally shows a curved surface A, bounded by the closed curve C. Decomposing A into surface elements of, e.g. triangular shape, and replacing $(\operatorname{curl}\boldsymbol{v})_3$ by $\boldsymbol{\nu}\cdot\operatorname{curl}\boldsymbol{v}$, where $\boldsymbol{\nu}$ is now the unit surface

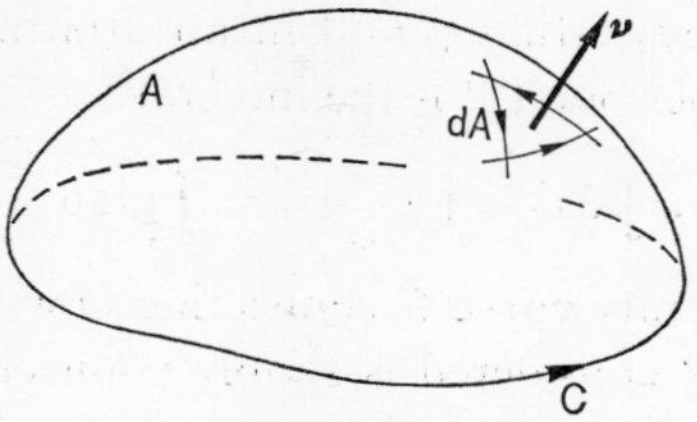

Fig. 1.5 Theorem of Stokes

normal forming a right-hand screw together with the sense of circulation around the element, we apply (1.84) to each one of these elements. Adding the results and observing that the contributions of all interior line segments cancel, we obtain

$$\int \boldsymbol{\nu}\cdot\operatorname{curl} \boldsymbol{v}\, \mathrm{d}A = \int \boldsymbol{v}\cdot \mathrm{d}\boldsymbol{r}, \tag{1.85}$$

where the integrals extend over the area A and its boundary C respectively, and the screw rule is still in force. The result is the so-called *theorem of Stokes*. It connects a surface and a line integral and implies, in particular, that the surface integral is the same for different surfaces bounded by the same curve.

On account of (1.79), the volume integral

$$\int \varphi\psi_{,ii}\, \mathrm{d}V = \int [(\varphi\psi_{,i})_{,i} - \varphi_{,i}\psi_{,i}]\, \mathrm{d}V \tag{1.86}$$

may be written

$$\int \varphi\psi_{,ii}\, \mathrm{d}V = \int \varphi\psi_{,i}\nu_i\, \mathrm{d}A - \int \varphi_{,i}\psi_{,i}\, \mathrm{d}V. \tag{1.87}$$

In symbolic notation, we thus obatin the relation

$$\int (\varphi\, \Delta\psi + \operatorname{grad}\varphi\, \operatorname{grad}\psi)\, \mathrm{d}V = \int \varphi\boldsymbol{\nu}\cdot\operatorname{grad}\psi\, \mathrm{d}A = \int \varphi \frac{\partial\psi}{\partial\nu}\, \mathrm{d}A, \tag{1.88}$$

called *Green's first identity*. Exchanging the roles of φ and ψ and subtracting the result from (1.88), we obtain *Green's second identity*

$$\int (\varphi\, \Delta\psi - \psi\, \Delta\varphi)\, \mathrm{d}V = \int \left(\varphi \frac{\partial\psi}{\partial\nu} - \psi \frac{\partial\varphi}{\partial\nu}\right) \mathrm{d}A. \tag{1.89}$$

We have noted that the field of a gradient is always vortex-free. Conversely, $\operatorname{curl} \boldsymbol{v} = 0$, written in components, supplies the integrability conditions necessary and sufficient for $\boldsymbol{v}$ to be the gradient of a function φ called its *potential*. The function φ is only determined within an additive constant which may be fixed by setting $\varphi = 0$ in an arbitrary point O (Fig. 1.6). The potential in another point P is the integral

$$\varphi = \int d\varphi = \int \varphi_{,i}\, dx_i = \int \operatorname{grad} \varphi \cdot d\boldsymbol{r} \tag{1.90}$$

extended over an arbitrary curve C connecting O with P; it is single-valued provided the region R considered is simply connected. In fact, under this

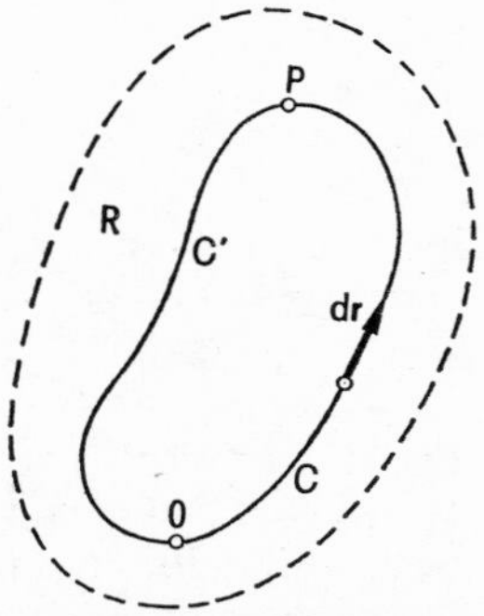

Fig. 1.6 Single-valuedness of the potential

condition, the left-hand side of (1.85) is zero for the area bounded by C and any other curve C' leading back within R from P to O; the integral (1.90) is therefore the same for all curves connecting O with P.

Let us finally show that, in a simply connected region V with regular boundary A, a vector field $\boldsymbol{v}$ is uniquely determined by its curl, its divergence, and by its normal component on A, provided div $\boldsymbol{v}$ in V and $\boldsymbol{v}\cdot\boldsymbol{\nu}$ on A are chosen so as to satisfy the theorem of Gauss $(1.81)_1$. If we tentatively assume that the problem has two solutions $\boldsymbol{v}^{(1)}$ and $\boldsymbol{v}^{(2)}$, the difference $\boldsymbol{v} = \boldsymbol{v}^{(1)} - \boldsymbol{v}^{(2)}$ satisfies the conditions curl $\boldsymbol{v} = 0$ and div $\boldsymbol{v} = 0$ in V and $\boldsymbol{v}\cdot\boldsymbol{\nu} = 0$ on A. On account of the first two conditions, $\boldsymbol{v}$ is the gradient of a single-valued potential φ satisfying *Laplace's equation* $\Delta\varphi = 0$ in V, and the last condition yields $\boldsymbol{\nu}\cdot \operatorname{grad} \varphi = \partial\varphi/\partial\nu = 0$ on A. Green's first identity (1.88), formulated for $\psi = \varphi$, thus reduces to

$$\int (\operatorname{grad} \varphi)^2\, dV = 0 \tag{1.91}$$

and hence requires that $\boldsymbol{v} = \operatorname{grad} \varphi$ be identically zero, i.e. that $\boldsymbol{v}^{(1)} = \boldsymbol{v}^{(2)}$ everywhere in V. Incidentally, the condition that V be simply connected is essential. In certain applications where this condition is not satisfied (Section 6.3) the solution is not unique.

Problems

1. Verify the identities

$$\begin{aligned} \operatorname{div}(\varphi \boldsymbol{v}) &= \varphi \operatorname{div} \boldsymbol{v} + \boldsymbol{v} \cdot \operatorname{grad} \varphi, \\ \operatorname{div}(\boldsymbol{u} \times \boldsymbol{v}) &= \boldsymbol{v} \cdot \operatorname{curl} \boldsymbol{u} - \boldsymbol{u} \cdot \operatorname{curl} \boldsymbol{v}, \\ \operatorname{curl}(\varphi \boldsymbol{v}) &= \varphi \cdot \operatorname{curl} \boldsymbol{v} - \boldsymbol{v} \times \operatorname{grad} \varphi. \end{aligned}$$

2. Verify the identities

$$\begin{aligned} \operatorname{div} \operatorname{curl} \boldsymbol{v} &= 0 \\ \operatorname{curl} \operatorname{curl} \boldsymbol{v} &= \operatorname{grad} \operatorname{div} \boldsymbol{v} - \Delta \boldsymbol{v}. \end{aligned}$$

3. The instantaneous velocity field $v_i(x_k)$ of a rigid body may be written $v_i = v_i^{(0)} + e_{ijk}\omega_j x_k$, where $v_i^{(0)}$ and ω_j are constant vectors. Show that the divergence of the field is zero and that its curl is $2\omega_i$.

4. Let a body of volume V and regular surface A be completely immersed in a liquid of constant specific weight γ, and use a coordinate system the origin O of which coincides with the center of gravity of the body, the axis x_3 pointing vertically upwards. According to the basic law of hydrostatics, the force acting on a surface element is $dF_i = -p\nu_i \, dA$, where ν_i is the exterior unit normal, $p = p_0 - \gamma x_3$ the hydrostatic pressure, and p_0 the value of p for $x_3 = 0$. Use the theorem of Gauss to prove the *principle of Archimedes*, stating that the surface forces reduce to a single force $K_3 = \gamma V$ in O.

CHAPTER 2

KINEMATICS

Kinematics describes the *motion* of an arbitrary body with respect to a reference frame. It requires a *time* scale, which in classical mechanics is assumed to be the same for all possible reference frames. In a purely kinematical description the forces responsible for the motion are disregarded. The motion is known as soon as the configuration of the body is specified as a function of time. If the motion is known for an infinitesimal time interval $t \ldots t+\mathrm{d}t$, i.e. if the instantaneous change of configuration is prescribed, we say that we know the *state of motion* at time t. In this chapter, we will study these concepts for a continuum.

2.1. The state of motion

In continuum mechanics the atomistic structure of matter is disregarded and the body is assumed to occupy a certain region in space in a continuous manner. This is justified for practical purposes as long as the body contains a sufficiently large number of atoms. We will see, however, that for the explanation of certain phenomena (crystal elasticity, thermal effects, etc.) the molecular structure has to be taken, at least temporarily, into account.

We will not specify at present whether the *continuum* considered is a gas, a liquid, or a solid; in fact, these terms will not be defined until Chapter 5. We will assume, however, that the body is deformable, in contrast to the rigid body treated in elementary mechanics. Referring a continuum to a cartesian coordinate system, we distinguish between *spatial points*, fixed in the reference system, and *material points* or *particles*, considered to be elements of the continuum and thus taking part in its motion. In a similar manner we distinguish between spatial and material curves, surfaces, and volumes.

For an arbitrary time t the state of motion of a continuum is described by a *velocity field* $v_k(x_j)$. It specifies the velocities of all material points at time t and will be assumed to be continuous and differentiable. The field lines of the velocity field, defined as curves on which v_k is tangential everywhere, are referred to as *streamlines*. The entire motion in a given time interval is known once the velocity is prescribed as a function $v_k(x_j, t)$ of position and time. The curve on which a single particle moves is called its *trajectory*. Obviously, the velocity of a particle is always tangential to its trajectory. The properties of streamlines and trajectories are thus similar but not identical: the streamlines are defined by the velocities at the same time, the trajectories by velocities at consecutive times. It follows that in general the trajectories do not coincide with the streamlines.

Let us concentrate now on the state of motion at a fixed time t and let us consider a material point P with coordinates x_j and velocity $v_k(x_j)$. To study the state of motion of an infinitesimal vicinity of P (Fig. 2.1), we consider an arbitrary point P′ in this vicinity, with coordinates $x_j + dx_j$. It is convenient to introduce, besides the coordinate system x_j, the system accompanying P

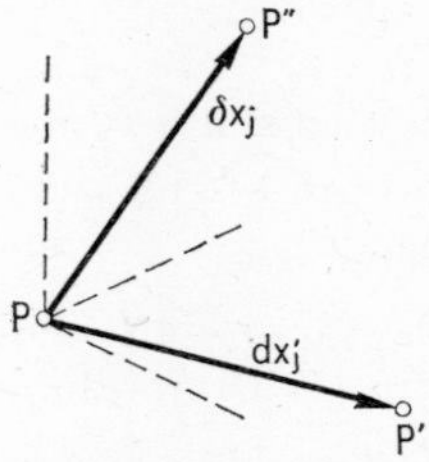

Fig. 2.1 Vicinity of a point P

during its motion, indicated in Figure 2.1 by dotted lines and defined by requiring that its origin be always at P and that its axes be always parallel to the axes x_j of the original reference system. This auxiliary coordinate system will be referred to as the *accompanying system*. The rate of change of any tensor $t_{ij\ldots l}$ for an observer at rest in the accompanying system will be called its *material derivative* $\dot{t}_{ij\ldots l}$ and will be denoted by a dot.

The velocity of P′ is given by

$$v'_k = v_k + dv_k = v_k + v_{k,j}\, dx_j. \tag{2.1}$$

The first term on the right, v_k, is the velocity of P and hence the velocity at which the accompanying system moves; it represents a *translation* of the vicinity of P. The second term,

$$\mathrm{d}v_k = v_{k,j}\,\mathrm{d}x_j, \tag{2.2}$$

represents the state of motion of the vicinity of P relative to the accompanying coordinate system. For a given point P′, (2.2) denotes its velocity for an observer in the accompanying system, and since the radius vector of P′ for this observer is $\mathrm{d}x_k$, the velocity $\mathrm{d}v_k$ is the material derivative

$$\mathrm{d}v_k = (\mathrm{d}x_k)^{\bullet}. \tag{2.3}$$

The tensor $v_{k,j}$ in (2.2) is the *velocity gradient*. It may be decomposed, according to

$$v_{k,j} = v_{(k,j)} + v_{[k,j]}, \tag{2.4}$$

into its symmetric and antimetric parts. The corresponding decomposition of $\mathrm{d}v_k$ reads

$$\mathrm{d}v_k = \mathrm{d}v_k^* + \mathrm{d}v_k^{**}, \tag{2.5}$$

where

$$\mathrm{d}v_k^* = v_{(k,j)}\,\mathrm{d}x_j, \qquad \mathrm{d}v_k^{**} = v_{[k,j]}\,\mathrm{d}x_j. \tag{2.6}$$

The two equations (2.6) define two states of motion which, after superposition, represent the state of motion of the vicinity of P with respect to the accompanying coordinate system.

In order to interpret (2.6), let us start with the antimetric part of the velocity gradient. According to (1.30), its dual vector is

$$w_i = \tfrac{1}{2} e_{ijk} v_{[k,j]} = \tfrac{1}{2} e_{ijk} v_{k,j} \tag{2.7}$$

or symbolically

$$\boldsymbol{w} = \tfrac{1}{2}\,\mathrm{curl}\,\boldsymbol{v}. \tag{2.8}$$

Conversely, (1.33) yields

$$v_{[k,j]} = e_{jki} w_i. \tag{2.9}$$

Inserting this in $(2.6)_2$, we obtain

$$\mathrm{d}v_k^{**} = e_{kij} w_i\,\mathrm{d}x_j \tag{2.10}$$

or

$$\mathrm{d}\boldsymbol{v}^{**} = \boldsymbol{w} \times \mathrm{d}\boldsymbol{r}, \tag{2.11}$$

where $\mathrm{d}\boldsymbol{r}$ is the vector connecting P and P′. We have seen at the end of Section 1.2 that the velocity field (2.11) corresponds to a *rotation* of the vicinity of P with angular velocity $\boldsymbol{w}$ about P. This angular velocity, obtained, according to (2.8), as half the curl of $\boldsymbol{v}$, is called the *vorticity* of the velocity field. The corresponding field lines are referred to as *vortex lines*.

The results obtained so far imply that, provided the symmetric part of the velocity gradient is zero, the vicinity of P moves as a rigid body. It is to be expected therefore that $v_{(k,j)}$ describes a deformation. To confirm this, we consider a second point P″ in the vicinity of P (Fig. 2.1), with coordinates $x_j + \delta x_j$. Its velocity relative to the accompanying system is obtained in the same manner as $\mathrm{d}v_k$ in (2.2) and (2.3) and hence is

$$\delta v_k = v_{k,j}\, \delta x_j = (\delta x_k)^{\bullet}. \tag{2.12}$$

The material rate of change of the scalar product $\mathrm{d}x_k\, \delta x_k$ is given by

$$\begin{aligned}(\mathrm{d}x_k\, \delta x_k)^{\bullet} &= \mathrm{d}x_k (\delta x_k)^{\bullet} + (\mathrm{d}x_k)^{\bullet}\, \delta x_k = v_{k,j}(\delta x_j\, \mathrm{d}x_k + \mathrm{d}x_j\, \delta x_k)\\ &= (v_{k,j} + v_{j,k})\, \mathrm{d}x_j\, \delta x_k = 2v_{(k,j)}\, \mathrm{d}x_j\, \delta x_k .\end{aligned} \tag{2.13}$$

The expression on the right vanishes for any choice of the vectors $\mathrm{d}x_j$ and δx_j if and only if $v_{(k,j)}$ is zero; thus, $(2.6)_1$ describes in fact a *deformation*.

The state of motion of the vicinity of P now appears decomposed into three parts: a translation with the velocity v_k of P, a rotation about P with an angular velocity determined by $v_{[k,j]}$, and a deformation determined by $v_{(k,j)}$.

To study the deformation in more detail, let us refer to the symmetric tensor

$$d_{jk} = v_{(k,j)} \tag{2.14}$$

as the *rate of deformation* and let us write (2.13) in the form

$$(\mathrm{d}x_k\, \delta x_k)^{\bullet} = 2d_{jk}\, \mathrm{d}x_j\, \delta x_k = 2d_{jk}\mu_j \nu_k\, \mathrm{d}s\, \delta s, \tag{2.15}$$

where $\mathrm{d}s$ and δs are the magnitudes of the vectors $\mathrm{d}x_k$, δx_k respectively and μ_k, ν_k their unit vectors. If ϑ denotes the angle between $\mathrm{d}x_k$ and δx_k, we also have

$$(\mathrm{d}x_k\, \delta x_k)^{\bullet} = (\mathrm{d}s\, \delta s \cos\vartheta)^{\bullet} = \left\{\left[\frac{(\mathrm{d}s)^{\bullet}}{\mathrm{d}s} + \frac{(\delta s)^{\bullet}}{\delta s}\right] \cos\vartheta - \sin\vartheta\dot{\vartheta}\right\} \mathrm{d}s\, \delta s, \tag{2.16}$$

and by comparing (2.15) with (2.16) we obtain

$$2d_{jk}\mu_j \nu_k = \left[\frac{(\mathrm{d}s)^{\bullet}}{\mathrm{d}s} + \frac{(\delta s)^{\bullet}}{\delta s}\right] \cos\vartheta - \sin\vartheta\dot{\vartheta}. \tag{2.17}$$

This equation provides a geometric interpretation of the rate of strain tensor d_{jk}.

If (Fig. 2.1) we let the two vectorial line elements dx_j and δx_j coincide, we have $\delta s = ds$, $\mu_j = \nu_j$, and $\vartheta = 0$. It follows from (2.17) that

$$\frac{(ds)^{\cdot}}{ds} = d_{jk}\mu_j\mu_k \tag{2.18}$$

independently of the magnitude of the (infinitesimal) vector dx_j. The left-hand side represents the relative increase in the length of dx_j per unit time and is called the *rate of extension* in the direction μ_j. Applying (2.18) to elements in the directions x_j, we see that the components $d_{11}, \ldots$ of the rate of deformation tensor are the rates of extension in the directions of the coordinate axes.

If, on the other hand, dx_j and δx_j are orthogonal, we have $\vartheta = \pi/2$ and hence

$$-\tfrac{1}{2}\dot{\vartheta} = d_{jk}\mu_j\nu_k . \tag{2.19}$$

The left-hand side is half the rate at which the angle ϑ between the two line elements decreases. It is called the *rate of shear* between the directions μ_j and ν_j. (In engineering texts, it is customary to denote $-\dot{\vartheta}$ as the rate of shear). Applying (2.19) to pairs of elements in the directions x_j, we see that $d_{23}, \ldots$ are the rates of shear between directions parallel to the axes.

Let us add a third line element Δx_j to the ones in Figure 2.1, so that dx_j, δx_j, and Δx_j form, in this sequence, a right-handed system. As we have seen in connection with (1.27) and (1.28), the volume of the block formed by the three elements is

$$dV = e_{ijk}\, dx_i\, \delta x_j\, \Delta x_k . \tag{2.20}$$

Its material derivative is given by

$$(dV)^{\cdot} = e_{ijk}[(dx_i)^{\cdot}\, \delta x_j\, \Delta x_k + dx_i(\delta x_j)^{\cdot}\, \Delta x_k + dx_i\, \delta x_j(\Delta x_k)^{\cdot}] \tag{2.21}$$

since e_{ijk} is a constant tensor. On account of (2.2) and (2.3), the contribution of the first term in the square bracket becomes

$$e_{ijk}v_{i,p}\, dx_p\, \delta x_j\, \Delta x_k = e_{pjk}v_{p,i}\, dx_i\, \delta x_j\, \Delta x_k , \tag{2.22}$$

and if the other contributions are treated similarly, (2.21) takes the form

$$(dV)^{\cdot} = (e_{pjk}v_{p,i} + e_{ipk}v_{p,j} + e_{ijp}v_{p,k})\, dx_i\, \delta x_j\, \Delta x_k . \tag{2.23}$$

It is easy to verify (P2) that the expression in parentheses is different from zero only if i, j, k is a permutation of 1, 2, 3 and that the contribution is $v_{p,p}$ for even permutations and $-v_{p,p}$ for odd ones. We thus have

$$(\mathrm{d}V)^{\cdot} = v_{p,p} e_{ijk}\, \mathrm{d}x_i\, \delta x_j\, \Delta x_k = v_{p,p}\, \mathrm{d}V \tag{2.24}$$

and hence

$$\frac{(\mathrm{d}V)^{\cdot}}{\mathrm{d}V} = v_{k,k} = d_{kk}\,. \tag{2.25}$$

The left-hand side is the relative increase of volume per unit time and is called the *rate of dilatation* at P. It is independent of the choice of the (infinitesimal) block and is obtained as the divergence of the velocity vector or, equivalently, as the trace of the rate of deformation tensor.

Since d_{jk} is a symmetric second-order tensor, the results of Section 1.3 are applicable: in any point P of the continuum the tensor d_{jk} has at least one system of principal directions. The corresponding *principal rates of extension* $d_{\mathrm{I}}, \ldots$ are stationary, and the corresponding rates of shear are zero. Moreover, if μ_j is a unit vector in one of the principal directions, it follows from $(2.6)_1$, (2.14), and (1.40) that

$$\mathrm{d}v_k^* = d_{jk}\, \mathrm{d}x_j = d_{jk}\mu_j\, \mathrm{d}s = d\,.\,\mu_k\, \mathrm{d}s, \tag{2.26}$$

where d is the corresponding principal value of d_{jk}. The vectors $\mathrm{d}v_k^*$ and μ_k are thus collinear: in a pure deformation the principal directions are not rotated.

Applying (1.71) to d_{ij}, we obtain

$$d_{ij} = d'_{ij} + \tfrac{1}{3} d_{kk}\delta_{ij}\,, \tag{2.27}$$

where the first term on the right is a deviator, the second one an isotropic tensor. By means of the first basic invariant $d_{(1)}$ of d_{ij}, the factor $d_{kk}/3$ might be written $d_{(1)}/3$; it is called the *mean extension rate* and is an invariant. Equation (2.27) implies that the instantaneous deformation may be considered as the result of two simpler ones. Since the trace of d'_{ij} is zero, the first one is a change of shape without change of volume, called a *distorsion.* For the second term on the right-hand side of (2.27), considered by itself, $(2.6)_1$ yields

$$\mathrm{d}v_k^* = d_{jk}\, \mathrm{d}x_j = \tfrac{1}{3} d_{ii}\delta_{jk}\, \mathrm{d}x_j = \tfrac{1}{3} d_{ii}\, \mathrm{d}x_k\,. \tag{2.28}$$

Thus, the elements emanating from P retain their directions and undergo the same extension: the second deformation is a change of volume without change of shape, denoted as a *dilatation*.

If one of the principal extension rates is zero, we call the deformation rate *plane*; if two principal extension rates vanish, we call it *uniaxial*.

A velocity field is called *plane* if all velocity vectors are parallel to a given plane and equal on its normals. If, in particular, these normals have the direction of x_3, the field is of the form

$$v_1 = v_1(x_1, x_2), \qquad v_2 = v_2(x_1, x_2), \qquad v_3 = 0, \tag{2.29}$$

and the velocity gradient is

$$v_{k,j} = \begin{pmatrix} v_{1,1} & v_{2,1} & 0 \\ v_{1,2} & v_{2,2} & 0 \\ 0 & 0 & 0 \end{pmatrix}. \tag{2.30}$$

Considering its symmetric part, we see that x_3 is a principal axis of the deformation rate and that the latter is plane since $d_{\text{III}} = 0$. This result cannot be inverted, for it is possible that the deformation rate is plane everywhere while its plane is different from point to point.

A velocity field is called *uniaxial* if all velocities are parallel to a given axis and equal on its normal planes, in particular, if

$$v_1 = v_1(x_1), \qquad v_2 = v_3 = 0. \tag{2.31}$$

Here, $v_{1,1}$ is the only non-vanishing component of the velocity gradient; the axes x_1, x_2, x_3 are therefore principal axes of the deformation rate, and the latter is uniaxial since $d_{\text{II}} = d_{\text{III}} = 0$. Again, the result cannot be inverted.

Problems

1. Show that in an arbitrary point of a continuum there is always a material direction which is unaffected by the instantaneous motion.
2. Verify equation (2.24).
3. Show that the deformation rate d_{ij} is uniaxial if and only if the basic invariants $d_{(2)}$ and $d_{(3)}$ vanish. What is the corresponding necessary and sufficient condition for d_{ij} to be plane?
4. The velocity field $v_1 = \lambda x_2$, $v_2 = v_3 = 0$ describes a state of motion called simple shear. Find its vorticity, the principal directions of the rate of deformation, and the principal extension rates.

2.2. Small displacements

In Section 2.1 we have studied the state of motion at a fixed time t. If, for the present, we denote the coordinates by y_j instead of x_j, the state of motion is described by the velocity field $v_k(y_j)$. It may be considered locally as the result of a translation, a rotation, and a deformation.

In the infinitesimal time interval $t \ldots t+dt$ the displacements of the various particles, referred to their configurations y_k at time t, are

$$du_k(y_j) = v_k(y_j)\, dt. \tag{2.32}$$

For an infinitesimal vicinity of the point P these displacements may again be interpreted as the result of a translation, a rotation, and a deformation. If v_k is the velocity of P, the velocity gradient $v_{k,j}$ is obtained by partial differentiation with respect to y_j. The infinitesimal translation during the time interval dt is $v_k\, dt$; the infinitesimal angle of rotation is determined by $v_{[k,j]}\, dt$, and the deformation by $v_{(k,j)}\, dt$.

Once the velocity field is specified as a function $v_k(y_j, t)$ of time, the entire motion of the continuum is known. This representation is particularly suited for the study of fluids in motion. In certain areas of continuum mechanics, e.g. for the treatment of solids, it is more convenient to describe the motion by its *displacement field* $u_k(x_j, t)$. Here, the x_j are not the instantaneous coordinates of the particles but their coordinates in a given reference configuration, e.g. in the configuration at time $t = 0$. The u_k are the displacement vectors connecting the initial configurations x_k of the particles with their configurations y_k at time t. The velocities are obviously the material derivatives $\dot{u}_k$ of the displacements.

The exact theory of the displacement field will be dealt with in Chapter 13. It sometimes happens, however, that the displacements are small compared to the dimensions of the body. In this case, the theory may be simplified. In the first instance the displacements may be treated, by way of approximation, as if they were infinitesimal. In the second instance the y_j, as arguments of any function, may be replaced by the x_j. A glance at (2.32) then shows that all the results obtained in Section 2.1 for the velocity field, $v_k(y_j)$ in the present notation, may be transferred to the displacement field $u_k(x_j)$. The velocity gradient $v_{k,j}$ is now to be replaced by the displacement gradient $u_{k,j}$, and the displacement, referred to the initial configuration, of an infinitesimal vicinity of the point P may be interpreted as the result of a translation with the displacement vector u_k of P, a rotation determined by $u_{[k,j]}$, and a deformation determined by $u_{(k,j)}$.

The symmetric tensor

$$\varepsilon_{jk} = u_{(k,j)} \tag{2.33}$$

is called the *strain tensor*. Its components $\varepsilon_{11}, \ldots$, referred to as *extensions*, are the relative increments in length of segments parallel to the coordinate axes. The $\varepsilon_{23}, \ldots$, referred to as *shear strains*, are half the decreases of the originally right angles between these segments. (In engineering texts, the total angles $\gamma_{23} = 2\varepsilon_{23}, \ldots$ are usually called shear strains). The trace of the strain tensor, $\varepsilon_{kk} = \varepsilon_{(1)}$, represents the relative increase in volume and is called *dilatation*. Finally, the material derivative $\dot{\varepsilon}_{jk}$ of the strain tensor is the rate of deformation d_{jk}.

Even if the displacements u_k are not small, the deformation of an element is determined by the displacement gradient $u_{k,j}$. Its symmetric part, however, has not the simple geometric significance just discussed for small deformations. For large displacements (Chapter 13) it is convenient to use a more general strain tensor, which may be considered as an extension of (2.33) and tends to (2.33) when the deformations become small. The linearized version (2.33) of this more general strain is sufficiently accurate for many practical purposes and is used, e.g. throughout the classical elasticity theory.

2.3. Material derivatives

Let us return to Section 2.1, where the motion of a continuum was described by a velocity field $v_k(x_j, t)$, and let us assume that, within this field, an arbitrary tensor $t_{kl\ldots n}(x_j, t)$ is defined as a function of position and time.

The *local change* of the tensor $t_{kl\ldots n}$ in the time element dt, i.e. its increment at a given spatial point P during dt, is given by $(\partial/\partial t)t_{kl\ldots n}\,\mathrm{d}t$ and is hence determined by the partial derivative of $t_{kl\ldots n}$ with respect to time. We will denote partial time-derivatives by an index zero, preceded by a comma, writing

$$\frac{\partial}{\partial t} t_{kl\ldots n}\,\mathrm{d}t = t_{kl\ldots n,0}\,\mathrm{d}t. \tag{2.34}$$

It is true that this rule is not quite consistent with the general tensor notation since the index zero does not increase the order of the tensor. However, the notation is convenient and certainly admissible if we limit the actual tensor indices to letters and avoid the use of the letter o.

The *instantaneous distribution* of the tensor $t_{kl\dots n}$ in the vicinity of P is described by its gradient and given by

$$\frac{\partial}{\partial x_j} t_{kl\dots n}\, \mathrm{d}x_j = t_{kl\dots n,j}\, \mathrm{d}x_j . \tag{2.35}$$

For an observer displacing himself by $\mathrm{d}x_j$ in the time element $\mathrm{d}t$, the change of the tensor is the sum of (2.34) and (2.35),

$$t_{kl\dots n,0}\, \mathrm{d}t + t_{kl\dots n,j}\, \mathrm{d}x_j . \tag{2.36}$$

If we define the *material change* of $t_{kl\dots n}$ as the increment on a given particle, i.e. the change for an observer moving with the accompanying coordinate system, we have $\mathrm{d}x_j = v_j\, \mathrm{d}t$ in (2.36) and hence

$$\mathrm{d}t_{kl\dots n} = t_{kl\dots n,0}\, \mathrm{d}t + t_{kl\dots n,j} v_j\, \mathrm{d}t. \tag{2.37}$$

The first term on the right is the local change; the second one is due to the displacement of the observer with the particle and is referred to as the *convective change* of $t_{kl\dots n}$. Dividing (2.37) by $\mathrm{d}t$, we obtain the *material derivative* of the tensor $t_{kl\dots n}$ as defined in Section 2.1:

$$\dot{t}_{kl\dots n} = t_{kl\dots n,0} + t_{kl\dots n,j} v_j . \tag{2.38}$$

It is composed of the *local* and the so-called *convective derivative.*

Applying (2.38) to the *density* of a continuum, i.e. to the mass per unit volume, $\varrho = \mathrm{d}m/\mathrm{d}V$, we obtain its material derivative

$$\dot{\varrho} = \varrho_{,0} + \varrho_{,j} v_j . \tag{2.39}$$

Another example is the *acceleration* of a particle, defined as the material derivative of its velocity. It is given by

$$a_k = \dot{v}_k = v_{k,0} + v_{k,j} v_j . \tag{2.40}$$

For certain purposes it is useful to write (2.40) in a slightly different manner. We have

$$v_{k,j} v_j = (v_{k,j} - v_{j,k}) v_j + v_{j,k} v_j = 2 v_{[k,j]} v_j + (\tfrac{1}{2} v_j v_j)_{,k} . \tag{2.41}$$

Inserting this in (2.40) and making use of (2.9), we obtain

$$a_k = v_{k,0} + 2 e_{kij} w_i v_j + (\tfrac{1}{2} v_j v_j)_{,k} \tag{2.42}$$

or, in symbolic notation,

$$\boldsymbol{a} = \frac{\partial \boldsymbol{v}}{\partial t} + 2\boldsymbol{w} \times \boldsymbol{v} + \operatorname{grad} \frac{v^2}{2}, \tag{2.43}$$

where $\boldsymbol{w}$ is the vorticity (2.8) of the velocity field.

So far, we have restricted ourselves to quantities defined at material points. Let us now consider quantities defined as volume integrals, extended over a certain material region V of the continnum. A simple example is the mass

$$m = \int \varrho(x_j, t)\, \mathrm{d}V. \tag{2.44}$$

The general expression of such an integral is

$$T_{kl\ldots n} = \int t_{kl\ldots n}(x_j, t)\, \mathrm{d}V, \tag{2.45}$$

and it is obvious that the $T_{kl\ldots n}$ are the components of a cartesian tensor since, in a rotation of the coordinate system, they transform as the $t_{kl\ldots n}$.

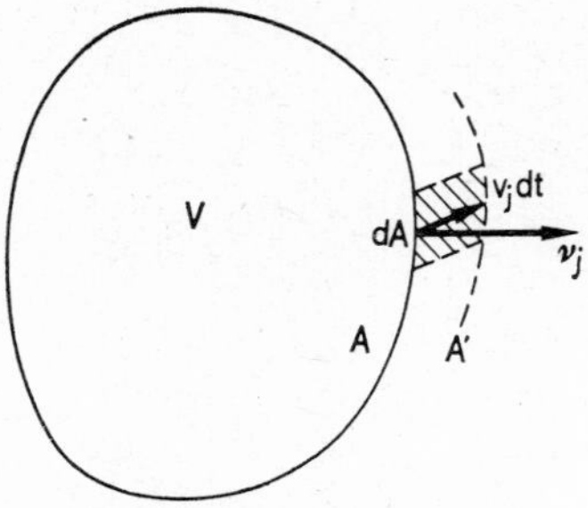

Fig. 2.2 Material volume in motion

Let the material portion V of the continuum (Fig. 2.2) be bounded by the regular surface A. Since the velocity field is assumed to be continuous, the particles forming the surface A at time t define another closed surface A' at time $t+\mathrm{d}t$, and this surface A' encloses the same particles as A. We now define the material derivative $\dot{T}_{kl\ldots n}$ as the change of $T_{kl\ldots n}$ per unit time within the material volume V.

A material point with coordinates x_j at time t moves to $x_j' = x_j + v_j\,\mathrm{d}t$ at time $t' = t+\mathrm{d}t$. The corresponding change of $T_{kl\ldots n}$ is

$$dT_{kl\ldots n} = \dot{T}_{kl\ldots n}\, \mathrm{d}t = \int t_{kl\ldots n}(x_j', t')\, \mathrm{d}V' - \int t_{kl\ldots n}(x_j, t)\, \mathrm{d}V, \tag{2.46}$$

where V' is the material volume at time t'. To calculate the right-hand side, let us first consider the volume elements contained in both volumes V and V'. Their contributions are

$$t_{kl\ldots n,0}\,\mathrm{d}t\,\mathrm{d}V. \tag{2.47}$$

The volume elements lying (as the shaded one in Figure 2.2) inside V' but not in V may be written $\mathrm{d}V = v_j\,\mathrm{d}t\nu_j\,\mathrm{d}A$; their contributions are

$$t_{kl\ldots n}v_j\nu_j\,\mathrm{d}t\,\mathrm{d}A. \tag{2.48}$$

It is easy to see that the contributions of the elements contained in V but not in V' are also given by (2.48). We thus have

$$\dot{T}_{kl\ldots n} = \int t_{kl\ldots n,0}\,\mathrm{d}V + \int t_{kl\ldots n}\,v_j\nu_j\,\mathrm{d}A. \tag{2.49}$$

According to the theorem of Gauss (1.79), the result may be written in the form

$$\dot{T}_{kl\ldots n} = \int [t_{kl\ldots n,0} + (t_{kl\ldots n}v_j)_{,j}]\,\mathrm{d}V, \tag{2.50}$$

and on account of (2.38) we finally obtain

$$\dot{T}_{kl\ldots n} = \int (\dot{t}_{kl\ldots n} + t_{kl\ldots n}v_{j,j})\,\mathrm{d}V. \tag{2.51}$$

If often happens that a tensor $T_{kl\ldots n}$ satisfies the *conservation condition*

$$\dot{T}_{kl\ldots n} = 0, \tag{2.52}$$

stating that $T_{kl\ldots n}$ remains constant for an observer moving with the volume V. In this case (2.49) yields

$$\int t_{kl\ldots n,0}\,\mathrm{d}V = -\int t_{kl\ldots n}v_j\nu_j\,\mathrm{d}A. \tag{2.53}$$

This is the conservation condition from the point of view of an observer at rest. In fact, if A is interpreted as a spatial surface, the integral on the left is the increase of $T_{kl\ldots n}$ within A per unit time, and the right-hand side is the supply from outside by flow through A.

Equations (2.52) and (2.53) are global forms of the conservation condition, valid for an arbitrary volume V and its boundary A. Two other global forms of this condition are obtained from (2.50) and (2.51) by equating the right-hand sides to zero. Since the two results apply for any volume V, we have

$$t_{kl\ldots n,0} + (t_{kl\ldots n}v_j)_{,j} = 0, \tag{2.54}$$

$$\dot{t}_{kl\ldots n} + t_{kl\ldots n}v_{j,j} = 0. \tag{2.55}$$

These are two local forms of the conservation condition.

The simplest example of the tensor $t_{kl\ldots n}$ is the scalar $t = 1$. Here the integral (2.45) is the volume

$$T = \int \mathrm{d}V = V, \tag{2.56}$$

and (2.51) yields

$$\dot{V} = \int v_{j,j}\, \mathrm{d}V. \tag{2.57}$$

The result confirms the information already contained in (2.25) that $v_{j,j}$ is the rate of dilatation. If, in particular, the motion of the continuum is such that volumes are conserved (e.g. if the continuum is incompressible), $v_{j,j}$ is zero. According to (2.51), this is the only case where material differentiation of $t_{kl\ldots n}$ and integration over V are interchangeable.

So far we have considered volume integrals of the type (2.45). Integrals extended over areas or curves can be treated similarly. As an example of practical importance, let us assume that $u_k(x_j, t)$ is a single-valued function, and let us consider the integral

$$L = \int u_k(x_j, t)\, \mathrm{d}x_k \tag{2.58}$$

extended over a closed material curve C. In place of (2.46) we now have

$$\mathrm{d}L = \dot{L}\, \mathrm{d}t = \int u_k(x_j', t')\, \mathrm{d}x_k' - \int u_k(x_j, t)\, \mathrm{d}x_k, \tag{2.59}$$

where

$$u_k(x_j', t') = u_k(x_j, t) + \dot{u}_k(x_j, t)\, \mathrm{d}t \tag{2.60}$$

and, on account of (2.3) and (2.2),

$$\mathrm{d}x_k' = \mathrm{d}x_k + (\mathrm{d}x_k)^{\bullet}\, \mathrm{d}t = \mathrm{d}x_k + v_{k,j}\, \mathrm{d}x_j\, \mathrm{d}t. \tag{2.61}$$

Inserting (2.60) and (2.61) in (2.59) and neglecting the second-order term in $\mathrm{d}t$, we obtain

$$\dot{L} = \int (\dot{u}_k\, \mathrm{d}x_k + u_k v_{k,j}\, \mathrm{d}x_j) = \int \dot{u}_k\, \mathrm{d}x_k + \int (u_k v_k)_{,j}\, \mathrm{d}x_j - \int u_{k,j} v_k\, \mathrm{d}x_j. \tag{2.62}$$

Since the integrand of the second integral on the right is the gradient of a single-valued function, the integral is zero, and (2.62) reduces to

$$\dot{L} = \int \dot{u}_k\, \mathrm{d}x_k - \int u_{k,j} v_k\, \mathrm{d}x_j. \tag{2.63}$$

If the field $u_k(x_j, t)$ is, in particular, the velocity field, the integral (2.58),

$$L = \int v_k\, \mathrm{d}x_k = \Gamma \tag{2.64}$$

is called the *circulation* of the flow around the closed curve C, denoted in hydrodynamics by Γ. In this case the second integrand in (2.63) is the gradient of the single-valued function $(v_k v_k)/2$. Thus, (2.63) reduces to

$$\dot{\Gamma} = \int \dot{v}_k \, dx_k . \tag{2.65}$$

Here material differentiation and integration over C are obviously interchangeable. In the special case where the acceleration $a_k = \dot{v}_k$ is the gradient of a single-valued function, the material derivative of the circulation vanishes for any closed curve C.

Problem

Write the condition of conservation of mass in the forms (2.52) through (2.55) and interpret the results.

2.4. Continuity

Figure 2.3 shows a partial volume V of a continuum, bounded by the regular surface A. An element of this volume is denoted by dV, its mass by

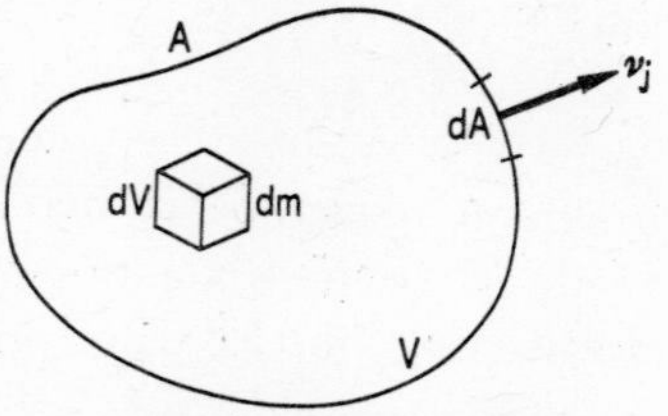

Fig. 2.3 Portion of a continuum

dm; dA is a surface element with exterior unit normal ν_j. The density is $dm/dV = \varrho(x_j, t)$, and the total mass contained in V, already given by (2.44), is

$$m = \int \varrho \, dV. \tag{2.66}$$

Classical mechanics is based on the *principle of conservation of mass*. The simplest global form of this principle is

$$\dot{m} = 0. \tag{2.67}$$

This equation is a special case of (2.52) and reflects the point of view of an observer moving with V. Another global form follows from (2.53).

It reads

$$\int \varrho_{,0}\, dV = -\int \varrho v_j \nu_j\, dA \tag{2.68}$$

and represents the point of view of an observer at rest. In fact, if A is interpreted as a spatial surface, the integral on the left is the increase of mass within A per unit time, and the right-hand side is the mass supply from outside by flow through A. Equations (2.54) and (2.55) provide two local forms of the conservation principle,

$$\varrho_{,0} + (\varrho v_j)_{,j} = 0 \tag{2.69}$$

and

$$\dot{\varrho} + \varrho v_{j,j} = 0, \tag{2.70}$$

referred to as *continuity equations.*

If the material elements retain their volumes dV during any motion, the continuum is called *incompressible.* Conservation of mass and volume imply conservation of density so that $\dot{\varrho} = 0$. However, the particles passing a given spatial point in the course of time may be of different density; incompressibility therefore does not imply that $\varrho_{,0}$ be zero. The simpler form of the continuity equation for this particular case is (2.70). It reduces to $v_{j,j} = 0$, i.e. to the statement that the divergence of the velocity field is identically zero.

If the fields describing the motion of a continuum are locally constant, the flow is called *steady* or *stationary*. Since here, in particular, $\varrho_{,0} = 0$, the appropriate form of the continuity equation is (2.69); it reduces to $(\varrho v_j)_{,j} = 0$. From the global equation (2.68) we obtain

$$\int \varrho v_j \nu_j\, dA = 0. \tag{2.71}$$

It follows from the definitions of streamlines and trajectories in Section 2.1 that in steady flow the two families of curves coincide.

A *stream tube* is defined by the streamlines passing through the points of a closed curve. In steady flow, its position and shape are fixed. A *stream filament* is a stream tube of infinitesimal cross section. Figure 2.4 shows a segment of a stream filament with end sections dA', dA''. The velocity is

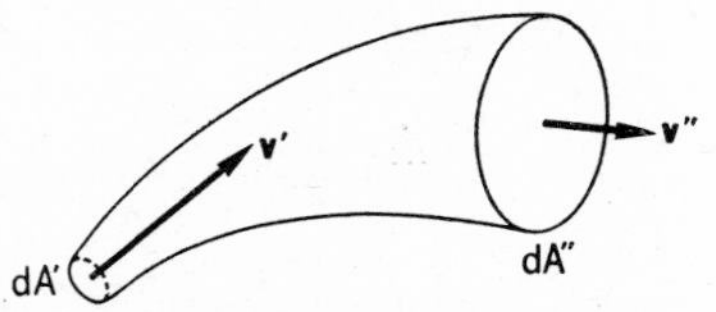

Fig. 2.4 Segment of a stream filament

everywhere tangential to the filament and is denoted by $\boldsymbol{v}'$, $\boldsymbol{v}''$ in the respective end sections. Equation (2.71), applied to the segment, yields

$$\varrho''v''\,\mathrm{d}A'' = \varrho'v'\,\mathrm{d}A'\,. \tag{2.72}$$

If an incompressible continuum moves in steady flow, the continuity equation reduces to $\varrho_{,j}v_j = 0$: the convective derivative of the density is zero. It follows that the density is constant along streamlines so that, in particular, (2.72) reduces to

$$v''\,\mathrm{d}A'' = v'\,\mathrm{d}A'\,. \tag{2.73}$$

The velocity is thus inversely proportional to the cross section of the stream filament.

If a flow satisfies the condition curl $\boldsymbol{v} = 0$, we conclude from (2.8) and (2.11) that the vicinity of any material point P does not rotate but performs a translation on which a pure deformation is superposed. The flow is therefore called *irrotational*. The velocity may be written $\boldsymbol{v} = \operatorname{grad}\varphi$, where $\varphi(x_j, t)$ is the so-called *velocity potential*. The motion is called a *potential flow*, and the continuity equation (2.70), written in terms of φ, takes the form $\dot{\varrho}+\varrho\varphi_{,jj} = 0$. In an incompressible continuum it reduces to the Laplace equation $\varphi_{,jj} = 0$.

In Section 2.3 we have considered integrals of the type (2.45) extended over material portions of the continuum. In many applications the density ϱ appears as a factor of the integrand, so that (2.45) takes the form

$$S_{kl\ldots n} = \int \varrho s_{kl\ldots n}\,\mathrm{d}V. \tag{2.74}$$

According to (2.50), the material derivative of $S_{kl\ldots n}$ is

$$\begin{aligned}\dot{S}_{kl\ldots n} &= \int [(\varrho s_{kl\ldots n})^{\boldsymbol{\cdot}} + \varrho s_{kl\ldots n}v_{j,j}]\,\mathrm{d}V\\ &= \int [(\dot{\varrho}+\varrho v_{j,j})s_{kl\ldots n} + \varrho\dot{s}_{kl\ldots n}]\,\mathrm{d}V.\end{aligned} \tag{2.75}$$

On account of the continuity equation (2.70), this reduces to

$$\dot{S}_{kl\ldots n} = \left(\int \varrho s_{kl\ldots n}\,\mathrm{d}V\right)^{\boldsymbol{\cdot}} = \int \varrho\dot{s}_{kl\ldots n}\,\mathrm{d}V \tag{2.76}$$

as was to be expected, since in material differentiation $\mathrm{d}m = \varrho\,\mathrm{d}V$ is constant.

The vortex field of a flow $v_k(x_j, t)$ is determined by (2.8); its divergence (P2 of Section 1.4) is identically zero. It follows from the theorem of Gauss in the form $(1.81)_1$ that

$$\int w_j v_j\,\mathrm{d}A = 0 \tag{2.77}$$

for every closed regular surface. The vortex lines have been defined in Section 2.1 as the field lines of $w_k(x_j, t)$. A *vortex tube* is formed by the vortex lines passing through the points of a closed curve. A *vortex filament* is a vortex tube of infinitesimal cross section. Let us consider a segment of a vortex filament (similar to the stream filament segment of Fig. 2.4) with end sections $\mathrm{d}A'$, $\mathrm{d}A''$. The vector $\boldsymbol{w}$ is everywhere tangential to the filament and may be denoted by $\boldsymbol{w}'$, $\boldsymbol{w}''$ in the respective end sections. Equation (2.77), applied to the segment, yields

$$w''\,\mathrm{d}A'' = w'\,\mathrm{d}A'\,. \tag{2.78}$$

The magnitude w of the vorticity is thus inversely proportional to the cross section of the vortex filament. It follows that vortex lines cannot end within the flow field. If they do not extend from boundary to boundary, they must be closed. A typical example is the well-known smoke ring.

According to (2.8) and the theorem of Stokes (1.85), the two sides of (2.78) are half the circulations of $\boldsymbol{v}$ along the boundaries of the two end sections. Since these sections are arbitrary, we see that, at a given time, the circulation has the same value for all normal sections of the vortex filament. The value of this circulation, $2w\,\mathrm{d}A$, is called the *vortex strength* of the filament at time t.

Vortex lines passing through the points of an arbitrary curve which itself is not a vortex line form a *vortex surface*. The vector $\boldsymbol{w}$ is everywhere tangential to this surface; thus, the vortex flow through the surface is zero. On account of Stokes' theorem (1.85), the circulation of $\boldsymbol{v}$ vanishes for the boundary of any simply-connected portion of a vortex surface. Finally, the intersection of two vortex surfaces is a vortex line.

Problem

Apply (2.76) to the scalar $s = 1$ and discuss the result.

CHAPTER 3

KINETICS

Kinetics deals with the forces acting on a body and with the manner these forces influence its motion. We know from particle mechanics that, to a certain extent, the connection between forces and motion depends on the choice (or, to be more precise, on the state of motion) of the reference system. This ambiguity, however, disappears if we restrict ourselves to inertial systems as reference frames.

3.1. The momentum theorems

In Figure 3.1 V is a partial volume of a continuum, bounded by the regular surface A. The volume element dV contains the mass dm, and dA is a surface element with exterior normal $\boldsymbol{\nu}$. The external forces acting on V are, by definition, the forces the reactions of which are acting outside V; conversely, the internal forces are characterized by reactions acting within V.

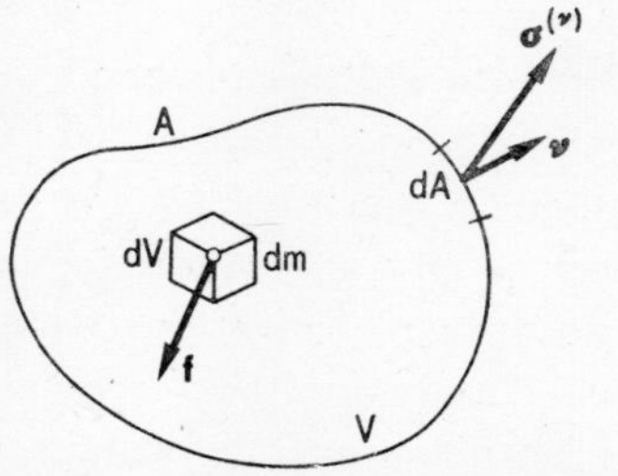

Fig. 3.1 External forces acting on a portion of a continuum

There are two types of external forces. One of them is the body force, acting on the volume elements and distributed over the whole volume V. It is convenient to describe it by the *specific body force* $\boldsymbol{f}$, referred to the unit of mass. The most common example is the specific weight, respresented by the vector $\boldsymbol{g}$ of the gravity acceleration. We assume that $\boldsymbol{f}$ is a continuous function, and we exclude couples acting on the volume elements. (In exceptional cases such couples may occur as a result of the molecular structure; they are taken into consideration in the so-called Cosserat theory [3]). The second type of external force is distributed over the surface A. It is convenient to refer this force to the unit area and to describe it by the *stress vector* $\boldsymbol{\sigma}^{(\nu)}$, dependent not only on position but also on the orientation of the element $\mathrm{d}A$, i.e. on the direction of its unit normal vector $\boldsymbol{\nu}$. We assume that $\boldsymbol{\sigma}^{(\nu)}$ is continuous, and we exclude surface couples (which are again taken care of in the Cosserat theory).

The internal forces, acting between the elements of the body V, need not be considered at present. On the other hand, we need the inertia force. It is distributed, like $\boldsymbol{f}$, over the entire body and will be represented by the *specific inertia force*, referred to the unit mass and equal to the negative acceleration, $-\dot{\boldsymbol{v}}$.

The general laws of kinetics may be derived from two principles: the principle of virtual rate of work and the reaction principle.

In order to formulate the first of these two principles, let us recall a few concepts used throughout mechanics. The *power* or *rate of work* of a force is the scalar product of the force vector and the velocity of its point of application. A state of motion is called *admissible* if it does not violate the constraints of the system in consideration. In the case of a continuum it is represented by a continuous velocity field satisfying the boundary conditions. The so-called *virtual power* of the forces acting in a system is always calculated from the actual forces present at a certain instant t and by means of an arbitrary admissible velocity field which is independent of and hence, in general, not identical with, the actual velocity field at time t. The *principle of virtual rate of work* asserts that, at any time t, the total virtual power of the external, internal, and inertia forces is zero in any admissible virtual state of motion.

The second of the two principles mentioned above is usually stated in a very special manner, valid only for the forces acting between a pair of particles. To obtain an entirely general formulation, let us recall (see for instance ([4], p. 72) that a system of forces satisfying the principle of virtual power for arbitrary admissible states of motion is said to be in *equilibrium*. Now, it is always possible to detach a body from its surroundings if, at the same time, the external reactions are introduced. The body may then be moved in particular as if it were rigid, and if the principle of virtual rate of work is satisfied at least for the corresponding "rigid" velocity fields, the forces acting on it are said to satisfy the equilibrium conditions of a rigid body. In its most general form, the *reaction principle* asserts that in an arbitrary system and, in particular, in a continuum, the internal forces satisfy the equilibrium conditions of a rigid-body, i.e. their virtual power vanishes in all rigid-body motions.

Let us assume now that a finite material portion of a continuum (Fig. 3.1) is detached from its surroundings, and let us apply the principle of virtual power to an arbitrary rigid state of motion, described by the velocity field $\boldsymbol{v}^*$. The external forces are the body and the surface forces. Their rates of virtual work are

$$\int \varrho f_k \overset{*}{v}_k \, \mathrm{d}V, \qquad \int \sigma_k^{(\nu)} \overset{*}{v}_k \, \mathrm{d}A \tag{3.1}$$

respectively. On account of the reaction principle, the rate of work of the internal forces in the rigid state of motion $\overset{*}{v}_k$ is zero. The power of the inertia forces is described by $(3.1)_1$ if f_k is replaced by $-\dot{v}_k$. Thus, the principle of virtual rate of work is expressed by the equation

$$\int \varrho f_k \overset{*}{v}_k \, \mathrm{d}V + \int \sigma_k^{(\nu)} \overset{*}{v}_k \, \mathrm{d}A - \int \varrho \dot{v}_k \overset{*}{v}_k \, \mathrm{d}V = 0, \tag{3.2}$$

where v_k is the real velocity and $\overset{*}{v}_k$ the virtual one.

Equation (3.2) holds for arbitrary states of rigid virtual motion. If we apply it to translations, the velocity $\overset{*}{v}_k$ is the same everywhere and can be dropped since it is arbitrary. We thus obtain

$$\int \varrho \dot{v}_k \, \mathrm{d}V = \int \varrho f_k \, \mathrm{d}V + \int \sigma_k^{(\nu)} \, \mathrm{d}A. \tag{3.3}$$

The linear momentum $\mathrm{d}B_k$ of the material volume element $\mathrm{d}V$ is the product of its mass $\varrho \, \mathrm{d}V$ and its velocity v_k. Defining the linear momentum of the body V as the sum of the linear momenta of its elements, we have

$$B_k = \int \varrho v_k \, \mathrm{d}V. \tag{3.4}$$

On account of (2.76), the material derivative of (3.4) is equal to the left-hand side of (3.3). Thus, (3.3) is the analytical form of the *theorem of linear momentum*, formulated for the material body V: the material derivative of the linear momentum is equal to the sum of the external forces or, equivalently, to the resultant force obtained by reduction of the external forces to an arbitrary point O. In exactly the same form, the theorem of linear momentum is known to hold in elementary mechanics, i.e. for arbitrary systems of particles and rigid bodies.

Let us apply the theorem just obtained to an infinitesimal tetrahedron (Fig. 3.2) three edges of which are parallel to the coordinate axes. If dA is the area of the oblique face with external unit normal $\boldsymbol{\nu}$, the areas of the remaining faces are dA_j = d$A\nu_j$. We denote the stress vector acting on dA by $\boldsymbol{\sigma}^{(\nu)}$ and the stress vectors on the other faces by $-\boldsymbol{\sigma}_j$. In this way, the stress

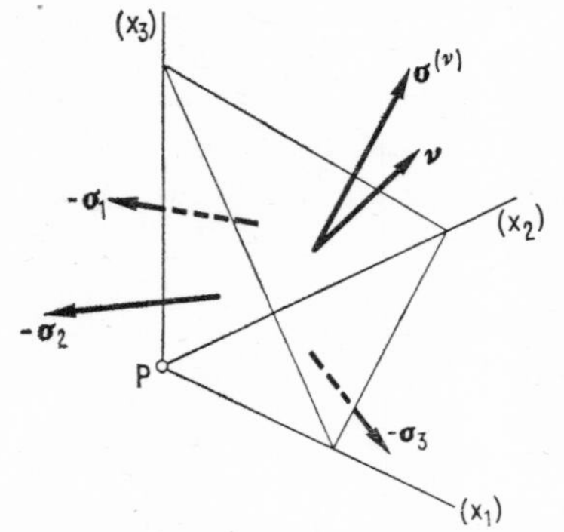

Fig. 3.2 Stresses acting on an infinitesimal tetrahedron

vector of a surface element the external unit vector of which points in the positive direction x_j is $\boldsymbol{\sigma}_j$, and its components σ_{ij} are the *normal stress* ($i = j$) and the *shear stresses* ($i \neq j$) acting on the element.

The stresses give rise to second-order forces, whereas the body and the inertia forces acting on the element of Figure 3.2 are of the third order and hence may be disregarded (as well as the stress difference between the oblique face and the surface element parallel to it passing through P). The linear momentum theorem thus reduces to the equilibrium condition for the surface forces,

$$\sigma_i^{(\nu)}\,\mathrm{d}A - \sigma_{ij}\,\mathrm{d}A\nu_j = 0, \tag{3.5}$$

which yields

$$\sigma_i^{(\nu)} = \sigma_{ij}\nu_j\,. \tag{3.6}$$

Once the stress components σ_{ij} that act on surface elements passing through P and parallel to the coordinate planes are known, (3.6) associates a stress vector $\boldsymbol{\sigma}^{(\nu)}$ with each direction $\boldsymbol{\nu}$ and thus supplies the stress acting on an arbitrary oblique element. Besides, equation (3.6) may be interpreted as the quotient law (Section 1.2) for the components σ_{ij}; it asserts that the σ_{ij}, which might be interpreted as normal and shear stresses acting on an infinitesimal cuboid (Fig. 3.3), are the components of a second-order tensor, called the *stress tensor*.

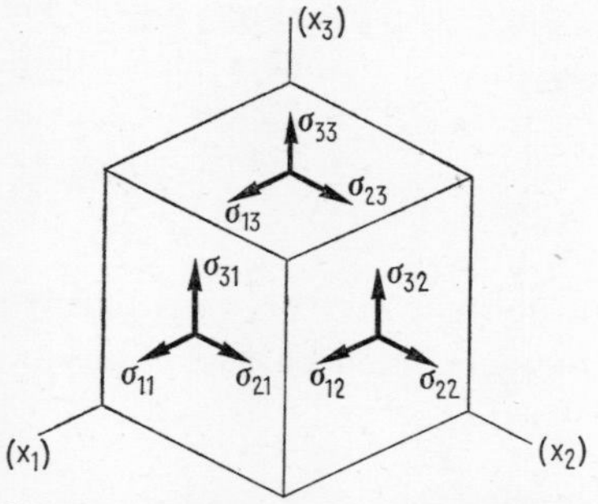

Fig. 3.3 Stresses acting on an infinitesimal cuboid

By means of (3.6) and the theorem of Gauss, (1.79), the global form (3.3) of the theorem of linear momentum may be written

$$\int \varrho \dot{v}_k \, \mathrm{d}V = \int \varrho f_k \, \mathrm{d}V + \int \sigma_{kl} \nu_l \, \mathrm{d}A = \int (\varrho f_k + \sigma_{kl,\,l}) \, \mathrm{d}V. \tag{3.7}$$

Subjecting the first equality (3.7) to (2.76), (2.50), and (1.79), we obtain an alternate global form of the theorem,

$$\int (\varrho v_k)_{,0} \, \mathrm{d}V = \int \varrho f_k \, \mathrm{d}V + \int (\sigma_{kl} - \varrho v_k v_l) \nu_l \, \mathrm{d}A. \tag{3.8}$$

This is the linear momentum theorem stated from the point of view of an observer at rest. In fact, if A is interpreted as a spatial surface, the integral on the left is the increase of linear momentum within A per unit time, and the various terms on the right represent the sum of the external forces and the momentum supply from outside by flow through A.

Observing that (3.7) holds for any material volume V, we obtain the local form

$$\varrho a_k = \varrho \dot{v}_k = \varrho f_k + \sigma_{kl,\,l} \tag{3.9}$$

of the linear momentum theorem. In the special case of a continuum at rest, it reduces to the equilibrium condition

$$\sigma_{kl,\,l} + \varrho f_k = 0. \tag{3.10}$$

Let us return to (3.2), the expression of the principle of virtual work for an arbitrary state of rigid virtual motion. Having exploited (3.2) for translations, we now apply it to an arbitrary rotation about the origin O of the coordinate system. Here the virtual velocity is

$$v_k^* = e_{kij} \omega_i x_j, \tag{3.11}$$

where ω_i denotes the angular velocity. Inserting (3.11) in (3.2) and observing that ω_i may be dropped since it is independent of the position of dV and arbitrary, we obtain

$$\int e_{ijk}\varrho x_j\dot{v}_k \,dV = \int e_{ijk}x_j\varrho f_k \,dV + \int e_{ijk}x_j\sigma_k^{(\nu)} \,dA. \tag{3.12}$$

The second integral on the right may be rewritten by means of (3.6). In the integral on the left, $x_j\dot{v}_k$ may be replaced by $(x_jv_k)^{\cdot}$ since $e_{ijk}\dot{x}_jv_k = e_{ijk}v_jv_k = 0$. We thus obtain

$$\int e_{ijk}\varrho(x_jv_k)^{\cdot} \,dV = \int e_{ijk}x_j\varrho f_k \,dV + \int e_{ijk}x_j\sigma_{kl}\nu_l \,dA. \tag{3.13}$$

The angular momentum dD_i, referred to O, of the material volume element dV is the moment of the momentum $\varrho v_k \,dV$ and hence is given by $e_{ijk}x_j\varrho v_k \,dV$. Defining the angular momentum of the body V as the sum of the angular momenta of its elements, we have

$$D_i = \int e_{ijk}\varrho x_jv_k \,dV. \tag{3.14}$$

On account of (2.76) and since e_{ijk} is constant, the material derivative of (3.14) is equal to the left-hand side of (3.13). Thus, (3.13) is the analytical form of the *theorem of angular momentum*, formulated for the material body V and referred to the origin O which, incidentally, is arbitrary: the material derivative of the linear momentum with respect to O is equal to the sum of the moments of the external forces with respect to O or, equivalently, to the resulting moment obtained by reduction of the external forces to O. In this form, the theorem of angular momentum is known to hold in elementary mechanics, i.e. for arbitrary systems of particles and rigid bodies.

We might reformulate the theorem just obtained, like the linear momentum theorem, from the point of view of an observer at rest. Instead of doing this, let us apply the theorem of Gauss (1.79) to obtain, in place of (3.12) or (3.13),

$$\int e_{ijk}\varrho x_j\dot{v}_k \,dV = \int e_{ijk}[x_j\varrho f_k + (x_j\sigma_{kl})_{,l}] \,dV \tag{3.15}$$

or, on account of (3.9) and of the identity $x_{j,l} = \delta_{jl}$,

$$\int e_{ijk}x_{j,l}\sigma_{kl} \,dV = \int e_{ijk}\sigma_{kj} \,dV = 0. \tag{3.16}$$

Since V is arbitrary, (3.16) yields the local form

$$e_{ijk}\sigma_{kj} = 0 \tag{3.17}$$

of the angular momentum theorem. Writing it out for $i = 1, 2, 3$, we find that the stress tensor σ_{ij} is symmetric.

Problems

1. In its common form the reaction principle states that the forces acting between two particles (action and reaction) are equal and opposite and have the same line of action. Show that this statement is a special case of the general reaction principle formulated in this section.

2. Show that the linear momentum, defined by (3.4), of a deformable body V may be written $m\dot{x}_k^{(C)}$, where

$$x_k^{(C)} = \frac{1}{m} \int \varrho x_k \, \mathrm{d}V$$

is the radius vector of the centroid C of V.

3. Show that the angular momentum of V, defined by (3.14), may be written

$$D_i = e_{ijk} x_j^{(C)} B_k + D_i' ,$$

where D_i' is the angular momentum with respect to the centroid C considered as the origin of its accompanying coordinate system.

3.2. The state of stress

According to Section 3.1, the state of stress within a continuum is described by the field of the stress tensor. Once this field is known, the stress vector acting on any surface element in an arbitrary point is given by (3.6). Since the stress tensor is symmetric, it is subject to the results obtained in Section 1.3. Thus, there exists at least one system of orthogonal *principal directions* in each point P. The surface elements normal to these directions are called *principal elements*. They are free of shear stresses, and the corresponding normal stresses $\sigma_{\mathrm{I}}, \ldots$, referred to as *principal stresses*, are stationary.

If the principal stresses $\sigma_{\mathrm{I}}, \ldots$ at a given point P are different, there exists a unique system of principal axes at this point. One of the principal stresses is the maximum, another one the minimum of all normal stresses acting on surface elements passing through P. If $\sigma_{\mathrm{I}} = \sigma_{\mathrm{II}} \neq \sigma_{\mathrm{III}}$, any cartesian coordinate system containing the third principal axis defines a principal system. If $\sigma_{\mathrm{I}} = \sigma_{\mathrm{II}} = \sigma_{\mathrm{III}}$, any cartesian system with the origin P is a principal system.

We have just noted that the extreme values of the normal stress in a given point are always principal stresses. For certain purposes it is important to know also the extreme values of the shear stress. We proceed to calculate them, using a coordinate system coinciding with the principal axes in the given point. In this system equation (3.6) for the stress vector acting on an arbitrary surface element dA with unit normal ν_i reduces to

$$\sigma_1^{(\nu)} = \sigma_{\mathrm{I}}\nu_1, \; \ldots \; . \tag{3.18}$$

The normal stress on dA,

$$\sigma^{(n)} = \boldsymbol{\sigma}^{(\nu)}\cdot\boldsymbol{\nu} = \sigma_{ij}\nu_i\nu_j, \tag{3.19}$$

thus assumes the form

$$\sigma^{(n)} = \sigma_{\mathrm{I}}\nu_1^2 + \ldots, \tag{3.20}$$

and the resultant shear stress is obtained from

$$\sigma^{(s)2} = \sigma_i^{(\nu)}\sigma_i^{(\nu)} - \sigma^{(n)2} = \sigma_{\mathrm{I}}^2\nu_1^2 + \ldots - (\sigma_{\mathrm{I}}\nu_1^2 + \ldots)^2. \tag{3.21}$$

In these relations ν_j appears only as a square. The stresses $\sigma^{(n)}$ and $\sigma^{(s)}$ are therefore equal for surface elements obtained by reflection along the principal planes. Since ν_j is a unit vector,

$$\nu_3^2 = 1 - \nu_1^2 - \nu_2^2. \tag{3.22}$$

Inserting this in (3.21), we have

$$\begin{aligned}\sigma^{(s)2} = {} & (\sigma_{\mathrm{I}}^2 - \sigma_{\mathrm{III}}^2)\nu_1^2 + (\sigma_{\mathrm{II}}^2 - \sigma_{\mathrm{III}}^2)\nu_2^2 + \sigma_{\mathrm{III}}^2 \\ & - [(\sigma_{\mathrm{I}} - \sigma_{\mathrm{III}})\nu_1^2 + (\sigma_{\mathrm{II}} - \sigma_{\mathrm{III}})\nu_2^2 + \sigma_{\mathrm{III}}]^2,\end{aligned} \tag{3.23}$$

where ν_1 and ν_2 now are independent variables. The extrema of $\sigma^{(s)2}$ are obtained by equating the partial derivatives of (3.23) with respect to ν_1 and ν_2 to zero. This yields

$$\frac{\partial\sigma^{(s)2}}{\partial\nu_1} = 2(\sigma_{\mathrm{I}}^2 - \sigma_{\mathrm{III}}^2)\nu_1 - 4[(\sigma_{\mathrm{I}} - \sigma_{\mathrm{III}})\nu_1^2 + (\sigma_{\mathrm{II}} - \sigma_{\mathrm{III}})\nu_2^2 + \sigma_{\mathrm{III}}](\sigma_{\mathrm{I}} - \sigma_{\mathrm{III}})\nu_1 = 0 \tag{3.24}$$

plus an analogous equation for ν_2. The two equations may be written

$$\begin{aligned}(\sigma_{\mathrm{I}} - \sigma_{\mathrm{III}})\,\nu_1\{\sigma_{\mathrm{I}} - \sigma_{\mathrm{III}} - 2[(\sigma_{\mathrm{I}} - \sigma_{\mathrm{III}})\nu_1^2 + (\sigma_{\mathrm{II}} - \sigma_{\mathrm{III}})\nu_2^2]\} &= 0, \\ (\sigma_{\mathrm{II}} - \sigma_{\mathrm{III}})\,\nu_2\{\sigma_{\mathrm{II}} - \sigma_{\mathrm{III}} - 2[(\sigma_{\mathrm{I}} - \sigma_{\mathrm{III}})\nu_1^2 + (\sigma_{\mathrm{II}} - \sigma_{\mathrm{III}})\nu_2^2]\} &= 0.\end{aligned} \tag{3.25}$$

In the general case where the three principal stresses are different, so are the two expressions between braces. A first solution of (3.25) then is $\nu_1 = \nu_2 = 0$ and, on account of (3.22), $\nu_3^2 = 1$. The corresponding dA is the third principal element, and the shear stress obtained from (3.23) is zero as expected. By cyclic permutation we obtain two more solutions of this type; they obviously correspond to minima of $\sigma^{(s)}$. Another type of solution is obtained by assuming that, e.g., $\nu_1 = 0$ and that the expression between braces in $(3.25)_2$ vanishes. The last condition, together with (3.22), yields $\nu_2^2 = \nu_3^2 = \frac{1}{2}$, and (3.23) supplies the shear stress $\sigma^{(s)} = \frac{1}{2}|\sigma_{II}-\sigma_{III}|$. By cyclic permutation we again obtain two similar solutions. The corresponding surface elements contain one of the principal axes each and bisect the angle of the other ones. The shear stresses are half the magnitudes of the differences between two principal stresses; they are called *principal shear stresses*. One of them is the maximal shear stress:

$$\sigma_{max}^{(s)} = \tfrac{1}{2}\max(|\sigma_{II}-\sigma_{III}|, \ldots). \tag{3.26}$$

The last result also applies in the event that two or more principal stresses are equal; these cases may be treated similarly (P1).

Like any symmetric tensor of the second order (Section 1.3) the stress tensor may be decomposed by

$$\sigma_{ij} = \sigma'_{ij}+\tfrac{1}{3}\sigma_{kk}\delta_{ij} \tag{3.27}$$

into a deviator and an isotropic tensor. The invariant $\sigma_{kk}/3 = \sigma_{(1)}/3$ is referred to as the *mean normal stress.*

A state of stress is called *plane* in a given point P if one of the principal stresses, e.g. σ_{III} is zero. The surface element perpendicular to the corresponding principal axis x_3 is then stress-free. According to (3.18), the stress vector acting on an arbitrary surface element passing through P has the components $\sigma_I\nu_1$, $\sigma_{II}\nu_2$, 0, and hence is parallel to the element which is free of stresses. If two principal stresses, e.g. σ_{II} and σ_{III} vanish in P, the state of stress is called *uniaxial.* Here the components of the stress vector acting on an arbitrary surface element are $\sigma_I\nu_1$, 0, 0: the vector is parallel to the axis of the stress state. Any element containing this axis is stress-free since $\nu_1 = 0$ and hence is a principal element. This was to be expected on account of a statement in the second alinea of this section.

Problems

1. Discuss the principal shear stresses and the corresponding surface elements for the cases where two or more principal stresses are equal.

2. Express the basic invariants $\sigma'_{(1)}, \sigma'_{(2)}, \sigma'_{(3)}$ of the stress deviator in terms of the invariants $\sigma_{(1)}, \sigma_{(2)}, \sigma_{(3)}$ of the stress tensor.

3. Consider a surface element at equal angles with the principal axes. For obvious reasons, the shear stress acting on this element is referred to as the *octahedral shear stress*. Express this stress in terms of the second basic invariant of the stress deviator, $\sigma'_{(2)}$.

3.3. The energy theorem

The momentum theorems have been obtained in Section 3.1 by applying the principle of virtual power to rigid translations and rotations of a finite portion V of the continuum. They are straightforward generalizations of the corresponding theorems in particle and rigid body mechanics. Guided by this analogy we now try to derive a further theorem, to be obtained as a consequence of the linear momentum theorem. If we multiply both sides of the second equality (3.9) by v_k and integrate over the volume V, we obtain

$$\begin{aligned}\int \varrho v_k \dot{v}_k \,\mathrm{d}V &= \int \varrho f_k v_k \,\mathrm{d}V + \int \sigma_{kl,l} v_k \,\mathrm{d}V \\ &= \int \varrho f_k v_k \,\mathrm{d}V + \int (\sigma_{kl} v_k)_{,l} \,\mathrm{d}V - \int \sigma_{kl} v_{k,l} \,\mathrm{d}V. \end{aligned} \tag{3.28}$$

On account of the symmetry of σ_{kl} and of (2.14), the velocity gradient $v_{k,l}$ may be replaced by the deformation rate d_{kl}. Besides, we can apply the theorem of Gauss (1.79) to the second integral in the second line. Thus, (3.28) assumes the form

$$\int \varrho v_k \dot{v}_k \,\mathrm{d}V = \int \varrho f_k v_k \,\mathrm{d}V + \int \sigma_{kl} v_k \nu_l \,\mathrm{d}A - \int \sigma_{kl} d_{kl} \,\mathrm{d}V. \tag{3.29}$$

It is obvious that (3.29) may be interpreted as the principle of virtual rate of work, applied to the real state of motion as a particular case of a virtual motion. Since V does not remain rigid in the actual motion, the power of the internal forces is not zero, and this accounts for the difference between (3.29) and the equation obtained from (3.2) by setting $v_k^* = v_k$. In fact, except for the sign, the integral on the left represents the rate of work or power of the inertia forces in the real motion. The first two integrals on the right are the rate of work of the external forces,

$$L^{(e)} = \int \varrho f_k v_k \,\mathrm{d}V + \int \sigma_{kl} v_k \nu_l \,\mathrm{d}A, \tag{3.30}$$

and it follows that the last term in (3.29),

$$L^{(i)} = -\int \sigma_{kl} d_{kl} \,\mathrm{d}V, \tag{3.31}$$

is the actual rate of work of the internal forces.

Intuitively, it seems obvious that the integral (3.31), interpreted as the total rate of work of the stresses on the deformation rates, represents the power of the internal forces. It is not quite as simple, however, to understand the minus sign. In fact, if we decompose the volume V into its elements in the shape of cuboids (Fig. 3.3) and if we consider the rate of work of the stress components on the corresponding deformation rates, we are tempted to conclude that the minus sign in (3.31) is wrong. However, by analogy with a discrete model it is not difficult to see that it is the *negative* stress tensor which represents the internal forces. Consider, e.g. a system of particles connected by massless springs. Applying the principle of virtual power to the virtual motion of a single particle, we note that, instead of the spring forces, their reactions, namely the forces acting on the particles, have to be considered, and it is obvious that these reactions correspond to the negative stresses in a continuum. In fact, in Figure 3.3 the cuboid corresponds to the spring and the surfaces separating the cuboids correspond to the particles; these surfaces, however, are acted upon by the reactions of the stresses shown in the figure, i.e. by the negative stresses.

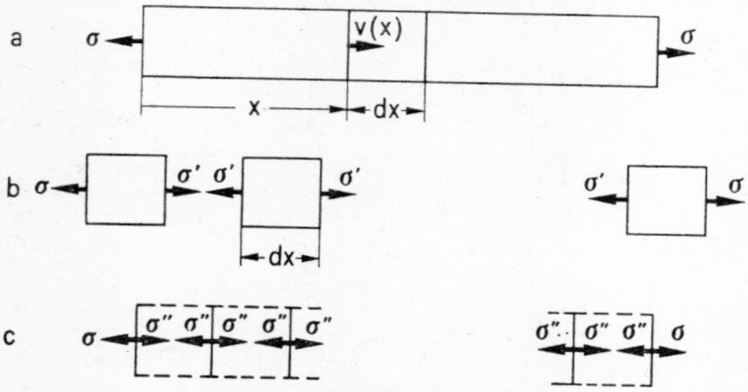

Fig. 3.4 Internal forces

As an example let us consider a rod of unit cross section (Fig. 3.4a) at rest under a given tension σ. Let $v(x)$ describe an admissible virtual state of motion corresponding, e.g. to an extension of the rod. According to the principle of virtual power, applied to the actual forces and to the virtual state of motion $v(x)$, the total virtual rate of work of the external and internal forces must be zero. If we decompose the rod into its elements according to Figure 3.4b, it is clear that the stresses σ acting in the two end sections represent the external forces, and we are tempted to interpret the stresses σ' $(= \sigma)$ acting between the elements as internal forces. However, since in an admissible state of motion the contact between the elements is maintained, the stresses σ' do not contribute to the virtual rate of work, whereas the contribution of the stresses σ in the end sections is different from zero. Thus, the interpretation of the internal forces suggested by Figure 3.4b is inconsistent with the principle of virtual power. If on the other hand the reactions σ'' $(=\sigma)$ of the stresses σ', i.e. the stresses acting on the partitions between the elements are considered as the internal forces, it follows immediately from Figure 3.4c that, as the principle of virtual power requires, their total virtual rate of work is the same, except for the sign, as that of the external forces σ.

The kinetic energy dT of the material volume element dV is half the product of its mass $\varrho\, dV$ and the square $v_k v_k$ of its velocity. Defining the kinetic energy of the body V as the sum of the kinetic energies of its elements, we have

$$T = \tfrac{1}{2} \int \varrho v_k v_k \, dV. \tag{3.32}$$

On account of (2.76), the material derivative of (3.32) is equal to the left-hand side of (3.29). Thus, (3.29) may be interpreted as the global form of the *energy theorem:* the material derivative of the kinetic energy is equal to the total rate of work of the external and internal forces. This is the form in which the energy theorem is also known to hold in elementary mechanics, i.e. for systems of particles and rigid bodies.

There is an essential difference between the momentum theorems derived in Section 3.1 and the energy theorem: in the momentum theorems the internal forces are absent; in the energy theorem they play an essential role. We will presently encounter other differences.

Solving (3.29) for the last integral

$$\int \sigma_{kl} d_{kl} \, dV, \tag{3.33}$$

we obtain a third interpretation of this equation: (3.33) is the part of the power of the external forces that is not converted into kinetic energy of the body V. We call it tentatively the *internal energy*, and we are immediately tempted to interpret it as the potential energy stored within V. However, to justify this interpretation, we need some knowledge of the structure of matter, and this knowledge is not available within the framework of continuum mechanics. It is clearly necessary to supplement, at least temporarily, the continuum concept or, in other words, the phenomenological viewpoint, by a more refined investigation into the microscopic structure of the element dV.

It is well-known that the element dV, which in this context may be referred to as the *macrosystem*, actually consists of a great number of elementary particles (atoms or molecules) forming the so-called *microsystem*, and it seems reasonable, in a first attempt, to interpret the internal energy as the total potential of all forces acting between these particles. Since in a static microsystem of this type the relative positions of the elementary particles are determined by the strain components ε_{kl} of dV, this interpretation implies that dV behaves as a conservative system, the stress components σ_{kl} depending on a potential which is a function of the ε_{kl} alone. We know, however, of no single body with this property, and the physical reason is obvious: accepting the notion of a microsystem, we are bound to accept also the possibility of relative motion between the elementary particles, and it is clear that the corresponding micro-velocities are not described by the macro-velocity v_k but rather superposed on it. The corresponding kinetic energy is part of the internal energy which thus is to be re-interpreted as the total (potential and kinetic) energy of the microsystem.

With this modified interpretation, it becomes clear that the internal energy cannot be determined by the ε_{kl} alone. The addition of one or more independent variables, however, leads us right into thermodynamics. There is yet another fact pointing in this direction. The element dV may be considered as a cuboid (Fig. 3.3) within a massless frame, containing the elementary particles connected among themselves and possibly to the frame by conservative forces. Any change of a strain component ε_{kl} corresponds to a change in the shape of the frame. The rate of work of the stresses, $\sigma_{kl}\dot{\varepsilon}_{kl} = \sigma_{kl}d_{kl}$, is equivalent to an increase or decrease of the internal energy. However, on account of the interaction between the elementary particles within the system and those of its surroundings or as a result of a micro-motion of the frame itself, the internal energy may also be changed even while the frame remains stationary for a macroscopic observer. This implies an energy exchange with the surroundings which, from the phenomenological viewpoint, is not explained by the power of the σ_{kl}; thermodynamically, it is interpreted as a *heat flow* through the boundary of the system.

Since there is no equivalent for the heat flow in (3.29), we conclude that the mechanical energy theorem does not describe the energy balance in a continuum in the phenomenological sense. It can be shown that it describes the energy balance for the microsystem; for the macrosystem, however, it must be replaced by the first fundamental law of thermodynamics (Chapter 4). Yet, (3.29) has been obtained as a consequence of the theorem of linear momentum; the question therefore arises whether the micro-motion does not invalidate the usual interpretation of the momentum theorems of Section 3.1. The answer is negative, and the reason is obvious. In the derivation of the momentum and energy theorems, the statistical micro-motion and the corresponding fluctuations of the stresses along the boundary have been disregarded. This is without consequence for the momentum theorems. The momenta are linear functions of the velocities, and the statistical fluctuations cancel out in the determination of the velocity and the angular velocity of the macroscopic element. The surface forces are linear in the stresses and hence also independent of their fluctuations. The kinetic energy, however, and the energy supply are quadratic expressions; it is therefore possible that a considerable part of the energy is due to the micro-motion and that this part is considerably influenced by the stress fluctuations along the boundary.

In the next chapter we will proceed to the thermodynamic investigation of the continuum. Before doing so, let us note for further reference that, decomposing the tensors σ_{kl} and d_{kl} into their deviatoric and isotropic parts, we may write the integrand in (3.33) in the form

$$\sigma_{kl}d_{kl} = (\sigma'_{kl}+\tfrac{1}{2}\sigma_{ii}\delta_{kl})(d'_{kl}+\tfrac{1}{2}d_{jj}\delta_{kl}) = \sigma'_{kl}d'_{kl}+\tfrac{1}{3}\sigma_{kk}d_{kk}. \qquad (3.34)$$

Problem

Motivate the negative sign in (3.31) by a discrete model in the form of a truss.

CHAPTER 4

THERMODYNAMICS

The field theory of mechanics, treated in the two preceding chapters, dates back for fluids to Euler (1755) and for solids to Cauchy (1822). Compared with mechanics as a whole, thermodynamics, established by Carnot (1824), Mayer (1844), and Clausius (1850), is a young science, and it is therefore surprising that until recently it has not been conceived as a field theory.

The object of thermodynamics has long been and, in almost the entire literature still is, a finite portion of matter, usually a gas, and as a rule it is tacitly assumed that the state is the same throughout the entire volume. In mechanics, this attitude would imply that only homogeneous states of strain and stress are considered. This would prevent us from solving even the simplest problems in strength of materials: in a bar subjected to flexure or torsion, e.g., strain and stress are functions of the position.

The lack of a thermodynamic field theory even at the end of the first half of the present century seems the more surprising as, on the other hand, the microscopic viewpoint, culminating in the statistical mechanics of Gibbs (1901), has been explored rather early. Between the two extremes of statistical mechanics and the phenomenological "thermostatics" of uniform states, the field theory in the sense of our last chapters has been missing, and it is not surprising that this fact had its far-reaching consequences.

In a finite portion V of a continuum the mechanical and the thermodynamic states generally differ from point to point. If V is isolated from its surroundings and left to itself, the state usually changes, and at least the temperature tends to become uniform in the course of time. This process is accompanied by an increase of entropy and hence is irreversible. To keep it negligible, classical thermodynamics was forced to restrict itself to infinitely slow processes or, in other words, to states in the immediate vicinity of equilibrium. Under this condition, most processes are reversible, and as a consequence certain texts on thermodynamics claim that every sufficiently slow process is reversible. Even this statement is incorrect: the deformation of a plastic body, e.g., is always an irreversible process no matter how slow it is.

In a thermodynamic field theory restriction to infinitely slow processes is as unnecessary as the exclusion of irreversible ones. Here the primary object is not a finite continuum, but its element, assumed to be so small that, within it, the state variables may be considered to be uniform although they differ from element to element. The analogy with continuum mechanics (strain and stress) is obvious. It is true that for very fast processes this approach is questionable on account of the fact that here the phenomenological treatment itself becomes inadequate. However, such processes are rare (see, e.g. the modern treatment of thermodynamics [5] by Traupel) and, in any case, they lie outside the scope of this book.

We have seen at the end of Section 3.3 that continuum mechanics cannot be separated from thermodynamics. A unified treatment, however, requires that the laws of thermodynamics be formulated in terms of concepts corresponding to those of the preceding chapters, i.e. as a field theory. It is the scope of the present chapter to establish the necessary thermodynamic field concepts, starting from the classical representation with which the reader is assumed to be familiar.

4.1. The classical theory

Classical thermodynamics deals with systems which, from a phenomenological point of view, are at rest. Let the state of such a system be described by a set of mutually independent *kinematical coordinates* or *parameters* $a_k (k = 1, 2, \ldots, n)$ and by the absolute *temperature*, which will be denoted by ϑ to distinguish it from the kinetic energy. As an example which will be of particular importance for the applications we have in mind, let us think of an element of a continuum in the form of a cuboid and let us consider the simple case where the kinematical state is completely described by the strain components ε_{ij}. Here the ε_{ij}, i.e. the components of a symmetric second-order tensor, play the roles of the a_k. The tensorial character of the a_k will become important at a later stage; in the present context, however, it is irrelevant and need not be discussed.

It is true that the basic concepts just defined are slightly more general than the ones we sometimes encounter in thermodynamics. There the object of investigation is usually an inviscid gas, and in this simple instance the volume V (e.g. of a mole) is the only kinematical state variable. Thermomechanics, however, is not restricted to this example. In viscous gases and in more complicated media V must be replaced by the ε_{ij} as kinematical coordinates. Sometimes we need even more kinematical parameters, and if chemical processes within the element are admitted, the concentrations of the constituents have to be added. In any event, our basis must be sufficiently large to cover all possible applications.

We have assumed that the a_k and ϑ completely describe the state of the system; we call these quantities *independent state variables*. Any function of them will be referred to as a *dependent state variable* or a *state function*. If the state is altered by the infinitesimal increments $\mathrm{d}a_k$, $\mathrm{d}\vartheta$, the *elementary work* done on the system is of the form

$$\mathrm{d}W = A_k\,\mathrm{d}a_k, \tag{4.1}$$

where use has been made of the summation convention. In the case of an inviscid gas, e.g., (4.1) has the simple form $\mathrm{d}W = -p\,\mathrm{d}V$, where p is the hydrostatic pressure. The coefficients A_k in (4.1) are called the *forces* corresponding to the a_k. In general, a change in the state of the system also implies that heat is supplied or extracted. Let $\mathrm{d}Q$ be the *heat supply* corresponding to the infinitesimal change of state. It is important to note that neither $\mathrm{d}W$ nor $\mathrm{d}Q$ are the total differentials of state functions. The symbol $\mathrm{d}W$ denotes the elementary work done on the system; the total work done in a finite time interval may be obtained by adding the $\mathrm{d}W$, and it may be denoted by W; however, the work done in a finite change of state depends not alone on the initial and the final states but also on the manner in which the transition takes place; thus W is not a state function. The same is true for $\mathrm{d}Q$ and the total heat supply Q; the term "heat" has no physical meaning.

The *first fundamental law* of thermodynamics states that there exists a state function $U(a_k, \vartheta)$, called the *internal energy*, such that

$$\mathrm{d}U = \mathrm{d}W + \mathrm{d}Q = A_k\,\mathrm{d}a_k + \mathrm{d}Q. \tag{4.2}$$

According to this theorem, the total differential of the internal energy is the sum of the elementary work done on the system and the elementary heat supply. If a process is such that $\mathrm{d}Q = 0$, i.e. if no heat is exchanged with the surroundings, the process is called *adiabatic*. If, on the other hand, all of the kinematical state variables remain constant, (4.1) yields $\mathrm{d}W = 0$, and we refer to the process as one of *pure heating* (or cooling). The first law may be considered as a balance equation; one of its well-known implications is the non-existence of a perpetuum mobile of the first kind.

The *second fundamental law* of thermodynamics states that there exists another state function $S(a_k, \vartheta)$, called *entropy*, such that

$$\vartheta \, dS \geqslant dQ. \tag{4.3}$$

If (4.3) holds with the equality sign, the process is referred to as *reversible*, otherwise as *irreversible*. This is the original and also the simplest form of the second law; the well-known assertions about cyclic processes and the non-existence of a perpetuum mobile of the second kind are consequences of it. The theorem can also be stated in the form

$$dS = d^{(r)}S + d^{(i)}S, \tag{4.4}$$

where

$$d^{(r)}S = \frac{dQ}{\vartheta} \tag{4.5}$$

is the reversible increment of S, called the *entropy supply* from outside, whereas

$$d^{(i)}S \geqslant 0 \tag{4.6}$$

is the irreversible increment, referred to as the *entropy production* inside the system. A reversible process is characterized by the absence of entropy production; in an irreversible process the entropy production within the system is always positive. If $dS = 0$, the entropy remains constant, and the process is called *isentropic*. According to (4.5) and (4.6), an adiabatic process is isentropic if and only if it is reversible. It should be noted that, unlike dS, the increments $d^{(r)}S$ and $d^{()}S$ are not total differentials. Finally, a process taking place at constant temperature is called *isothermal*.

We now proceed to deduce a few general consequences, important for the next sections, of the fundamental laws. By means of (4.2), (4.5), and (4.4), the elementary work may be written

$$dW = dU - dQ = dU - \vartheta \, d^{(r)}S = dU - \vartheta \, dS + \vartheta \, d^{(i)}S. \tag{4.7}$$

On account of (4.1) and the fact that U and S are state functions, (4.7) may be replaced by the relation

$$A_k \, da_k = \left(\frac{\partial U}{\partial a_k} - \vartheta \frac{\partial S}{\partial a_k}\right) da_k + \left(\frac{\partial U}{\partial \vartheta} - \vartheta \frac{\partial S}{\partial \vartheta}\right) d\vartheta + \vartheta \, d^{(i)}S. \tag{4.8}$$

For pure heating (4.8) reduces to

$$\left(\frac{\partial U}{\partial \vartheta} - \vartheta \frac{\partial S}{\partial \vartheta}\right) d\vartheta + \vartheta\, d^{(i)}S = 0. \tag{4.9}$$

According to (4.6), the second term is non-negative for $\vartheta > 0$, whereas the quantity between parentheses is a state function and hence is independent of $d\vartheta$. Since (4.9) must hold for positive and negative values of $d\vartheta$, it follows that

$$\frac{\partial U}{\partial \vartheta} - \vartheta \frac{\partial S}{\partial \vartheta} = 0. \tag{4.10}$$

Even though we have obtained (4.10) by considering a special process, the result is generally valid, for the left-hand side of (4.10) is a state function and as such is independent of the type of process. Equation (4.9), on the other hand, only holds for pure heating. Combining (4.9) and (4.10), we find that, in this special case, $d^{(i)}S = 0$, i.e. that the process of pure heating is reversible as expected.

Returning to the general case and inserting (4.10) in (4.8), we obtain

$$A_k\, da_k = \left(\frac{\partial U}{\partial a_k} - \vartheta \frac{\partial S}{\partial a_k}\right) da_k + \vartheta\, d^{(i)}S. \tag{4.11}$$

It follows that $\vartheta\, d^{(i)}S$ has the form of an elementary work and may be written

$$\vartheta\, d^{(i)}S = A_k^{(d)} da_k = dW^{(d)}, \tag{4.12}$$

where

$$A_k^{(d)} = A_k - \frac{\partial U}{\partial a_k} + \vartheta \frac{\partial S}{\partial a_k}. \tag{4.13}$$

We call the $A_k^{(d)}$ the *dissipative forces* of the system and refer to $dW^{(d)}$ as the *dissipative elementary work*. Besides, we note that in classical thermodynamics, where negative temperatures are excluded, (4.12), together with (4.6), imply

$$dW^{(d)} = A_k^{(d)}\, da_k \geqslant 0. \tag{4.14}$$

It is convenient to consider the $A_k^{(d)}$ defined by (4.13) as parts of the forces A_k, writing

$$A_k = A_k^{(d)} + A_k^{(q)}, \tag{4.15}$$

where the remainders $A_k^{(q)}$ are given by

$$A_k^{(q)} = \frac{\partial U}{\partial a_k} - \vartheta \frac{\partial S}{\partial a_k} \tag{4.16}$$

and hence are state functions. Conversely, the $A_k^{(d)}$, provided they are different from zero, cannot be state functions, since the inequality sign in (4.14) holds for arbitrary signs of $\mathrm{d}a_k$. We call the $A_k^{(q)}$ the *quasiconservative forces* of the system and note that, by (4.15), the actual forces appear decomposed into their dissipative and quasiconservative parts.

The results just obtained can be simplified if we introduce another state function, the so-called *free energy* of the system, defined by

$$\Psi = U - \vartheta S. \tag{4.17}$$

Its differential is

$$\mathrm{d}\Psi = \mathrm{d}U - \vartheta\,\mathrm{d}S - S\,\mathrm{d}\vartheta. \tag{4.18}$$

Writing it out by means of the partial derivatives of Ψ with respect to ϑ and the a_k and combining the result with (4.10), we obtain

$$S = -\frac{\partial \Psi}{\partial \vartheta}\,; \tag{4.19}$$

combination with (4.16) yields

$$A_k^{(q)} = \frac{\partial \Psi}{\partial a_k}\,. \tag{4.20}$$

It follows that the free energy $\Psi(a_k, \vartheta)$ plays the role of a potential: its partial derivatives with respect to the kinematical parameters and to the temperature are the quasiconservative forces and the negative entropy respectively. We note in particular that (4.20) justifies the attribute "quasiconservative" for the $A_k^{(q)}$: they depend on a potential, but this potential is not a function of the a_k alone but also dependent on ϑ.

In the preceding considerations we have used the language familiar from texts on thermodynamics, expressing everything in terms of differentials or other infinitesimal quantities. This language is quite adequate as long as one limits oneself, as classical thermodynamics did, to the study of extremely slow processes in the vicinity of an equilibrium state. In a thermodynamic field theory this restriction is unnecessary. Here it is more convenient to deal with finite quantities. This is achieved by dividing all terms in the differential relations obtained above by the time element $\mathrm{d}t$, replacing in this manner the differentials by time derivatives and the remaining elementary quantities by time rates. We thus obtain, in place of (4.1), the equation

$$L = A_k \dot{a}_k \tag{4.21}$$

for the power of the forces A_k. The first fundamental law (4.2) assumes the form

$$\dot{U} = L+Q^* = A_k\dot{a}_k+Q^*, \tag{4.22}$$

where Q^* denotes the heat supply per unit time. Similarly, the second fundamental law (4.3) may be written

$$\vartheta\dot{S} \geqslant Q^*, \tag{4.23}$$

and in place of (4.4) through (4.6) we obtain

$$\dot{S} = \dot{S}^{(r)}+\dot{S}^{(i)}, \tag{4.24}$$

where

$$\dot{S}^{(r)} = \frac{Q^*}{\vartheta} \quad \text{and} \quad \dot{S}^{(i)} \geqslant 0. \tag{4.25}$$

Let us note here that there exist no state functions $S^{(r)}$ or $S^{(i)}$; the left-hand sides in (4.25) are separate contributions to $\dot{S}$, the time derivative of the state function $S(a_k, \vartheta)$.

On account of (4.15), an alternate form of the first fundamental law (4.22) is

$$\dot{U} = A_k^{(q)}\dot{a}_k+A_k^{(d)}\dot{a}_k+Q^*. \tag{4.26}$$

Using (4.12) and (4.5), we derive from (4.26)

$$\dot{U} = A_k^{(q)}\dot{a}_k+\vartheta\dot{S}^{(i)}+\vartheta\dot{S}^{(r)} = A_k^{(q)}\dot{a}_k+\vartheta\dot{S}. \tag{4.27}$$

In its last form, this is the famous equation of Gibbs [6]. The term $\vartheta\dot{S}^{(r)}$ in the first equality (4.27) is equal to the heat supply Q^* per unit time. The term $\vartheta\dot{S}^{(i)}$, sometimes called *dissipation rate*, is given by

$$\vartheta\dot{S}^{(i)} = A_k^{(d)}\dot{a}^k = L^{(d)} \geqslant 0 \tag{4.28}$$

and hence is equal to the *power of dissipation*, i.e. the rate of work $L^{(d)}$ of the dissipative forces. Like Q^*, it is determined by the state and its instantaneous change, i.e. by the state variables and their derivatives. We thus have

$$\vartheta\dot{S}^{(i)} = \Phi(\dot{a}_k, \dot{\vartheta}, a_k, \vartheta) \geqslant 0, \tag{4.29}$$

where Φ is called the *dissipation function*. According to (4.28), the dissipative forces depend on the same arguments as Φ. Since Φ is non-negative and, on account of (4.28), is zero for pure heating, the dissipation function is at least positive semidefinite in the $\dot{a}_k$. Using (4.28) and (4.29), we obtain a final form of the first fundamental law (4.26),

$$\dot{U} = A_k^{(q)}\dot{a}_k + \Phi + Q^*. \tag{4.30}$$

So far, we have considered the a_k and ϑ as independent state variables and we will adhere to this rule for most of the remainder of this book. Once the choice of the independent variables is made, the meaning of the partial derivatives is clear, and it is not necessary to indicate by subscripts —the way it is usually done in thermodynamics—which variables are to be kept constant in the derivation. There are cases, however, where it is convenient to exchange the roles of ϑ and S, using the a_k and S as independent state variables. If this happens, we will avoid the cumbersome subscript notation by means of primes indicating that the corresponding quantity is to be considered as a function of the a_k and of S, e.g.

$$U'(a_k, S) = U(a_k, \vartheta). \tag{4.31}$$

The time derivative of U' is

$$\dot{U}' = \frac{\partial U'}{\partial a_k}\dot{a}_k + \frac{\partial U'}{\partial S}\dot{S}. \tag{4.32}$$

Comparison with (4.27) shows that

$$A_k^{(q)} = \frac{\partial U'}{\partial a_k}, \qquad \vartheta = \frac{\partial U'}{\partial S}. \tag{4.33}$$

These equations are analogous to (4.20) and (4.19). They show that, in the new independent variables, it is the function U' in place of Ψ that assumes the role of a potential: its partial derivatives with respect to the kinematical parameters and to the entropy are the quasiconservative forces and the temperature respectively.

Problem

Discuss the relative significance of the potentials introduced in this section for isothermal and isentropic processes respectively.

4.2. State variables

It has been pointed out in Section 4.1 that, to arrive at a thermodynamic field theory, we have to start from the element of a continuum. Since the conservation condition holds for the mass of a material element whereas the volume of the element is variable, it is clear that the object to be considered is the element of mass. We assume that its shape in a given reference configuration, e.g. at time $t = 0$, is a cuboid, and we know from Section 2.2 that its shape at time t is determined by the strain tensor ε_{ij}. For large displacements the strain tensor will be discussed in Chapter 13. For the present we will restrict ourselves to small displacements. Here the geometric interpretation of the ε_{ij} is straightforward: they represent the extensions and the shear strains defining the shape of the element at an arbitrary time.

It is obvious that, in general, the six strain components ε_{ij} appear as independent kinematical state variables a_k in the sense of Section 4.1, and it is equally obvious that the temperature $\vartheta > 0$ is another independent state variable. Provided these are the only independent parameters, the state functions are completely determined by them. It is convenient to refer them to the unit mass and to denote the *specific internal energy* by $u(\varepsilon_{ij}, \vartheta)$, the *specific entropy* by $s(\varepsilon_{ij}, \vartheta)$, and to define the *specific free energy* in analogy to (4.17) by

$$\psi = u - \vartheta s. \tag{4.34}$$

Since the object we are dealing with is the element of mass, the time derivatives of Section 4.1 are now to be interpreted as material derivatives. For small displacements, the material derivative $\dot{\varepsilon}_{ij}$ of the strain tensor (Section 2.2) is equal to the rate of deformation d_{ij}. According to Section 3.3, the rate of work done on the unit of volume is $\sigma_{ij} d_{ij}$; thus, the *specific rate of work* (referred to the unit mass) is

$$l = \frac{1}{\varrho} \sigma_{ij} d_{ij}. \tag{4.35}$$

Comparing this to (4.21), we conclude that the forces corresponding to the kinematical state variables ε_{ij} are the quotients σ_{ij}/ϱ. Decomposing these forces according to (4.15), i.e. by means of

$$\sigma_{ij} = \sigma_{ij}^{(q)} + \sigma_{ij}^{(d)}, \tag{4.36}$$

into their quasiconservative and dissipative parts, we obtain, in analogy to (4.20) and (4.19),

$$\sigma_{ij}^{(q)} = \varrho \frac{\partial \psi}{\partial \varepsilon_{ij}}, \qquad s = -\frac{\partial \psi}{\partial \vartheta}. \tag{4.37}$$

The *specific power of dissipation* is given, on account of (4.28) and (4.35), by

$$l^{(d_1)} = \frac{1}{\varrho} \sigma_{ij}^{(d)} d_{ij}. \tag{4.38}$$

In many materials the independent state variables ε_{ij} and ϑ suffice to determine the state. There are other continua, however, where the state at a given time t is not a function of the instantaneous values of ε_{ij} and ϑ alone but also depends on the previous history of these parameters. Such cases may be dealt with in various manners. One of these considers the dependent state variables (internal energy, entropy, etc.) as *functionals* of ε_{ij} and ϑ, depending not only on their instantaneous values but also on their values at all previous times. Another approach, which will be preferred throughout most of this book, consists in the introduction, at least temporarily, of *internal parameters.* To illustrate these concepts, let us consider a very simple one-dimensional model.

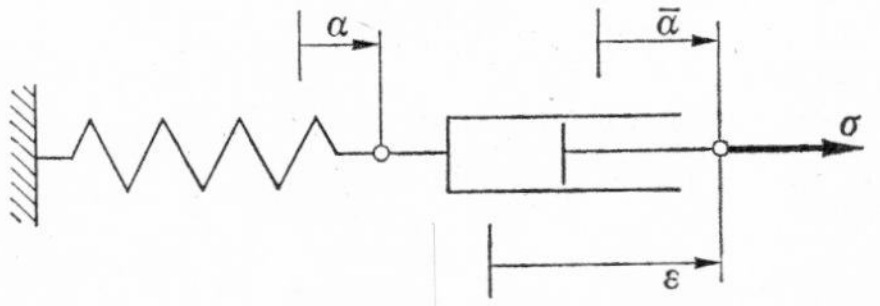

Fig. 4.1 Spring and dashpot

Figure 4.1 shows a system consisting of a spring and a dashpot. The force acting in the spring is a function of its elongation α; the force in the dashpot depends on its rate of elongation $\dot{\bar{\alpha}}$. Let us assume that the motion under the force σ acting at the free end of the system is so slow that the temperature ϑ is the same all over the system and that the inertia forces may be neglected. The forces in either element, spring and dashpot, are then equal to σ. The only external kinematical parameter is the elongation ε of the entire system, and it is clear that the force σ is not determined by ε alone (nor by ε in connection with ϑ). In fact, if ε is kept constant, σ tends to zero while the dashpot allows the spring to return to its natural length. It is intuitively obvious, however, that the force σ at time t is determined by the history of the elongation ε, i.e. if the function $\varepsilon(\tau)$ is known for all times τ between $\tau = -\infty$ and $\tau = t$. Thus, $\sigma(t)$ may be considered as a functional, i.e. as a function of the function $\varepsilon(\tau)$ which, incidentally, will be calculated for linear elements in Chapter 11. On the other hand, the value of σ at time t is also known if, beside ε, either α or $\bar{\alpha}$ are given at time t. Since $\alpha + \bar{\alpha} = \varepsilon$, we need only one of the two quantities, and we refer to it as an internal parameter.

In Chapter 11 we will consider more complicated models, characterized by more than one internal parameter. Needless to mention that systems of this type may be poor models for the behavior of a material element under general states of strain and stress and that they are to be handled with caution. However, they clearly illustrate the usefulness of internal parameters for certain continua. Let us denote them generally by α_{kl} and let us add them to the independent state variables ε_{ij} and ϑ. The specific internal energy is now of the form $u(\varepsilon_{ij}, \alpha_{kl}, \vartheta)$; the specific entropy becomes $s(\varepsilon_{ij}, \alpha_{kl}, \vartheta)$, and the specific free energy is again defined by (4.34). The forces corresponding to the internal parameters α_{kl} may be tentatively denoted by β_{kl}/ϱ and will be referred to, in a sense slightly different from the one used at the beginning of Section 3.1, as *internal forces*. Their quasi-conservative parts are defined by the equation

$$\beta_{kl}^{(q)} = \varrho \frac{\partial \psi}{\partial \alpha_{kl}}, \tag{4.39}$$

which is analogous to $(4.37)_1$. They are state functions, dependent on ε_{ij}, α_{kl}, and ϑ. The dissipative parts $\beta_{kl}^{(d)}$ of the internal forces also depend on d_{ij}, $\dot{\alpha}_{kl}$, and $\dot{\vartheta}$; their specific power of dissipation is

$$l^{(d2)} = \frac{1}{\varrho} \beta_{kl}^{(d)} \dot{\alpha}_{kl}; \tag{4.40}$$

it is to be added to (4.38).

The internal forces do not appear in the expression (4.35) for the rate of external work done on the unit of mass. Thus, in an arbitrary process,

$$\frac{1}{\varrho} \beta_{kl} \dot{\alpha}_{kl} = \frac{1}{\varrho} (\beta_{kl}^{(q)} + \beta_{kl}^{(d)}) \dot{\alpha}_{kl} = 0. \tag{4.41}$$

We conclude that the internal forces are zero and that, therefore,

$$\beta_{kl}^{(d)} = -\beta_{kl}^{(q)}, \tag{4.42}$$

provided we neglect an arbitrary gyroscopic contribution to $\beta_{kl}^{(d)}$, i.e. a set of internal forces dependent in such a way on the $\dot{\alpha}_{kl}$ that their power is always zero. Such a gyroscopic contribution is irrelevant at present and will be dealt with in Chapter 14.

So far, the internal parameters have been denoted by α_{kl} irrespective of their tensorial character. Guided by the model of Figure 4.1, we are tempted to presume that they are symmetric second-order tensors, representing auxiliary states of strain. This assumption will be justified in the applications to be considered in this book, and this is the reason why we have used two subscripts to distinguish the various internal parameters. It should be noted, however, that there are exceptions. If, e.g. chemical processes are admitted in the material under consideration, the concentrations of the various constituents, as far as they are independent, have the properties of internal parameters [7]; they may be represented by a vector in an n-dimensional space, and the corresponding quasiconservative (or dissipative) force is a similar vector determined by the various chemical potentials.

Let us close this section with a remark concerning the formalism of our presentation. Equation $(4.37)_1$, representing the quasiconservative stress tensor in terms of the partial derivatives of the free energy with respect to the strain components, is open to criticism since, on account of the symmetry of the strain tensor, its nine components are interdependent. A similar remark holds in the case of (4.39) provided α_{kl} is a symmetric tensor. Confining ourselves strictly to cyclic permutations indicated by dots, and introducing the engineering shear strains

$$\gamma_{23} = 2\varepsilon_{23} = 2\varepsilon_{32}, \ldots, \tag{4.43}$$

we note that the specific free energy may be written in the two equivalent forms

$$\psi^*(\varepsilon_{11}, \ldots, \gamma_{23}, \ldots) = \psi(\varepsilon_{11}, \ldots, \varepsilon_{23}, \ldots, \varepsilon_{32}, \ldots), \tag{4.44}$$

where we have assumed for simplicity that the α_{kl} and ϑ are fixed. The six arguments of ψ^* are independent; for the nine arguments of ψ this is not true. Recalling the signification of a partial differentiation, we observe that, on account of (4.43),

$$\frac{\partial \psi^*}{\partial \gamma_{23}} = \frac{\partial \psi}{\partial \varepsilon_{23}} \frac{\partial \varepsilon_{23}}{\partial \gamma_{23}} + \frac{\partial \psi}{\partial \varepsilon_{32}} \frac{\partial \varepsilon_{32}}{\partial \gamma_{23}} = \frac{1}{2}\left(\frac{\partial \psi}{\partial \varepsilon_{23}} + \frac{\partial \psi}{\partial \varepsilon_{32}}\right), \ldots. \tag{4.45}$$

Provided we write ψ as a symmetric function of the arguments ε_{23} and $\varepsilon_{32}, \ldots$, the two terms between parentheses are equal, and we obtain

$$\frac{\partial \psi^*}{\partial \gamma_{23}} = \frac{\partial \psi}{\partial \varepsilon_{23}}, \ldots. \tag{4.46}$$

We thus have the choice of two approaches:

(a) We write the specific free energy as a function ψ^* of the six independent engineering strain components $\varepsilon_{11}, \ldots, \gamma_{23}, \ldots$. The quasiconservative stresses are then given by

$$\sigma_{11}^{(q)} = \varrho \frac{\partial \psi^*}{\partial \varepsilon_{11}}, \quad \ldots, \quad \sigma_{23}^{(q)} = \varrho \frac{\partial \psi^*}{\partial \gamma_{23}}, \quad \ldots . \tag{4.47}$$

(b) We consider ψ as a function of the nine strain components $\varepsilon_{11}, \ldots$, $\varepsilon_{23}, \ldots, \varepsilon_{32}, \ldots$, written symmetrically in the corresponding shear strains. Differentiating formally, i.e. neglecting the interdependence of the arguments, we then obtain

$$\sigma_{11}^{(q)} = \varrho \frac{\partial \psi}{\partial \varepsilon_{11}}, \quad \ldots, \quad \sigma_{23}^{(q)} = \varrho \frac{\partial \psi}{\partial \varepsilon_{23}}, \quad \ldots, \quad \sigma_{32}^{(q)} = \varrho \frac{\partial \psi}{\partial \varepsilon_{32}}, \ldots, \tag{4.48}$$

and the symmetry of ψ results automatically in a symmetric stress tensor.

If, in particular, the quasiconservative stress tensor is isotropic (P of Section 1.1), we have

$$\sigma_{11}^{(q)} = \ldots = -p, \qquad \sigma_{23}^{(q)} = \ldots = 0. \tag{4.49}$$

Using (4.48), we conclude that ψ may be considered as a function in the 3-dimensional space $\varepsilon_{11}, \ldots$ and that it is constant on the planes

$$\varepsilon_{11} + \ldots = \text{const} \tag{4.50}$$

and hence is of the form $\psi(\varepsilon_{ii}, \alpha_{kl}, \vartheta)$, if the arguments α_{kl} and ϑ are re-introduced. It follows that

$$\frac{\partial \psi}{\partial \varepsilon_{11}} = \ldots = \frac{\partial \psi}{\partial \varepsilon_{ii}} \tag{4.51}$$

and that, on account of (4.49) and (4.48),

$$p = -\sigma_{11}^{(q)} = \ldots = -\varrho \frac{\partial \psi}{\partial \varepsilon_{ii}}, \tag{4.52}$$

where $\varepsilon_{ii} = \varepsilon_{(1)}$, according to Section 2.2, is the dilatation.

Problems

1. Consider the model of Figure 11.3, where a spring and a dashpot are arranged side by side and hence have the same elongation ε. Assume that both elements are linear and let E denote the spring constant and F the viscosity constant (force/velocity) of the dashpot. Find the quasiconservative and dissipative forces, the free energy, and the dissipation function.

2. Establish the differential equation connecting ε and σ in the model of Figure 11.3.

3. Consider α as the internal parameter and solve P1 and P2 for the model of Figure 4.1.

4.3. The field theory

With the help of the state variables introduced in Section 4.2 we are now in a position to formulate the fundamental laws of thermodynamics in terms of a field theory. We proceed in the same manner as in the deduction of the momentum theorems in Section 3.1, considering a partial volume V of a continuum (Fig. 3.1), bounded by the regular surface A.

The internal energy contained in V is obtained by integrating the specific internal energy $u(\varepsilon_{ij}, \alpha_{kl}, \vartheta)$ over the entire mass and hence is given by

$$U = \int \varrho u \, \mathrm{d}V. \tag{4.53}$$

Integrating the specific entropy $s(\varepsilon_{ij}, \alpha_{kl}, \vartheta)$ in the same manner, we obtain the total entropy

$$S = \int \varrho s \, \mathrm{d}V. \tag{4.54}$$

On account of (2.76), the material derivatives of U and S are

$$\dot{U} = \int \varrho \dot{u} \, \mathrm{d}V, \qquad \dot{S} = \int \varrho \dot{s} \, \mathrm{d}V \tag{4.55}$$

respectively.

The rate of work $L^{(e)}$ of the external forces is given by (3.30). If we exclude any heat exchange by radiation, the heat supply Q^* per unit time is due alone to convection across the surface A. To give it a precise form, let us define a local *heat flow vector* q_k by means of the condition that (Fig. 4.2), given a

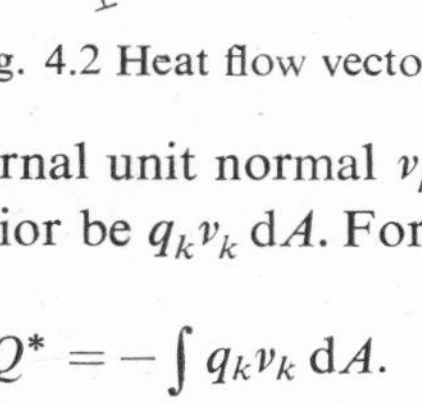

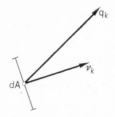

Fig. 4.2 Heat flow vector

surface element $\mathrm{d}A$ with external unit normal ν_k, the heat flow through $\mathrm{d}A$ from the interior to the exterior be $q_k \nu_k \, \mathrm{d}A$. For the entire body V, the heat supply per unit time is thus

$$Q^* = -\int q_k \nu_k \, \mathrm{d}A. \tag{4.56}$$

In view of $(4.25)_1$ it seems reasonable to define a local *entropy flow vector* by q_k/ϑ, where ϑ is the local temperature. The entire entropy supply across A per unit time is then

$$\dot{S}^{(r)} = -\int \frac{q_k}{\vartheta} \nu_k \, \mathrm{d}A. \tag{4.57}$$

In formulating the field version of the first fundamental law, we need to remember that, in general, the continuum is in motion. The kinetic energy of the portion we are considering is given by (3.32). Its material derivative,

$$\dot{T} = \int \varrho v_k \dot{v}_k \, \mathrm{d}V, \tag{4.58}$$

represents part of the energy increase of V and hence must be taken into account. Thus, the *first fundamental law* assumes the global form

$$\dot{T} + \dot{U} = L^{(e)} + Q^*. \tag{4.59}$$

It states that the total rate of increase of the kinetic and internal energies is equal to the rate of work of the external forces plus the heat supply per unit time. Inserting (4.58), $(4.55)_1$, (3.30), and (4.56) in (4.59), we obtain in more detail

$$\int \varrho(v_k \dot{v}_k + \dot{u}) \, \mathrm{d}V = \int \varrho f_k v_k \, \mathrm{d}V + \int (\sigma_{kl} v_k - q_l) \nu_l \, \mathrm{d}A. \tag{4.60}$$

This is the equation which takes the place of the mechanical energy theorem (3.28). It is interesting, however, to note that from a microscopic point of view (4.59) or (4.60) may still be interpreted as expressions of a mechanical energy theorem: on the left-hand side the kinetic energy T of the macro-motion is supplemented by the sum U of the potential energy of the internal forces and the kinetic energy of the micro-motion; on the right the power $L^{(e)}$ of the external macro-forces is supplemented by the power of the micro-stresses which, phenomenologically, appears as an extra-mechanical energy supply Q^*.

Applying the theorem of Gauss (1.79) to the first part of the last integral in (4.60), we obtain

$$\int \varrho(v_k \dot{v}_k + \dot{u}) \, \mathrm{d}V = \int (\varrho f_k v_k + \sigma_{kl} v_{k,l} + \sigma_{kl,l} v_k) \, \mathrm{d}V - \int q_k \nu_k \, \mathrm{d}A. \tag{4.61}$$

On account of the linear momentum theorem (3.9), this reduces to

$$\int \varrho \dot{u} \, \mathrm{d}V = \int \sigma_{kl} d_{kl} \, \mathrm{d}V - \int q_k \nu_k \, \mathrm{d}A, \tag{4.62}$$

where use has been made of (2.14) and of the symmetry of σ_{kl}. In this form, the first law states that the rate of increase of the internal energy is equal to the rate of heat supply minus the power of the internal forces.

The theorem of Gauss may be used once more to convert the surface integral in (4.62) into a volume integral. Observing that the result is valid for any volume V, we obtain the local form of the first fundamental law

$$\varrho \dot{u} = \sigma_{kl} d_{kl} - q_{k,k}. \tag{4.63}$$

It states that the material derivative of the internal energy, referred to the unit of volume, is equal to the power of the stress tensor plus the rate of heat supply, represented by the negative divergence of the heat flow. Note the similarity between (4.63) and (4.22) and the fact that the overall motion of the element does not appear any more in (4.63).

Proceeding to the *second fundamental law* we start from the global form

$$\dot{S} \geqslant \dot{S}^{(r)}, \tag{4.64}$$

stating that the rate of entropy increase in the material volume V is never less than the entropy supply per unit time. If (4.64) holds with the equality sign, the process is reversible, otherwise irreversible. Inserting $(4.55)_2$ and (4.57) in (4.64), we obtain

$$\int \varrho \dot{s} \, \mathrm{d}V \geqslant -\int \frac{q_k}{\vartheta} \nu_k \, \mathrm{d}A \tag{4.65}$$

or, using the theorem of Gauss (1.79),

$$\int \varrho \dot{s} \, \mathrm{d}V \geqslant -\int \left(\frac{q_k}{\vartheta}\right)_{,k} \mathrm{d}V = \int \left(\frac{q_k}{\vartheta^2} \vartheta_{,k} - \frac{q_{k,k}}{\vartheta}\right) \mathrm{d}V. \tag{4.66}$$

This inequality holds for any volume V and hence yields the local form of the second fundamental law,

$$\varrho \dot{s} \geqslant -\left(\frac{q_k}{\vartheta}\right)_{,k} = \frac{\vartheta_{,k}}{\vartheta^2} q_k - \frac{q_{k,k}}{\vartheta}. \tag{4.67}$$

The last inequality states that the rate of entropy increase per unit volume is never less than the rate of entropy supply. Depending on the classical approach, where the rate of entropy supply is represented by $(4.25)_1$, we might be tempted to question the first term on the right in (4.67), as $-q_{k,k}/\vartheta$ alone seems to correspond to Q^*/ϑ. The classical theory, however, is based on the fiction of a uniform temperature and hence on the assumption, untenable in a field theory, that $\vartheta_{,k} = 0$.

We know from Section 4.1 that the application of the fundamental laws is facilitated by the use of the free energy. In the present context we need the specific free energy, defined by (4.34). Its material derivative is given by

$$\dot{\psi} = \dot{u} - \vartheta\dot{s} - s\dot{\vartheta}. \tag{4.68}$$

Since ψ is a function of ε_{ij}, α_{kl}, and ϑ, we also have

$$\varrho\dot{\psi} = \varrho\frac{\partial\psi}{\partial\varepsilon_{ij}}d_{ij} + \varrho\frac{\partial\psi}{\partial\alpha_{kl}}\dot{\alpha}_{kl} + \varrho\frac{\partial\psi}{\partial\vartheta}\dot{\vartheta}. \tag{4.69}$$

Making use of (4.37) and (4.39), we obtain instead

$$\varrho\dot{\psi} = \sigma_{ij}^{(q)}d_{ij} + \beta_{kl}^{(q)}\dot{\alpha}_{kl} - \varrho s\dot{\vartheta} \tag{4.70}$$

or, on account of (4.68),

$$\varrho(\dot{\psi} + s\dot{\vartheta}) = \varrho(\dot{u} - \vartheta\dot{s}) = \sigma_{ij}^{(q)}d_{ij} + \beta_{kl}^{(q)}\dot{\alpha}_{kl}. \tag{4.71}$$

The dissipative stresses follow from (4.36). If, in addition, (4.42) is used, the second equality (4.71) becomes

$$\varrho(\dot{u} - \vartheta\dot{s}) = \sigma_{ij}d_{ij} - \sigma_{ij}^{(d)}d_{ij} - \beta_{kl}^{(d)}\dot{\alpha}_{kl}. \tag{4.72}$$

On account of the first fundamental law (4.63), this is equivalent to

$$\varrho\dot{s} = \frac{1}{\vartheta}(\sigma_{ij}^{(d)}d_{ij} + \beta_{kl}^{(d)}\dot{\alpha}_{kl} - q_{k,k}) \tag{4.73}$$

or

$$\varrho\dot{s} = \frac{1}{\vartheta}\sigma_{ij}^{(d)}d_{ij} + \frac{1}{\vartheta}\beta_{kl}^{(d)}\dot{\alpha}_{kl} - \frac{\vartheta_{,k}}{\vartheta^2}q_k - \left(\frac{q_k}{\vartheta}\right)_{,k}. \tag{4.74}$$

The significance of these transformations becomes evident as soon as we integrate (4.74) over the whole body V. On account of (2.76) and of the theorem of Gauss (1.79), applied to the last integral, we obtain

$$\dot{S} = \int \varrho\dot{s}\,\mathrm{d}V = \int \frac{1}{\vartheta}\sigma_{ij}^{(d)}d_{ij}\,\mathrm{d}V + \int \frac{1}{\vartheta}\beta_{kl}^{(d)}\dot{\alpha}_{kl}\,\mathrm{d}V - \int \frac{\vartheta_{,k}}{\vartheta^2}q_k\,\mathrm{d}V - \int \frac{q_k}{\vartheta}\nu_k\,\mathrm{d}A. \tag{4.75}$$

The first integral on the right is obviously the rate of entropy production due to the power of the dissipative stresses, which, referred to the unit mass, is given by (4.38). The second integral is the rate of entropy production due to the change of the internal parameters; it corresponds to (4.40). The third term (minus-sign included) is to be interpreted as the rate of entropy production due to heat exchange within V, for the last term is the entropy supply (4.57) across A. The entropy production within V has thus three different sources and is given by the first three terms. The second fundamental law (4.64) may now be stated in the global form

$$\dot{S}^{(i)} = \int \frac{1}{\vartheta} \sigma_{ij}^{(d)} d_{ij} \, \mathrm{d}V + \int \frac{1}{\vartheta} \beta_{kl}^{(d)} \dot{\alpha}_{kl} \, \mathrm{d}V - \int \frac{\vartheta_{,k}}{\vartheta^2} q_k \, \mathrm{d}V \geqslant 0, \tag{4.76}$$

which is far more transparent than (4.66).

Since V is arbitrary, we obtain from (4.76) the local form

$$\varrho \vartheta \dot{s}^{(i)} = \sigma_{ij}^{(d)} d_{ij} + \beta_{kl}^{(d)} \dot{\alpha}_{kl} - \frac{\vartheta_{,k}}{\vartheta} q_k \geqslant 0 \tag{4.77}$$

of the second law, sometimes referred to as the *Clausius–Duhem inequality*. It could have been derived directly by inserting (4.74) in (4.67). The first two terms are rates of work per unit volume, easily interpreted by means of (4.38) and (4.40). The third term reflects an additional entropy production, stemming from the heat flow across the element. The term itself has the dimension of a rate of work per unit volume, and it is obvious that q_k is to be interpreted as the corresponding velocity and

$$-\frac{\vartheta_{,k}}{\vartheta} = -(\ln \vartheta)_{,k} = \left(\ln \frac{1}{\vartheta}\right)_{,k}, \tag{4.78}$$

i.e. the negative gradient of the logarithmic temperature, as the corresponding force. In terms of a specific power of dissipation the last term in (4.77) yields

$$l^{(d3)} = -\frac{1}{\varrho} (\ln \vartheta)_{,k} q_k. \tag{4.79}$$

The second law requires that the left-hand side of (4.77) is non-negative as a whole. There are reasons to surmise that the inequality holds for the mechanical and the thermal terms separately. We will return to this question in Section 15.3. At present, let us just note that there is no experimental evidence to the contrary and that, in any event, the conditions

$$\sigma_{ij}^{(d)} d_{ij} + \beta_{kl}^{(d)} \dot{\alpha}_{kl} \geqslant 0, \qquad \vartheta_{,k} q_k \leqslant 0 \tag{4.80}$$

may be introduced as postulates restricting the field of continua to be treated in the next few chapters.

CHAPTER 5

MATERIAL PROPERTIES

The laws established in the preceding chapters are valid for arbitrary continua. There is a single exception: the inequalities (4.80) are stronger than second law in the form (4.77), and it is therefore possible that they exclude certain types of continua. On the other hand, it is clear that we need more than the general relations obtained so far if we want to treat specific problems of motion. We know that different materials behave differently in a given situation, and we conclude that our next task is to describe the various materials by means of relations connecting the stresses with their kinematic *response*, i.e. with the deformation of the body.

Equations or inequalities describing the response of a material are called *constitutive relations*. Each of them defines a certain class of materials, and by means of a sufficient number of such relations it is possible to describe a given material with any desired accuracy. It is the purpose of the present chapter to establish a preliminary classification, which will be refined in various directions in the following chapters. The first classification of this type has been given in terms of group theory by Noll [8]. Some of his concepts will be used here; others will be replaced by physical definitions which, thanks to the thermomechanical approach, will turn out to be surprisingly simple.

5.1. Basic concepts

Throughout this book we will restrict ourselves to *simple bodies*. They are defined by constitutive relations connecting local quantities, e.g. the strain tensor at a given point with the local stresses or the heat flow with the local temperature gradient. We thus exclude cases where the response of a given mass element is influenced by elements that are not in direct contact with it.

On the other hand, we do not exclude materials in which the response at a given time depends on the previous history, but we will describe this history by internal parameters.

In order to classify these simple bodies, we return to Section 4.2, noting that the stress tensor may be decomposed, according to (4.36), into its quasiconservative and dissipative parts. On account of $(4.37)_1$, the quasiconservative stress is $\sigma_{ij}^{(q)} = \varrho\, \partial\psi/\partial\varepsilon_{ij}$, where ϱ is the density and where the specific free energy is a state function $\psi\,(\varepsilon_{ij}, \alpha_{kl}, \vartheta)$. It follows in particular that $\sigma_{ij}^{(q)}$ is independent of the deformation rate.

If, on the other hand, we write (4.77) in terms of a *specific dissipation function* φ, we obtain, as in (4.29),

$$\vartheta \dot{s}^{(i)} = \varphi(d_{ij}, \dot{\alpha}_{kl}, q_m, \dot{\vartheta}, \varepsilon_{ij}, \alpha_{kl}, \vartheta) \geqslant 0. \tag{5.1}$$

The presence of the first three arguments in φ follows immediately from (4.77); the other ones are explained by the fact that φ, as a part of $\vartheta\dot{s}$, depends on the state variables and their material derivatives. The essential arguments in φ are d_{ij}, $\dot{\alpha}_{kl}$, and q_m; they are concerned with the irreversible process itself whereas ε_{ij}, α_{kl}, ϑ are mere state variables and $\dot{\vartheta}$, which might be replaced by $q_{m,\,m}$, may be given arbitrary values by superposing a reversible process of pure heating. In the simple applications to be treated starting with Chapter 9 the specific dissipation function depends on d_{ij}, $\dot{\alpha}_{kl}$, and q_m alone, and for the principal discussions of the last chapters it is always possible to assume that $\dot{\vartheta}$, ε_{ij}, α_{kl}, and ϑ are prescribed. The quantities $\sigma_{ij}^{(d)}$, $\beta_{kl}^{(d)}$, and $-\vartheta_{,k}$, which, in essence, represent the dissipative forces, depend on the same variables as φ, and the connection must be such that (4.77) is satisfied and that, provided the material obeys separate inequalities (4.80), the total rate of work of $\sigma_{ij}^{(d)}$ and $\beta_{ij}^{(d)}$, in particular, is non-negative.

Let us now define the *elastic body* by the identities

$$\alpha_{kl} = 0, \qquad \sigma_{ij}^{(d)} = 0, \tag{5.2}$$

i.e. by the conditions that there are no internal parameters nor dissipative stresses. It follows that the stress tensor is purely quasiconservative and that, on account of (4.37) and (4.34), the free energy and all other state functions including the stresses are functions of the strain and the temperature alone.

Conversely, the *purely viscous body* may be defined by the identities

$$\alpha_{kl} = 0, \qquad \sigma_{ij}^{(q)} = 0, \tag{5.3}$$

i.e. by the conditions that there are no internal parameters nor quasiconservative stresses. Here the stress tensor is purely dissipative, depending, like the dissipation function, at least on the deformation rate. On account of (4.37) and (4.34), the free energy, the entropy, and the internal energy are functions of the temperature alone.

The materials just mentioned are important limiting cases. In general, both parts $\sigma_{ij}^{(q)}$ and $\sigma_{ij}^{(d)}$ of the stress tensor are different from zero; besides, the description of many materials requires the use of internal parameters α_{kl}. Since in this case the α_{kl} are arguments of ψ in $(4.37)_1$, the subdivision of σ_{ij} into $\sigma_{ij}^{(q)}$ and $\sigma_{ij}^{(d)}$ depends on the choice of the internal parameters. We will consider this point in more detail in Section 11.3. For the definition of the solid and the fluid states, let us restrict ourselves provisionally to materials without internal parameters.

According to Section 1.3, the stress tensor can always be decomposed into an isotropic tensor $(\sigma_{kk}/3)\delta_{ij}$ and a deviator σ'_{ij}. Similar decompositions are possible for $\sigma_{ij}^{(q)}$ and $\sigma_{ij}^{(d)}$. In certain bodies one or the other of the four partial stress tensors obtained in this manner may be identically zero; all known materials without internal parameters, however, are capable of sustaining an isotropic quasiconservative stress, $(\sigma_{kk}^{(q)}/3)\delta_{ij}$. In the case of a gas, e.g., this stress is necessary to keep its volume finite; in other bodies it represents the tendency to restore the original volume after a dilatation.

A body without internal parameters capable of sustaining deviatoric stresses $\sigma_{ij}^{(q)\prime}$ of the quasiconservative type is called a *solid*. Here the tensor $\sigma_{ij}^{(q)}$ renders the material resistant against illimited deformations; besides, it tends to restore the original volume and shape after deformation. If on the other hand a body without internal parameters is incapable of sustaining deviatoric stresses of the quasiconservative type or, in other words, if $\sigma_{ij}^{(q)\prime}$ is identically zero, the body is referred to as a *fluid*. Here illimited distortions are possible, and the tendency to restore the original shape is missing. Since the quasiconservative stress is isotropic, the specific free energy, according to one of the last statements in Section 4.2, is of the form $\psi(\varepsilon_{ii}, \vartheta)$ and hence is independent of the strain deviator.

It has been mentioned in the introduction to this chapter that there are other ways of classifying the various continua. The one used here appears particularly simple and convenient. It will be necessary to elaborate it in more detail, distinguishing, within the class of fluids, e.g. between liquids and gases. No matter what classification is used, it always happens that the status of certain materials appears to differ from the one suggested by naive intuition. With the classification adopted here it will turn out, e.g. that a perfectly plastic body (Chapter 10) is to be considered as a liquid.

For its simplicity and its role played in practice, the class of *isotropic materials* is particularly important. Let us define it by the condition that the constitutive relations are independent of the orientation of the coordinate system with respect to the material element. If, in particular, a constitutive equation connects two symmetric second-order tensors (e.g. the strain or the deformation rate on one hand and the quasiconservative or dissipative stress on the other) and if the relationship admits a power series expansion, its general form is given by (1.68), where the properties of the material are represented by the tensors C_{ij}, C_{ijkl}, ... In general, the components of these tensors depend on the orientation of the coordinate system within the element, and it follows that (1.68) describes the response of an anisotropic body. In the case of an isotropic material the tensor function (1.68) itself must be isotropic in the sense of Section 1.3, i.e. of the special form (1.67), where f, g, and h are scalar functions of the basic invariants $t_{(1)}$, $t_{(2)}$, $t_{(3)}$, defining the properties of the material.

Developing the functions $f(t_{(1)}, t_{(2)}, t_{(3)})$, ... into power series of their arguments and noting that, according to (1.52), $t_{(1)}$, $t_{(2)}$, and $t_{(3)}$ are of the first, second, and third order in t_{ij} respectively, we are in a position to discuss approximations of (1.67), taking into account only terms up to a certain order. The simplest relevant approximation is of the order one. Here h is zero; g is to be replaced by its constant term, which may be denoted by 2μ, and f reduces to a linear function of $t_{(1)}$, say $\varkappa+\lambda t_{(1)}$, where $\varkappa$ and λ are constants. We thus obtain

$$s_{ij} = (\varkappa+\lambda t_{(1)})\delta_{ij}+2\mu t_{ij}. \tag{5.4}$$

It is generally possible to define the tensors s_{ij} and t_{ij} in such a way that

$$s_{ij} = 0 \text{ for } t_{kl} = 0. \tag{5.5}$$

It follows that $\varkappa = 0$ and that (5.4) therefore reduces to

$$s_{ij} = \lambda t_{kk}\delta_{ij}+2\mu t_{ij}, \tag{5.6}$$

where the constants λ and μ are typical for the body in consideration.

The functions describing the properties of a material or, in particular, the "constants" $\varkappa$, λ, μ, may still differ from element to element. If this is the case, the material is called *inhomogeneous*. We will restrict ourselves in this book to *homogeneous* bodies, characterized by properties that are the same for all material elements.

To a certain extent, the behavior of a material is determined by its specific free energy

$$\psi = \psi(\varepsilon_{ij}, \alpha_{kl}, \vartheta). \tag{5.7}$$

The special form that (5.7) assumes for a given continuum is called its *caloric equation of state*. Once it is known, the specific entropy follows from $(4.37)_2$ and the specific internal energy from (4.34). Equation $(4.37)_1$ is sometimes referred to as the *thermal equation of state*; it supplies the quasiconservative stresses $\sigma_{ij}^{(q)}$. By means of all these relations it is possible to treat reversible processes (e.g. the adiabatic behavior of elastic bodies to be discussed in Section 7.1).

The dissipative forces, as $\sigma_{ij}^{(d)}/\varrho$ or $\beta_{kl}^{(d)}/\varrho$, cannot be derived from the free energy. In general, we need more information to deal with irreversible processes. Incidentally, any process governed by a specific free energy that really depends on internal parameters α_{kl} is irreversible. In fact, the corresponding quasiconservative forces are given by (4.39), and from (4.42) we conclude that also the dissipative forces are non-zero.

Perhaps the simplest irreversible process is *heat conduction* in a continuum at rest. It is characterized by the heat flow vector q_k, dependent on the temperature gradient $\vartheta_{,k}$ and possibly on the other arguments of φ in (5.1) except d_{ij}. The simplest connection between the two vectors q_k and $\vartheta_{,k}$ is

$$q_k = -\lambda_{kl}\vartheta_{,l}, \tag{5.8}$$

where λ_{kl} is a second-order tensor the matrix of which, according to $(4.80)_3$, is positive definite. In general (5.8) describes an anisotropic body. In an isotropic material the tensor λ_{kl} itself must be isotropic and hence (P of Section 1.1) of the form $\lambda_{kl} = \lambda\delta_{kl}$. Thus, (5.8) reduces to *Fourier's law*

$$q_k = -\lambda\vartheta_{,k}. \qquad (\lambda > 0) \tag{5.9}$$

The scalar λ is the *thermal conductivity* of the material. It may depend on the state variables but is usually considered as a constant.

The material derivative of the internal energy $u(\varepsilon_{ij}, \alpha_{kl}, \vartheta)$ is

$$\dot{u} = \frac{\partial u}{\partial \varepsilon_{ij}} d_{ij} + \frac{\partial u}{\partial \alpha_{kl}} \dot{\alpha}_{kl} + \frac{\partial u}{\partial \vartheta} \dot{\vartheta}. \tag{5.10}$$

In a process of pure heating ($d_{ij} = 0$, $\dot{\alpha}_{kl} = 0$) it reduces to $\dot{u} = (\partial u/\partial\vartheta)\dot{\vartheta}$ or

$$\dot{u} = c\dot{\vartheta}, \quad \text{where} \quad c = \frac{\partial u}{\partial \vartheta} \tag{5.11}$$

is a state function called the *specific heat capacity* for constant strains and internal parameters. It follows that the specific internal energy may be written

$$u = \int c\,d\vartheta, \tag{5.12}$$

where the integration is to be carried out for constant values of ε_{ij} and α_{kl}.

In pure heat conduction the first fundamental law (4.63) reduces to

$$\varrho\dot{u} = -q_{k,k}. \tag{5.13}$$

Let us assume now that the material is isotropic and that its internal parameters are constant. Inserting (5.11) and (5.9) in (5.13), we obtain $\varrho c\dot{\vartheta} = \lambda\vartheta_{,kk}$ or, equivalently,

$$\varrho c\vartheta_{,0} = \lambda\vartheta_{,kk}. \tag{5.14}$$

This is the *differential equation of heat conduction* in an isotropic rigid material.

5.2. Fluids without internal parameters

The fluid without internal parameters has been defined in Section 5.1 by the absence of a quasiconservative stress deviator, i.e. by the conditions $\alpha_{kl} = 0$ and $\sigma_{ij}^{(q)\prime} = 0$. It follows that

$$\sigma_{ij}^{(q)} = -p\delta_{ij}, \tag{5.15}$$

where p is a scalar, called the *hydrostatic pressure*, generally dependent on the material, the deformation, and the temperature. Besides, p is positive provided we neglect effects like surface tension or capillarity.

Small deformations can be represented by the strain tensor ε_{ij} defined in Section 2.2. Here the hydrostatic pressure is given, according to (4.52), by

$$p = -\varrho\frac{\partial\psi}{\partial\varepsilon_{kk}}, \tag{5.16}$$

where $\psi(\varepsilon_{kk}, \vartheta)$ is the specific free energy. Since ε_{kk} and ϱ determine one another, p is a function of ε_{kk} and ϑ or of ϱ and ϑ. Provided (5.16) is the only constitutive equation (i.e. if the dissipative stresses vanish identically and no heat flow is present), the fluid is isotropic in the sense of Section 5.1.

If the dilatation ε_{kk}, referred to the stress-free state, is small compared to unity, the fluid is called a *liquid*. Here, the function $p(\varepsilon_{kk}, \vartheta)$ may be linearized with respect to ε_{kk}; the dependence on ϑ is usually negligible. We thus have

$$p = p_0 - K\varepsilon_{kk} \geqslant 0, \tag{5.17}$$

where ε_{kk} is now referred to the state under pressure p_0, and K is a constant called the *bulk modulus*. An important limiting case, often used as an idealization, is the *incompressible liquid*, defined by $K \to \infty$ so that $\varepsilon_{kk} \to 0$. Since the last term becomes indefinite, (5.17) is useless in this case. The hydrostatic pressure becomes a reaction (in the same manner as a spring force if the spring constant tends to infinity).

If, on the other hand, a fluid is capable of arbitrary dilatations, it is called a *gas*. Here it is convenient to express ψ and p as functions of ϱ and ϑ. The corresponding relations

$$\psi = \psi(\varrho, \vartheta) \quad \text{and} \quad p = p(\varrho, \vartheta) \tag{5.18}$$

are the caloric and the thermal equations of state.

Inviscid fluids are characterized by the absence of dissipative stresses:

$$\sigma_{ij}^{(d)} = 0. \tag{5.19}$$

It follows from (5.2) that inviscid fluids are elastic, provided no internal parameters are present. In the case of a liquid, (5.15), (5.17), and (5.19) are the constitutive equations; for gases, $(5.18)_2$ takes the place of (5.17).

Viscous fluids, on the other hand, are characterized by the presence of a dissipative stress tensor $\sigma_{ij}^{(d)}$, representing the influence of internal friction. In general $\sigma_{ij}^{(d)}$ depends on d_{ij} and possibly on the other arguments of φ in (5.1). For practical purposes it is usually sufficient to assume that

$$\sigma_{ij}^{(d)} = \sigma_{ij}^{(d)}(d_{kl}, \vartheta). \tag{5.20}$$

Equation (5.20) replaces (5.19) as a constitutive equation, and the stress tensor is obtained, according to (4.36), by adding the quasiconservative and disspiative stresses, (5.15) and (5.20) respectively.

It follows from Section 5.1 that in an *isotropic fluid* the function (5.20) itself must be isotropic, i.e. of the form (1.67). Hence,

$$\sigma_{ij}^{(d)} = f\delta_{ij} + gd_{ij} + hd_{ik}d_{kj}, \tag{5.21}$$

where f, g, and h are functions of the basic invariants $d_{(1)}, \ldots$, dependent on ϑ and the material. Developing the right-hand side, we find as in (5.6) that the simplest form of (5.21) satisfying the obvious condition

$$\sigma_{ij}^{(d)} = 0 \qquad \text{for } d_{kl} = 0 \tag{5.22}$$

is

$$\sigma_{ij}^{(d)} = \lambda d_{kk}\delta \; + 2\mu d_{ij}. \tag{5.23}$$

Incidentally, (5.22), provided the function (5,21) is continuous, is a consequence of the second law $(4.80)_1$. Equation (5.23) defines the so-called *Newtonian fluid.* Decomposing the deformation rate into an isotropic tensor and a deviator, we obtain

$$\sigma_{ij}^{(d)} = \left(\lambda + \frac{2}{3}\,\mu\right) d_{kk}\delta_{ij} + 2\mu d_{ij} \tag{5.24}$$

in place of (5.23). The so-called *viscosity coefficients* λ and μ depend on the material and on the teperature. They describe the internal friction, and it becomes clear from (5.24) that μ and $\lambda+\frac{2}{3}\mu$ are characteristic for the viscosity effects in the cases of distortion or dilatation respectively. On account of (5.15) and (5.24), the total mean normal stress is

$$\frac{1}{3}\,\sigma_{kk} = -p + \left(\lambda + \frac{2}{3}\,\mu\right) d_{kk}. \tag{5,25}$$

It is not identical with the negative hydrostatic pressure but also contains a viscous contribution.

In the special case where the so-called *bulk viscosity* $\lambda+\frac{2}{3}\,\mu$ is zero, (5.24) reduces to the simple constitutive equation

$$\sigma_{ij}^{(d)} = 2\mu d'_{ij} \tag{5.26}$$

of the so-called *Navier–Stokes fluid.* The same result is obtained if the Newtonian fluid is assumed incompressible. In this last case d'_{ij} might be replaced by d_{ij} in (5.26). Besides, p becomes a reaction, doing no work, and it follows that the quasiconservative stress tensor $-p\delta_{ij}$ may as well be interpreted as a dissipative stress. Adopting this point of view, we obtain $\sigma_{ij}^{(q)} = 0$, and the fluid becomes purely viscous in the sense of Section 5.1.

The foregoing classification is based on the constitutive equations of the various fluids. In Chapters 6 and 9 we will list all equations necessary to treat flows of inviscid and viscous fluids, especially liquids, and we will apply them to specific problems. Chapter 8 will be concerned with gases and Chapter 10 with the plastic body, a limiting case of the viscous liquid. In Chapter 11 the discussion will be extended to viscoelastic fluids, i.e. to fluids with internal parameters.

Problems

1. The ideal gas is a fluid without internal parameters, defined by the caloric equation of state

$$\psi = \int c^{(v)}\,\mathrm{d}\vartheta - \vartheta \int \frac{c^{(v)}}{\vartheta}\,\mathrm{d}\vartheta + R\vartheta \ln \varrho,$$

where $c^{(v)}(\vartheta)$ is the specific heat capacity for constant volume and R is the gas constant, referred to the unit mass. Find the specific entropy, the specific internal energy, and the thermal equation of state $p = p(\varrho, \vartheta)$.

2. Generalize (5.23) by taking second-order terms in (5.21) into account. Specialize the result for an incompressible liquid and make use of the fact that in this case $\sigma_{ij}^{(d)}$ is a deviator.

5.3. Elastic solids

The elastic body has been defined in Section 5.1 by the absence of internal parameters and dissipative stresses, i.e. by the conditions $\alpha_{kl} = 0$ and $\sigma_{ij}^{(d)} = 0$, the solid by the presence of a quasiconservative stress deviator, i.e. by $\sigma_{ij}^{(q)\prime} \neq 0$. Restricting ourselves to small deformations, we obtain the specific free energy in the form

$$\psi = \psi(\varepsilon_{ij}, \vartheta). \tag{5.27}$$

The specific entropy and the stress tensor $\sigma_{ij} = \sigma_{ij}^{(q)}$ follow from (4.37) and the specific internal energy from (4.34). Inserting $d_{ij} = \dot{\varepsilon}_{ij}$, $\dot{\alpha}_{kl} = 0$, and $\sigma_{ij}^{(q)} = \sigma_{ij}'$ in (4.71), we obtain

$$\sigma_{ij}\dot{\varepsilon}_{ij} = \varrho(\dot{\psi} + s\dot{\vartheta}) = \varrho(\dot{u} - \vartheta\dot{s}). \tag{5.28}$$

From the first equality (5.28) we deduce

$$\frac{1}{\varrho}\sigma_{ij}\,\mathrm{d}\varepsilon_{ij} - s\,\mathrm{d}\vartheta = \mathrm{d}\psi(\varepsilon_{ij}, \vartheta), \tag{5.29}$$

where the differentials denote material increments. It follows that

$$\frac{1}{\varrho}\sigma_{ij} = \frac{\partial\psi}{\partial\varepsilon_{ij}}, \qquad s = -\frac{\partial\psi}{\partial\vartheta} \tag{5.30}$$

in accordance with (4.37) for $\sigma_{ij} = \sigma_{ij}^{(q)}$. The specific free energy appears as a potential for the tensor $-\sigma_{ij}/\varrho$ representing the internal forces per unit mass, and for s. The representation (5.30) is generally valid but particularly suitable for *isothermal processes*, defined by $\vartheta = \text{const}$. Here ψ is a function of ε_{ij} alone. Referring the strains to the stress-free state and developing ψ in its vicinity, we have

$$\psi(\varepsilon_{ij}) = \psi_0 + \left(\frac{\partial \psi}{\partial \varepsilon_{ij}}\right)_0 \varepsilon_{ij} + \ldots, \tag{5.31}$$

where the suffix 0 denotes the stress-free state. On account of $(5.30)_1$, the coefficients of the linear terms are zero. It follows that $\psi - \psi_0$ is of the second order and that, therefore, in the identity

$$\frac{\partial}{\partial \varepsilon_{ij}}[\varrho(\psi - \psi_0)] = \varrho \frac{\partial \psi}{\partial \varepsilon_{ij}} + (\psi - \psi_0)\frac{\partial \varrho}{\partial \varepsilon_{ij}} \tag{5.32}$$

the last term is to be dropped in the present approximation. Thus, ϱ may be considered as a constant, and $(5.30)_1$ becomes equivalent to

$$\sigma_{ij} = \frac{\partial W}{\partial \varepsilon_{ij}}, \tag{5.33}$$

where

$$W(\varepsilon_{ij}) = \varrho(\psi - \psi_0) \tag{5.34}$$

may be interpreted as the *strain energy* per unit volume.

It has been pointed out at the end of Section 4.1 that it is sometimes convenient to replace the temperature as an independent state variable by the specific entropy. Indicating this transition to new arguments in the state functions by primes, we obtain, e.g.,

$$u = u'(\varepsilon_{ij}, s) \tag{5.35}$$

for the specific internal energy. The second equality (5.28) yields

$$\frac{1}{\varrho}\sigma_{ij}\,\mathrm{d}\varepsilon_{ij} + \vartheta\,\mathrm{d}s = \mathrm{d}u'(\varepsilon_{ij}, s). \tag{5.36}$$

It follows that

$$\frac{1}{\varrho}\sigma_{ij} = \frac{\partial u'}{\partial \varepsilon_{ij}}, \qquad \vartheta = \frac{\partial u'}{\partial s}. \tag{5.37}$$

The specific internal energy thus appears as a potential for the $-\sigma_{ij}/\varrho$ and for $-\vartheta$. The representation (5.37) is again generally valid but particularly suitable for *adiabatic processes*, defined by $q_k = 0$, since, on account of (4.73), they are also isentropic in the present case. Here u' is a function of ε_{ij} alone, and an argumentation similar to the one following the expansion (5.31) shows that the stresses are still given by (5.33) provided the strain energy is now defined by

$$W(\varepsilon_{ij}) = \varrho(u' - u_0') \tag{5.38}$$

in place of (5.34).

Let us assume now that the vicinity of an elastic solid is maintained at a constant temperature. On account of the heat exchange inside the body and with its surroundings, the temperature within and outside will be practically the same provided the deformations are sufficiently slow, and the process will be close to isothermal. If, on the other hand, the deformations are sufficiently fast, there is no time for an appreciable heat exchange, and the process will be practically adiabatic. In the theory of elasticity one usually restricts oneself to these two limiting cases. Here (5.33) is valid; the internal forces are conservative, and W is the potential of the negative stresses $-\sigma_{ij}$.

If, in addition, the body is *isotropic*, the strain energy W is a function of the basic strain invariants $\varepsilon_{(1)}, \ldots$ alone. Moreover, the stress tensor is an isotropic function of the strains, i.e. of the form (1.67). Hence,

$$\sigma_{ij} = f\delta_{ij} + g\varepsilon_{ij} + h\varepsilon_{ik}\varepsilon_{kj}, \tag{5.39}$$

where f, g, and h are functions of the invariants $\varepsilon_{(1)}, \ldots$, dependent on the material (and possibly on the temperature). On account of (5.33), however, the functions f, g, and h are not arbitrary but determined by the strain energy W.

Developing the right-hand side of (5.39), we find as in (5.6) that the simplest form satisfying the condition

$$\sigma_{ij} = 0 \text{ for } \varepsilon_{kl} = 0 \tag{5.40}$$

is

$$\sigma_{ij} = \lambda\varepsilon_{kk}\delta_{ij} + 2\mu\varepsilon_{ij}. \tag{5.41}$$

This is the general form of *Hooke's law*. The body with the constitutive equation (5.41) is called a *Hookean solid*; λ and μ are referred to as *Lamé's constants*. It is interesting to note that, with (5.41), the existence of a strain energy is automatically guaranteed; in fact, (5.41) follows from (5.33) with

$$W = \frac{\lambda}{2}\,\varepsilon_{ii}\varepsilon_{jj} + \mu\varepsilon_{ij}\varepsilon_{ij}. \tag{5.42}$$

In Chapter 7 the results obtained here will be applied to problems concerned with elastic solids, including anisotropic bodies and processes that are neither isothermal nor adiabatic. The extension of Hooke's law (5.41) to the nonlinear case (physical nonlinearity) is of little practical importance compared to the fact that, for large deformations, the strain tensor (2.33) cannot be used any more (geometric nonlinearity). Chapter 13 will deal with this problem. Inelastic solids or, more precisely, solids with internal parameters, will be discussed in Chapter 11.

Problems

1. Solve Hooke's law (5.41) for the strain components.
2. Express the strain energy W in terms of the stress components.

CHAPTER 6

IDEAL LIQUIDS

In the preceding chapters we have established all theorems relevant for the treatment of deformations and motions of continua. We now propose to apply these theorems to various materials and to the solution of specific problems. The present chapter and the following ones will be concerned with this task, beginning with simple situations and proceeding to more complicated ones.

The principal object of the present chapter will be the *ideal liquid*, i.e. the inviscid incompressible liquid without internal parameters. It is worthwhile, however, to include the inviscid gas in the first section in order to establish a few results which will remain valid in a more general context.

6.1. Basic equations

The constitutive equations of an *inviscid fluid without internal parameters* have been derived in Section 5.2; they are given by (5.15), $(5.18)_2$, and (5.19). The dissipative stress is identically zero; the quasiconservative stress reduces to the hydrostatic pressure, and the pressure is connected with the density and the temperature by the thermal equation of state, $p = p(\varrho, \vartheta)$. The fluid is mechanically isotropic.

In many practically important cases it is possible to eliminate the temperature from the thermal equation of state. In an isothermal process, e.g., ϑ is a constant. Besides, we will see in Section 8.1 that also in adiabatic processes ϑ can be eliminated. Let us restrict ourselves to cases where this is possible, i.e. where $p = p(\varrho)$, and let us refer to such cases as *barotropic*.

The motion of the fluid we are considering here is subject to the continuity equation in the form (2.69) or (2.70), to the theorem of linear momentum, (3.9), and to the constitutive equations enumerated above. The differential equations governing the motion are thus

$$\dot{\varrho}+\varrho v_{j,j} = \varrho_{,0}+(\varrho v_j)_{,j} = 0, \tag{6.1}$$

$$\varrho a_i = \varrho \dot{v}_i = \varrho(v_{i,0}+v_{i,j}v_j) = \varrho f_i+\sigma_{ij,j}, \tag{6.2}$$

and

$$\sigma_{ij} = -p\delta_{ij}, \quad \text{where} \quad \varrho = \varrho(p) \tag{6.3}$$

is the barotropic equation. There are altogether $1+3+6+1 = 11$ differential equations (of orders 0 through 1) for the 11 unknowns ϱ, p, v_i, and σ_{ij}, and it is clear that in any particular case they must be supplemented by appropriate boundary and initial conditions.

We note that the differential equations (6.1) through (6.3) do not contain any of the thermodynamic variables ϑ, q_i. The problem is thus purely mechanical. It is true that the material flow v_i may be accompanied by a heat flow q_i and by temperature differences. If the fields $q_i(x_j)$ and $\vartheta(x_j)$ are of interest, they can be obtained in a second step from Fourier's law, the first fundamental law of thermodynamics, and from the corresponding boundary and initial conditions. The entire problem thus resolves into a mechanical and a thermodynamic part.

In order to simplify the system (6.1) through (6.3), let us insert the stress tensor $(6.3)_1$ in (6.2). We thus obtain the equation

$$a_i = \dot{v}_i = v_{i,0}+v_{i,j}v_j = f_i-\frac{1}{\varrho}p_{,i}, \tag{6.4}$$

usually referred to as *Euler's differential equation.* Another simplification results from the introduction of the so-called *pressure function*

$$P(p) = \int_{p_0}^{p} \frac{dp'}{\varrho(p')}, \tag{6.5}$$

where the prime is used to distinguish the integration variable from the upper limit of the integral, and where p_0 is an arbitrary reference pressure. In the special case of an incompressible liquid, ϱ is constant, and the pressure function becomes

$$P(p) = \frac{p-p_0}{\varrho_0}. \tag{6.6}$$

In any event, the gradient of the pressure function is

$$P_{,i} = \frac{\mathrm{d}P}{\mathrm{d}p} p_{,i} = \frac{1}{\varrho} p_{,i}, \tag{6.7}$$

so that Euler's equation (6.4) assumes the form

$$a_i = \dot{v}_i = v_{i,0} + v_{i,j} v_j = f_i - P_{,i}. \tag{6.8}$$

Writing (6.8) in symbolic notation and making use of (2.43), we obtain

$$\boldsymbol{a} = \frac{\partial \boldsymbol{v}}{\partial t} + 2\boldsymbol{w} \times \boldsymbol{v} + \operatorname{grad} \frac{v^2}{2} = \boldsymbol{f} - \operatorname{grad} P, \tag{6.9}$$

where $\boldsymbol{w}$ is the vorticity (2.8) of the velocity field.

In the particular case of a fluid at rest, the continuity equation (6.1) is trivial, and Euler's equation (6.4) reduces to the *equilibrium condition*

$$\frac{1}{\varrho} p_{,i} = P_{,i} = f_i \tag{6.10}$$

or, in symbolic notation,

$$\frac{1}{\varrho} \operatorname{grad} p = \operatorname{grad} P = \boldsymbol{f}, \tag{6.11}$$

sometimes referred to as the *fundamental equation of hydrostatics.* In general, the specific body force is prescribed, and the problem is to find the density and pressure fields. Since (6.11) represents three scalar equations, the problem seems to be over-determined. It follows from the second equality (6.11), however, that $\operatorname{curl} \boldsymbol{f} = 0$. In other words: the fluid cannot be at rest unless the force field is irrotational. Thus, $\boldsymbol{f}$ has a potential U such that $f_i = -U_{,i}$. Equation (6.10) now yields $P_{,i} = -U_{,i}$ or, equivalently,

$$P + U = \text{const.} \tag{6.12}$$

Except for an additive constant, the field of the pressure function is determined by (6.12), and the fields of p and ϱ follow from $(6.3)_2$ and (6.5).

Returning to the case of a fluid in motion, let us still assume that the specific body force has a potential. Equation (6.8) then becomes

$$a_i = -(U+P)_{,i}, \tag{6.13}$$

and the function $U+P$ may be interpreted as an acceleration potential. The pressure function, on account of its definition (6.5), is single-valued.

Provided that also $U(x_j, t)$ is a single-valued function, the condition mentioned in the last sentence of Section 2.3 is satisfied, and in consequence the material derivative of the circulation Γ is zero for any closed curve. This is *Thomson's* (Lord Kelvin's) *vortex theorem*, stating that in an inviscid barotropic fluid the specific body force of which has a single-valued potential the circulation is constant along any closed material curve.

Helmholtz's vortex theorems, although established prior to the one of Thomson, may be derived as consequences of it. Consider two simply-

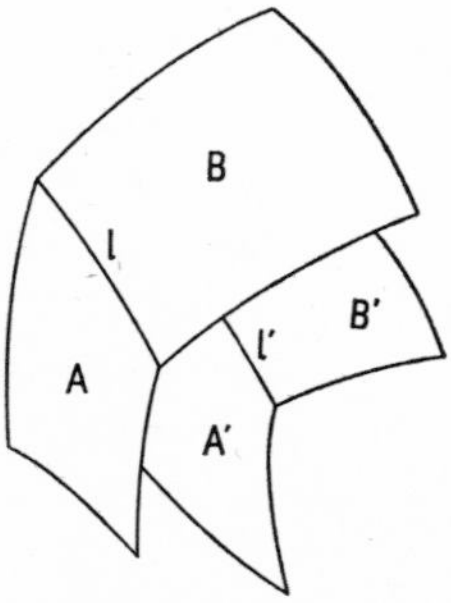

Fig. 6.1 Material vortex surfaces

connected material surfaces A and B (Fig. 6.1) which at time t consist of vortex lines and hence are vortex surfaces. Their material intersection l is a vortex line. At time t' the positions of the two surfaces and their intersection are A', B', and l' respectively. According to Thomson's theorem, the circulation along any closed curve on A' or B' is still zero; thus, A' and B' are still vortex surfaces and l' is still a vortex line. It follows that vortex surfaces may be considered as material surfaces and vortex lines as material curves. In a similar manner, vortex filaments may be conceived as material filaments. Their vortex strength is not only the same for all cross sections as shown at the end of Section 2.4 but also independent of time.

An essential consequence of these theorems is the fact that in an inviscid barotropic fluid with a specific body force that has a single-valued potential, a flow remains irrotational, provided it has been irrotational at any given time and in particular if it has developed from a state of rest. This property is responsible for the important role played by irrotational flows.

Inserting the potential U of $\boldsymbol{f}$ in (6.9), we obtain

$$\frac{\partial \boldsymbol{v}}{\partial t} + 2\boldsymbol{w} \times \boldsymbol{v} = -\operatorname{grad}\left(U + P + \frac{v^2}{2}\right). \tag{6.14}$$

Provided the flow is irrotational, the second term on the left is zero. At the same time the velocity may be derived in the form

$$\boldsymbol{v} = \operatorname{grad} \varphi, \tag{6.15}$$

from a velocity potential $\varphi(x_j, t)$, and the motion is therefore a potential flow. Since partial differentiation with respect to t and x_j is interchangeable, (6.14) and (6.15) yield

$$\operatorname{grad}\left(\frac{\partial \varphi}{\partial t} + U + P + \frac{v^2}{2}\right) = 0 \tag{6.16}$$

or

$$\frac{\partial \varphi}{\partial t} + U + P + \frac{v^2}{2} = f(t), \tag{6.17}$$

where f(t) is an arbitrary function and where v^2 might be expressed in terms of φ by means of (6.15). For the representation of the flow field, an additive time function in the velocity potential is irrelevant. We are thus allowed to drop the function $f(t)$ and to write

$$\varphi_{,0} + U + P + \frac{v^2}{2} = 0 \tag{6.18}$$

in place of (6.17).

6.2. Steady potential flows

We have restricted ourselves in Section 6.1 to barotropic inviscid fluids with specific body forces having a single-valued potential, and we have added the assumption that the flow be irrotational. Under these conditions the motion is governed by the continuity equation (6.1), Euler's equation in the integrated form (6.18) [$\boldsymbol{v}$ and φ being connected by (6.15)] and by the constitutive equations (6.3).

Let us now introduce three additional restrictions. In the first place, we limit ourselves to the case of an *incompressible* and hence *ideal liquid.* Here the material derivative of the density ϱ is zero, and it follows from the assumption of homogeneity that ϱ is constant with respect to place and time. We secondly neglect the body force, setting $U = 0$, and we finally assume that the flow is *steady*, i.e. independent of time. The continuity equation then reduces to

$$v_{j,j} = 0 \quad \text{or} \quad \operatorname{div} \boldsymbol{v} = 0. \tag{6.19}$$

In terms of the velocity potential, (6.19) yields the *Laplace equation*

$$\varphi_{,jj} = 0 \quad \text{or} \quad \Delta\varphi = 0. \tag{6.20}$$

In (6.18) the first two terms are zero, and since the pressure function assumes the simple form (6.6), we obtain the so-called *Bernoulli equation*

$$\frac{p}{\varrho} + \frac{v^2}{2} = \text{const} \tag{6.21}$$

or

$$\frac{p}{\varrho} + \frac{v^2}{2} = \frac{p_0}{\varrho} + \frac{v_0^2}{2}, \tag{6.22}$$

where p_0 and v_0 denote the pressure and the velocity respectively in an arbitrary reference point, e.g. at infinity. The constitutive equation $(6.3)_2$ is to be dropped, and p becomes a reaction.

In (6.22) the velocity might be expressed in terms of the potential φ. The problem is thus reduced to a pair of scalar differential equations, (6.20) and (6.22), for φ and p. They must be supplemented, not by initial conditions since the flow is steady, but by boundary conditions. On a free surface, e.g., the pressure is to be prescribed. Along a solid boundary at rest, on the other hand, the flow is tangential. The corresponding boundary condition reads

$$v_j \nu_j = \boldsymbol{v} \cdot \boldsymbol{\nu} = 0 \tag{6.23}$$

or, in terms of φ,

$$\varphi_{,j} \nu_j = \frac{\partial \varphi}{\partial \nu} = 0, \tag{6.24}$$

where ν_j is the unit normal of the boundary and $\partial\varphi/\partial\nu$ denotes the so-called normal derivative of φ.

Neither the Laplace equation (6.20) nor the boundary condition (6.24) contains the pressure p. It follows that in flows without a free boundary the problem resolves once more. The function $\varphi(x_j)$ is obtained from (6.20) and (6.24) as a solution of the second boundary value problem of potential theory. In the case of a region V with a regular boundary A, we have curl $\boldsymbol{v} = 0$ and div $\boldsymbol{v} = 0$ in V and $\boldsymbol{v} \cdot \boldsymbol{\nu} = 0$ on A, and for simply-connected regions V it has been shown at the end of Section 1.4 that the solution is unique. Once it is known, the velocity field follows from (6.15) and the pressure field from (6.22).

The simplest example is the *uniform flow*, e.g. in the direction x_1, defined by the velocity components $v_1 = u$, $v_2 = v_3 = 0$, where u is a constant.

The corresponding potential is $\varphi = ux_1$, and it is easy to verify that it satisfies the Laplace equation (6.20). The pressure obtained from (6.21) is a constant, $p = p_0$. The flow may be considered as a motion covering the entire three-dimensional space or as a flow inside a cylinder of arbitrary cross section with generators parallel to the axis x_1; in fact, in the last case the boundary condition (6.23) is satisfied on the whole boundary.

Another example is the uniform flow disturbed by a sphere at rest (Fig. 6.2) with center O and radius a. Here the potential is

$$\varphi = ux_1\left(1+\frac{1}{2}\frac{a^3}{r^3}\right), \quad \text{where} \quad r^2 = x_jx_j \tag{6.25}$$

is the square of the distance from O. A simple calculation (P1) shows that the Laplace equation (6.20) is satisfied everywhere outside the origin O. Since φ is rotationally symmetric, the flow occurs in planes containing the axis x_1, and the problem can be studied in the plane x_1, x_2. The velocity components could be obtained directly from (6.25). We may as well use polar coordinates r, ϑ in the plane x_1, x_2 and write (6.25) in the form

$$\varphi = u \cos\vartheta\left(r+\frac{1}{2}\frac{a^3}{r^2}\right). \tag{6.26}$$

A simple calculation (P2) yields the velocity components in polar coordinates,

$$v_r = \frac{\partial\varphi}{\partial r} = u\cos\vartheta\left(1-\frac{a^3}{r^3}\right), \qquad v_\vartheta = \frac{1}{r}\frac{\partial\varphi}{\partial\vartheta} = -u\sin\vartheta\left(1+\frac{1}{2}\frac{a^3}{r^3}\right). \tag{6.27}$$

From (6.27) it follows that $v_r = 0$ for $r = a$, i.e. that the flow does not cross the meridian of Figure 6.2. Thus, (6.27) may be interpreted as a motion either inside or outside the sphere. We are interested in the second case. Here the sphere may be considered as a solid body at rest, and since the potential (6.25) tends towards ux_1 with increasing distance from O, we are in fact dealing with what might be briefly, but not quite correctly, called a *uniform flow past a sphere*.

Evaluating (6.27) pointwise, we obtain the streamline pattern of Figure 6.2. Since $v_\vartheta = 0$ for $\vartheta = 0$ and $\vartheta = \pi$, the axis x_1, supplemented by the semi-circles *ACB* and *ADB*, is a streamline. In *A* and *B* both velocity com-

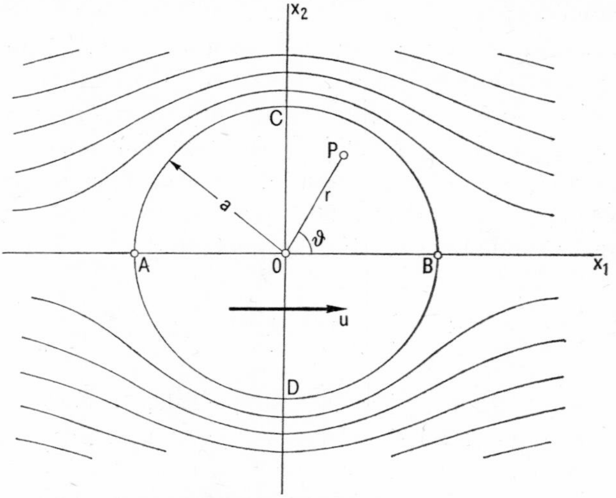

Fig. 6.2 Uniform flow past a sphere

ponents (6.27) are zero; the two points are therefore referred to as *stagnation points*. Along the sphere the magnitude of the velocity is

$$v = |v_\vartheta| = \frac{3}{2} u |\sin \vartheta| \qquad (r = a). \tag{6.28}$$

This confirms that $v = 0$ at A and B. On the other hand, the velocity at C or D turns out to be $\frac{3}{2}u$. In fact, since the flow through all planes perpendicular to the axis is the same, it is obvious that the velocities must exceed u in the equatorial plane x_2, x_3 of the sphere.

According to (6.22), the pressure distribution is given by

$$p_0 = p + \frac{\varrho}{2}(u^2 - v^2), \tag{6.29}$$

where p_0 denotes the pressure at infinity. Inserting (6.28) in (6.29), we obtain the pressure on the sphere,

$$p = p_0 + \frac{\varrho}{2} u^2 \left(1 - \frac{9}{4} \sin^2 \vartheta\right) \qquad (r = a). \tag{6.30}$$

In the two stagnation points the excess pressure over p_0 is $(\varrho/2)u^2$, whereas the pressure near C and D is less than p_0. The distribution (6.30) is symmetric with respect to the axes x_1 and x_2. It follows that, by reduction of the surface forces to O, we obtain neither a resultant force nor a moment; thus, the flow does not influence the sphere. Since the three-dimensional space outside the sphere is simply-connected, the solution obtained here is the only one. It can be shown to hold, as far as the absence of a resultant force is concerned, for finite bodies of arbitrary shape. This result, called *d'Alembert's paradox*, is not in accordance with experimental evidence and has long prevented the application of hydrodynamics to real flow problems, in particular to the problem of flight.

Problems

1. Verify that the potential (6.25) satisfies the Laplace equation (6.20).
2. Show that, in plane polar coordinates r, ϑ, the velocity components are $v_r = \partial\varphi/\partial r$, $v_\vartheta = (1/r)\,\partial\varphi/\partial\vartheta$.
3. Calculate the cartesian velocity components v_1, v_2 of the problem of Figure 6.2 directly from (6.25) and show that their polar components v_r, v_ϑ are the right-hand sides of (6.27).

6.3. The plane problem

A more realistic approach to the main problem of the last section, the explanation and calculation of the forces experienced by a body in an otherwise uniform flow, requires that one or more of the simplifying assumptions be dropped. The scope of this book does not allow us to go into details, but we will at least show that plane flows are not subject to d'Alembert's paradox.

Let us apply the basic equations of Section 6.2 to a plane flow defined by the velocity components $v_1(x_1, x_2)$, $v_2(x_1, x_2)$, $v_3 = 0$. It follows from Bernoulli's equation (6.22) that the pressure has the form $p(x_1, x_2)$. As far as the forces and amounts of flow are concerned, it is convenient to refer them to the unit of length in the direction x_3, i.e. to consider a layer of unit thickness extending along the plane x_1, x_2.

In Figure 6.3 P is an arbitrary point in the plane of flow, and C, C' are two arbitrary regular curves connecting the origin O with P; ds is a line element on C, and its unit normal ν_j is assumed to point to the right for an observer moving from O towards P. Let us denote the *flow* through C, i.e. the amount of liquid crossing C in the direction of ν_j per unit time, by ψ. The divergence (6.19) of the velocity is zero, and it follows from $(1.81)_1$ that the

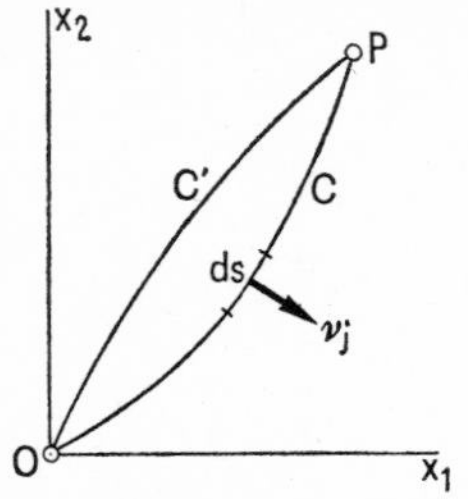

Fig. 6.3 Stream function

flow through the curves C and C' is the same (i.e. that the amount of liquid between C and C' remains constant). Since C and C' are arbitrary except for their end points, ψ is a function of P alone, called the *stream function* $\psi(x_1, x_2)$. It is given by the integral

$$\psi = \int v_j \nu_j \, ds = \int (v_1 \nu_1 + v_2 \nu_2) \, ds. \tag{6.31}$$

extended from O to P along an arbitrary curve C. From Figure 6.3 we obtain $\nu_1 = dx_2/ds$ and $\nu_2 = -dx_1/ds$ and hence

$$d\psi = v_1 \, dx_2 - v_2 \, dx_1. \tag{6.32}$$

We thus have

$$v_1 = \varphi_{,1} = \psi_{,2}, \qquad v_2 = \varphi_{,2} = -\psi_{,1} \tag{6.33}$$

and in consequence $\varphi_{,12} = \psi_{,22} = -\psi_{,11}$: the stream function ψ, like the potential φ, obeys the Laplace equation

$$\psi_{,jj} = 0 \quad \text{or} \quad \Delta\psi = 0. \tag{6.34}$$

According to (6.32), the curves $\psi = \text{const}$ consist of elements in the direction of the flow and hence are streamlines. They are perpendicular to the potential curves $\varphi = \text{const}$, the two families thus forming an orthogonal net.

It is convenient for many purposes to identify the plane of flow with the *complex plane*

$$z = x_1 + ix_2 = re^{i\vartheta} \tag{6.35}$$

and to define a *complex potential* by

$$\chi = \varphi + i\psi. \tag{6.36}$$

On account of (6.20) and (6.34), the function $\chi(z)$ is subject to the differential equation

$$\Delta\chi = 0. \tag{6.37}$$

The two relations (6.33) are the corresponding *Cauchy–Riemann equations*, ensuring that the function $\chi(z)$ is *analytic*, i.e. *differentiable*. Conversely, the real and imaginary parts of an arbitrary analytic function may be interpreted as the potential and the stream function of a potential flow.

The derivative of the function $\chi(z)$ in an arbitrary point z is

$$\chi' = \frac{\mathrm{d}\chi}{\mathrm{d}z} = \chi_{,1} = \varphi_{,1} + i\psi_{,1} = v_1 - iv_2, \tag{6.38}$$

where use has been made of (6.33). This result suggests the introduction of a *complex velocity*

$$v = \chi' = v_1 - iv_2, \tag{6.39}$$

to be interpreted geometrically as the reflection of the vector $\boldsymbol{v}$ on a parallel to the axis x_1 passing through z.

In Section 6.2 we have treated the uniform flow past a sphere. Let us now consider the corresponding plane problem: a uniform flow in the direction x_1, disturbed by a circle (actually, a circular cylinder with axis x_3) of radius a. The complex potential of this problem can be shown to be

$$\chi = u\left(z + \frac{a^2}{z}\right) + \frac{ic}{2\pi} \ln z, \tag{6.40}$$

where u and c are real constants. The function $\chi(z)$ is analytic outside the origin (P3); its imaginary part is the stream function

$$\psi = u\left(1 - \frac{a^2}{r^2}\right) x_2 + \frac{c}{2\pi} \ln r, \tag{6.41}$$

and the complex velocity is

$$v = \chi' = u\left(1 - \frac{a^2}{z^2}\right) + \frac{ic}{2\pi z}. \tag{6.42}$$

On account of (6.41) and (6.42), the circle of radius a is a streamline, and u is the velocity at infinity.

Since the constant c is still free, (6.40) represents a whole family of flows with the required characteristics. Provided $c = 0$, the stream function (6.41) vanishes for $x_2 = 0$; the axis x_1 is thus a streamline, and the flow is similar to the one shown in Figure 6.2. For $c \neq 0$, on the other hand, the flow may be considered as composed of two partial flows: the one just considered and another one with the velocity $v = ic/2\pi z$ or

$$v_1 = \frac{c}{2\pi} \frac{x_2}{r^2}, \qquad v_2 = -\frac{c}{2\pi} \frac{x_1}{r^2}. \tag{6.43}$$

The velocity of the last flow is perpendicular everywhere to the radius vector (Fig. 6.4), and its magnitude is proportional to $1/r$. The flow may be

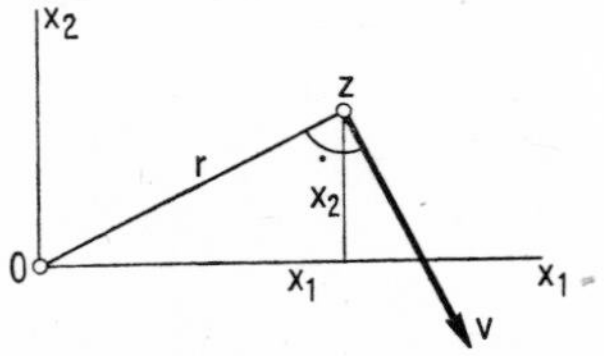

Fig. 6.4 Flow around an isolated vortex line

interpreted as the velocity field of an isolated vortex line (a vortex filament of vanishing cross section) coincident with the axis x_3 and having the arbitrary strength c. In fact, the streamlines are circles around this axis, and the circulation along any streamline is

$$\Gamma = 2\pi r \, |v| = c. \tag{6.44}$$

Once the two flows are superposed, the vortex is located inside the solid circle, and the flow outside the circle is irrotational. Since c is free, there is an infinity of possible solutions. This does not violate the uniqueness proof of Section 1.4, for the region outside the circle (or cylinder) is now doubly connected. Figure 6.5 shows the streamline pattern for a reasonably small positive value of c. Compared to Figure 6.2, the stagnation points A, B are lowered, and the distances between the streamlines are smaller above the circle than below it. This implies an increase in velocity above the circle and, on account of Bernoulli's equation (6.21), a decrease in pressure, whereas the opposite is true for the region below the circle. It follows that now the body is acted upon by a force in the direction x_2.

The problem just outlined is fundamental for the theory of flight. In a first approximation an airfoil (Fig. 6.6) may be considered as a cylinder (of infinite length). By conformal mapping it is possible to reduce the flow past the actual wing-profile to the flow past a circle, described by the complex potential (6.40). The constant c is determined by the fact that the actual

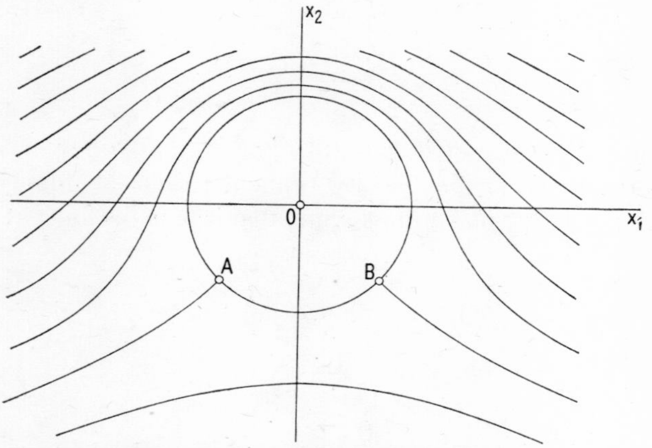

Fig. 6.5 Flows of Figures 6.2 and 6.4 superposed

Fig. 6.6 Airfoil in normal flight

profile has a sharp trailing edge which, in normal flight, corresponds to the stagnation point B. To the resultant force acting on the circle of Figure 6.5 there corresponds a force on the wing, perpendicular to the flow at infinity and hence representing a lift. Actually, the wing also experiences a drag. To explain it, it would be necessary to take viscosity effects into account.

Problems

1. Discuss the flows represented by the complex potential

$$\chi = \mu C z^{1/\mu},$$

where C is a real constant and μ assumes the values $\frac{1}{2}$, 1, $\frac{3}{2}$, and 2.

2. Consider a stagnation point connecting n branches of a streamline. Show that these branches form equal angles at the stagnation point.

3. Show that the complex potential (6.40) satisfies the differential equation (6.37).

4. Generalize the complex potential (6.40) for the case where the velocity at infinity is slightly inclined so that its complex value is $u = u_1 - iu_2$.

5. Show that the function

$$z = z^* + \frac{a^2}{z^*}$$

maps the flow of P4 past a circle in the plane z^* onto a flow past a straight contour in the plane z, the velocity being the same at infinity. Find the complex velocity in the plane z, assuming that it remains finite at the trailing edge.

CHAPTER 7

LINEAR ELASTICITY

In this chapter the general theorems established in Chapters 2 through 5 will be applied to the *elastic solid*. We will restrict ourselves for the present to small displacements and to the study of equilibrium states rather than motions.

In the first two sections the material will be assumed to be isotropic, and we will limit the investigation to cases of isothermal or adiabatic behavior. Section 7.3 will contain a brief outline of anisotropic response, and in Section 7.4 the elastic body will be studied under nonisothermal and nonadiabatic conditions.

7.1. Basic equations

For small displacements and isothermal or adiabatic behavior the constitutive equation of the isotropic elastic solid has been established in Section 5.3. It has the form of equation (5.41), where λ and μ are Lamé's constants and σ_{ij} is a purely quasiconservative stress.

For the treatment of equilibrium problems we also need the kinematic relation (2.33) and the equilibrium condition (3.10). The differential equations governing the problem are thus

$$\varepsilon_{ij} = u_{(j,i)}, \tag{7.1}$$

$$\sigma_{ij,j} + \varrho f_i = 0, \tag{7.2}$$

$$\sigma_{ij} = \lambda \varepsilon_{kk} \delta_{ij} + 2\mu \varepsilon_{ij}. \tag{7.3}$$

It has been shown in Section 5.3 that in the approximation considered here the density ϱ is to be treated as a constant. Incidentally, if we add the negative acceleration to the specific body force f_i, (7.2) becomes the local form of the linear momentum theorem, which is needed for the treatment of motions.

Relations (7.1) through (7.3) represent 15 scalar differential equations (of orders 0 through 1) for the unknowns u_i, ε_{ij}, and σ_{ij}. They do not contain thermodynamic variables, and if the same is true for the boundary (and initial) conditions, the problem is of a purely mechanical nature.

It is sometimes convenient to decompose the tensors ε_{ij} and σ_{ij} into their isotropic and deviatoric parts. Hooke's law (7.3) then resolves into a relation

$$\sigma'_{ij} = 2\mu\varepsilon'_{ij} \tag{7.4}$$

between the deviators, and another one,

$$\sigma_{kk} = (3\lambda+2\mu)\varepsilon_{kk}, \tag{7.5}$$

connecting the isotropic parts of the two tensors. The inversions of (7.4) and (7.5) are

$$\varepsilon'_{ij} = \frac{1}{2\mu}\sigma'_{ij}, \qquad \varepsilon_{kk} = \frac{1}{3\lambda+2\mu}\sigma_{kk}. \tag{7.6}$$

It follows from (7.6) that

$$\varepsilon_{ij} = \frac{1}{2\mu}(\sigma_{ij}-\tfrac{1}{3}\sigma_{kk}\delta_{ij})+\frac{1}{3(3\lambda+2\mu)}\sigma_{kk}\delta_{ij}, \tag{7.7}$$

and from (7.7) we obtain the inversion of Hooke's law (7.3):

$$\varepsilon_{ij} = \frac{1}{2\mu}\left(\sigma_{ij}-\frac{\lambda}{3\lambda+2\mu}\sigma_{kk}\delta_{ij}\right). \tag{7.8}$$

As a first example, let us consider the state of *uniaxial stress*, occurring in a slender rod subjected to tension or compression and defined by the condition that of all the stress components only a single normal stress, e.g. σ_{11}, is nonzero. Equation (7.8) yields the corresponding strain components

$$\varepsilon_{11} = \frac{\lambda+\mu}{\mu(3\lambda+2\mu)}\sigma_{11}, \qquad \varepsilon_{22} = \varepsilon_{33} = -\frac{\lambda}{2\mu(3\lambda+2\mu)}\sigma_{11}, \qquad \varepsilon_{23} = \ldots = 0. \tag{7.9}$$

In engineering literature, relations $(7.9)_1$ and $(7.9)_2$ are usually written in the forms

$$\varepsilon_{11} = \frac{1}{E}\sigma_{11}, \qquad \varepsilon_{22} = \varepsilon_{33} = -\frac{\nu}{E}\sigma_{11}, \tag{7.10}$$

and the constants

$$E = \frac{\mu(3\lambda+2\mu)}{\lambda+\mu}, \qquad \nu = \frac{\lambda}{2(\lambda+\mu)} \tag{7.11}$$

are referred to as *Young's modulus* and *Poisson's ratio* respectively.

Another simple case is *hydrostatic stress*, occurring in bodies immersed in liquids under compression and defined by $(6.3)_1$:

$$\sigma_{ij} = -p\delta_{ij}. \tag{7.12}$$

Here $\sigma'_{ij} = 0$ and $\sigma_{kk} = -3p$. It follows from (7.6) that $\varepsilon'_{ij} = 0$ and

$$\varepsilon_{kk} = -\frac{3}{3\lambda+2\mu}p \tag{7.13}$$

or

$$\varepsilon_{kk} = -\frac{1}{K}p, \quad \text{where} \quad K = \lambda+\tfrac{2}{3}\mu \tag{7.14}$$

is the so-called *bulk modulus*.

As a last example let us cite the state of *simple shearing stress* defined by the condition that $\sigma_{23} = \sigma_{32}$ is the only non-vanishing stress component. Here (7.8) yields

$$\varepsilon_{11} = \ldots = 0, \qquad \varepsilon_{23} = \frac{1}{2\mu}\sigma_{23}, \qquad \varepsilon_{31} = \varepsilon_{12} = 0. \tag{7.15}$$

The only strain component which is not zero corresponds to the engineering shear strain

$$\gamma_{23} = \frac{1}{G}\sigma_{23}, \quad \text{where} \quad G = \mu \tag{7.16}$$

is the so-called *shear modulus*.

Stability of the material in the unstressed state requires that in the last example ε_{23} is positive for $\sigma_{23} > 0$. Similarly, a hydrostatic pressure causes a negative dilatation, and a positive uniaxial stress results in a positive extension. In other words, the constants G, K, and E are to be positive; hence, on account of (7.16), (7.14), $(7.10)_1$, and $(7.11)_1$,

$$\mu > 0, \qquad 3\lambda+2\mu > 0, \qquad \lambda+\mu > 0. \tag{7.17}$$

If μ obeys the first of these inequalities, the last one is a consequence of the second and can be dropped. Figure 7.1 shows the region in the plane λ, μ

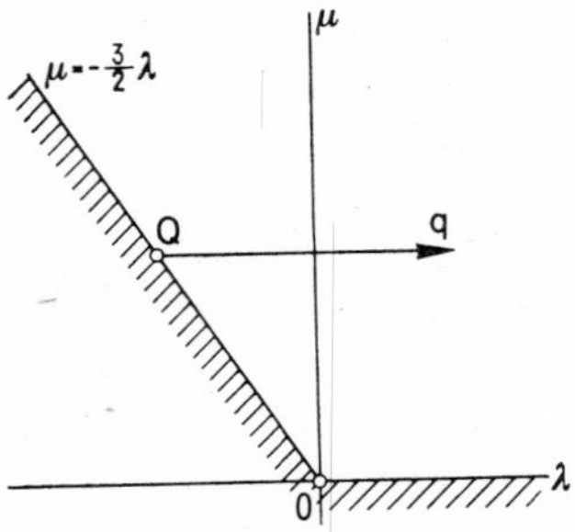

Fig. 7.1 Region where the inequalities (7.17) are satisfied

where the remaining inequalities are satisfied. On its entire left-hand boundary, Poisson's ratio ν $(7.11)_2$ assumes the value -1. Since

$$\frac{\partial \nu}{\partial \lambda} = \frac{\mu}{2(\lambda+\mu)^2} > 0, \tag{7.18}$$

ν increases monotonically on any parallel q to the axis λ, tending towards $\frac{1}{2}$ with increasing distance from Q. It follows that Poisson's ratio is confined to the interval

$$-1 \leqslant \nu \leqslant \tfrac{1}{2}. \tag{7.19}$$

According to $(7.10)_2$, the possibility of ν being negative implies that the cross section of a slender rod might increase in tension. There is no experimental evidence whatsoever for this effect, and we conclude therefore that in reality

$$0 \leqslant \nu \leqslant \tfrac{1}{2}. \tag{7.20}$$

The actual nonexistence of negative values of ν must be due to the microscopic structure of the elastic body and to the nature of the micro-forces. The situation is similar to the one encountered in connection with the mechanical energy theorem (Section 3.3): we have again reached the confines of pure continuum mechanics, and we have to cross them, adopting (7.20) in place of (7.19), in order to remain in agreement with the observed facts.

In the theory of elasticity one usually knows the shape of the body, its loading (by body and surface forces) and the way it is supported. The problem is to find the displacements and the strains and stresses inside the body. The differential equations for the unknowns are relations (7.1) through (7.3), where the last one may be replaced by its inversion (7.8). On surface areas that are loaded or free, the stress vector $\sigma_i^{(\nu)}$ is known, and from (3.6) we obtain the boundary condition

$$\sigma_{ij}\nu_j = \sigma_i^{(\nu)} \tag{7.21}$$

for the stresses. On fixed regions of the surface the boundary condition

$$u_i = 0 \tag{7.22}$$

for the displacements must be satisfied. If a surface element dA with external unit normal ν_j is simply supported and free of friction, we have a first boundary condition for the displacement,

$$u_j\nu_j = 0, \tag{7.23}$$

and since the shear stress $\sigma_i^{(\nu)} - \sigma_k^{(\nu)}\nu_k\nu_i$ vanishes, we obtain an additional boundary condition

$$(\sigma_{ik} - \sigma_{jk}\nu_i\nu_j)\nu_k = 0 \tag{7.24}$$

for the stress. It is easy to verify that in each of these cases three independent scalar boundary conditions are to be satisfied (P1). For surface elements with friction the problem is far more complicated and will not be discussed here.

The differential equations (7.1) through (7.3) connect three systems of unknowns: the displacements, strains, and stresses. Eliminating two of these systems, it is possible to reduce the number of differential equations in such a way that a single system of equations results for, e.g. the displacements or the stresses alone. Thus, the problem appears simplified as far as the differential equations are concerned. In general, however, this simplification is accompanied by a complication of the boundary conditions, since they usually contain more than one set of unknowns.

By way of an example, let us eliminate the strain components with the help of the kinematic relation (7.1). Hooke's law (7.3) then assumes the form

$$\sigma_{ij} = \lambda u_{k,k}\delta_{ij} + \mu(u_{j,i} + u_{i,j}). \tag{7.25}$$

Inserting (7.25) in the equilibrium condition (7.2), we eliminate the stresses, obtaining

$$\lambda u_{k,ki} + \mu(u_{j,ij} + u_{i,jj}) + \varrho f_i = 0 \tag{7.26}$$

or, in more compact form,

$$\mu u_{i,jj} + (\lambda + \mu)u_{j,ji} + \varrho f_i = 0. \tag{7.27}$$

This is *Navier's differential equation* for the displacement vector. In a similar manner a set of differential equations for the stresses alone may be obtained (P2).

In the theory of ordinary differential equations it is often possible to find the general solution and to adapt it to the boundary conditions by an appropriate choice of the integration constants. Here, however, we are confronted with 9 partial differential equations (7.1) and (7.2) apart from the equations (7.3), and it would be hopeless to search for the general solution. The exact solutions obtained so far are due to intuition and to the fact that, the differential equations being linear, superposition of simple solutions is legitimate and sometimes helps so solve more complicated problems. However, the question then arises whether a solution obtained in this manner is the only one. This question has been settled once and for all by *Kirchhoff's uniqueness proof*, which will be reproduced here under the assumption that the material is not idealized as incompressible, the constant λ, on account of (7.14), thus being finite.

Let us assume that the fields $\overset{*}{u}_i$, $\overset{*}{\varepsilon}_{ij}$, $\overset{*}{\sigma}_{ij}$, and $\overset{**}{u}_i$, $\overset{**}{\varepsilon}_{ij}$, $\overset{**}{\sigma}_{ij}$ are solutions of one and the same problem. In this case the fields $u_i = \overset{*}{u}_i - \overset{**}{u}_i$, $\varepsilon_{ij} = \overset{*}{\varepsilon}_{ij} - \overset{**}{\varepsilon}_{ij}$, $\sigma_{ij} = \overset{*}{\sigma}_{ij} - \overset{**}{\sigma}_{ij}$ satisfy the basic equations (7.1) through (7.3) for $f_i = 0$ and the boundary condition (7.21) with $\sigma_i^{(\nu)} = 0$ on the surface areas that in reality are either loaded or free. The new solution obviously corresponds to an unloaded body, and since the boundary conditions (7.22) through (7.24) are such that the reactions do not work, $\sigma_i^{(\nu)} u_i = \sigma_{ij} u_i \nu_j$ is zero on the whole supported surface area and hence on the entire surface. It follows that

$$\int \sigma_{ij} u_i \nu_j \, \mathrm{d}A = 0 \tag{7.28}$$

or, according to Gauss' theorem (1.79),

$$\int (\sigma_{ij} u_i)_{,j} \, \mathrm{d}V = \int \sigma_{ij} u_{i,j} \, \mathrm{d}V + \int \sigma_{ij,j} u_i \, \mathrm{d}V = 0. \tag{7.29}$$

The last integral vanishes on account of (7.2) written for $f_i = 0$. In the preceding integral, equations (7.1) and the symmetry of σ_{ij} allow us to replace $u_{i,j}$ by ε_{ij}. Making use of (7.3), we get

$$\int \sigma_{ij} \varepsilon_{ij} \, \mathrm{d}V = \int (\lambda \varepsilon_{ii} \varepsilon_{jj} + 2\mu \varepsilon_{ij} \varepsilon_{ij}) \, \mathrm{d}V = 0. \tag{7.30}$$

The last integrand is twice the strain energy per unit volume, (5.42), and as such is positive definite. In fact, it may be written

$$\lambda \varepsilon_{ii} \varepsilon_{jj} + 2\mu (\varepsilon'_{ij} + \tfrac{1}{3} \varepsilon_{kk} \delta_{ij})(\varepsilon'_{ij} + \tfrac{1}{3} \varepsilon_{ll} \delta_{ij}) = (\lambda + \tfrac{2}{3}\mu) \varepsilon_{ii} \varepsilon_{jj} + 2\mu \varepsilon'_{ij} \varepsilon'_{ij}. \tag{7.31}$$

Here the right-hand side is a sum of squares with coefficients that are positive according to (7.17). Thus, (7.30) requires that ε_{ii} and ε'_{ij}, in consequence ε_{ij} and on account of (7.3) also σ_{ij} are identically zero. It follows that $\varepsilon^*_{ij} = \varepsilon^{**}_{ij}$ and $\sigma^*_{ij} = \sigma^{**}_{ij}$, so that the difference $u_i = u^*_i - u^{**}_i$ of the displacement fields is either zero or a rigid displacement of the entire body. In most cases, such a rigid displacement is prevented by the supports.

Note that the compressibility of the body is essential for the proof. In an incompressible material ($\lambda = \infty$, $\varepsilon_{kk} = 0$) a hydrostatic stress remains indetermined in Hooke's law (7.3); the stress states σ^*_{ij} and σ^{**}_{ij} might thus differ in a hydrostatic stress even though the strains ε^*_{ij} and ε^{**}_{ij} are the same. Equally essential is the assumption that the displacements are small (or, more precisely, infinitesimal). In fact, in the linear theory we tacitly assume that (7.2) and (7.3) are satisfied in the undeformed configuration of the body, whereas the correct procedure would be to satisfy them in the deformed configuration. It is therefore obvious that the uniqueness proof is only valid for sufficiently small loads. Conversely, the loss of uniqueness may be used as a stability criterion: the buckling load of a slender rod, e.g., is sometimes defined as the smallest load for which the (straight) equilibrium configuration is not unique.

Problems

1. Show that equation (7.24) represents only two scalar boundary conditions.

2. Deduce the so-called *Beltrami–Michell equations*

$$\sigma_{ij,kk} + \frac{2(\lambda+\mu)}{3\lambda+2\mu}\sigma_{kk,ij} + \varrho(f_{i,j}+f_{j,i}) + \frac{\lambda}{\lambda+2\mu}\varrho f_{k,k}\delta_{ij} = 0$$

from (7.1) through (7.3). They can be shown to be interdependent and hence must be supplemented by the equilibrium conditions in order to form a complete set of differential equations for the stresses.

3. Verify that in an elastic solid without body forces the differential equations

$$\Delta\varepsilon_{ii} = \Delta\sigma_{ii} = \Delta\Delta u_i = \Delta\Delta\varepsilon_{ij} = \Delta\Delta\sigma_{ij} = 0$$

are satisfied.

4. Let $u_i(x_j)$ be the displacement field of an elastic solid which is in equilibrium when subjected to the surface stresses $\sigma^{(\nu)}_i$, and let $u^*_i(x_j)$ be the displacement field corresponding to the surface stresses $\sigma^{(\nu)*}_i$. Prove the reciprocity relation

$$\int \sigma^{(\nu)}_i u^*_i \, dA = \int \sigma^{(\nu)*}_i u_i \, dA.$$

7.2. Torsion

One of the simplest applications of the linear theory of elasticity is the torsion of a cylindrical shaft of length l and arbitrary but simply connected cross section A (Fig. 7.2). We refer it to a coordinate system with axes x_1 and x_2 in the lower base, and we assume that it is twisted by couples acting in the two end sections.

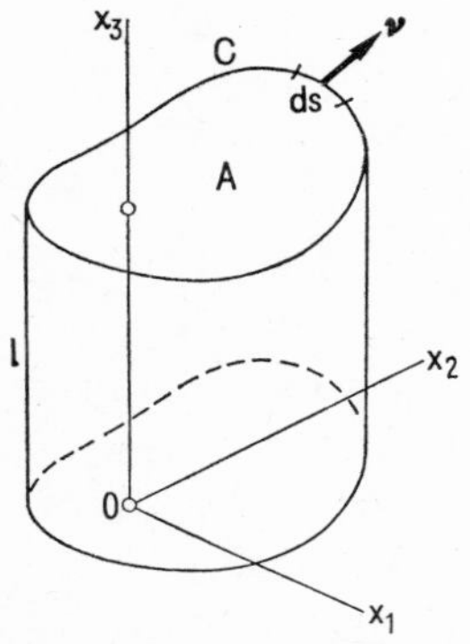

Fig. 7.2 Cylindrical shaft with simply connected cross section

Let us disregard the body force and let us assume that the end sections are free to warp, horizontal displacements however being prevented in the lower base and corresponding, in the upper base, to a rotation about the axis x_3. The curved surface is stress-free; the boundary condition for the stress in either end section is $\sigma_{33} = 0$; the boundary conditions for the displacement are

$$u_1 = u_2 = 0 \qquad (x_3 = 0) \tag{7.32}$$

$$u_1 = -\vartheta l x_2, \qquad u_2 = \vartheta l x_1, \quad (x_3 = l) \tag{7.33}$$

where ϑl is the total angle of twist and the displacements are assumed to be small.

The problem is clearly of the type discussed in Section 7.1. To solve it, let us try a displacement field of the form

$$u_1 = -\vartheta x_2 x_3, \qquad u_2 = \vartheta x_1 x_3, \qquad u_3 = w(x_1, x_2). \tag{7.34}$$

The first two equations suggest that the cross sections rotate about the axis x_3, the angle of twist being ϑx_3 for the segment $0 \ldots x_3$ of the shaft and hence amounting to ϑ per unit length. The third equation represents a warping of the cross section which is the same for all sections. The question now is whether the function $w(x_1, x_2)$ can be so determined that the displacement field (7.34) and the corresponding strain and stress fields satisfy all differential equations and boundary conditions.

Let us note that the displacement field (7.34) satisfies the boundary conditions (7.32) and (7.33). From the kinematic relation (7.1) we obtain the strain components $\varepsilon_{11} = \ldots = 0$, $\varepsilon_{12} = 0$, and

$$\varepsilon_{23} = \tfrac{1}{2}(w_{,2}+\vartheta x_1), \qquad \varepsilon_{31} = \tfrac{1}{2}(w_{,1}-\vartheta x_2). \tag{7.35}$$

Thus, the only nonvanishing strains are the shear components ε_{23} and ε_{31}; they are independent of x_3. Hooke's law (7.3) yields $\sigma_{11} = \ldots = 0$, $\sigma_{12} = 0$, and

$$\sigma_{23} = \mu(w_{,2}+\vartheta x_1), \qquad \sigma_{31} = \mu(w_{,1}-\vartheta x_2). \tag{7.36}$$

It follows that the only non-vanishing stress components are the shear stresses σ_{23} and σ_{31} in the cross section and in the corresponding surface elements perpendicular to it; they, too, are independent of x_3 and hence are the same in all cross sections. Since $\sigma_{33} = 0$, the boundary conditions for the stresses in the end sections are satisfied. The same is true for the first two equilibrium conditions (7.2); the last one reduces to

$$\sigma_{31,1}+\sigma_{23,2} = 0. \tag{7.37}$$

On substitution of (7.36) in (7.37), we would obtain a differential equation for w. Let us note, instead, that (7.37) is satisfied provided the two non-vanishing stress components can be derived from a *stress function* $\varphi(x_1, x_2)$ in accordance with

$$\sigma_{23} = -\varphi_{,1}, \qquad \sigma_{31} = \varphi_{,2}. \tag{7.38}$$

Conversely, it can be shown (P1) that any solution of (7.37) can be expressed in terms of such a stress function.

Equations (7.38) reduce the unknowns σ_{23} and σ_{31} to a single one. Introducing it in (7.36), we obtain

$$\varphi_{,1} = -\mu(w_{,2}+\vartheta x_1), \qquad \varphi_{,2} = \mu(w_{,1}-\vartheta x_2), \tag{7.39}$$

where φ and w are the only remaining unknowns. We eliminate the last one by differentiating these equations with respect to x_1 and x_2 respectively; adding the results, we get

$$\varphi_{,11}+\varphi_{,22} = \varphi_{,ii} = \Delta\varphi = -2\mu\vartheta. \tag{7.40}$$

This is *Poisson's differential equation* formulated for the stress function. The curved surface is free of stresses. The corresponding boundary condition is $\sigma_i^{(\nu)} = \sigma_{ij}\nu_j = 0$; it reduces to the condition

$$\sigma_{3j}\nu_j = \sigma_{31}\nu_1 + \sigma_{23}\nu_2 = 0, \tag{7.41}$$

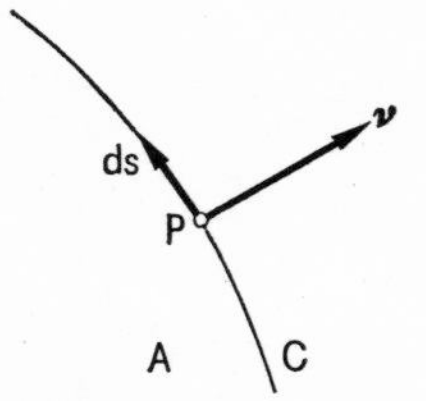

Fig. 7.3 Boundary of a cross section

to be satisfied on the boundary C of the cross section A (Fig. 7.3). Inserting (7.38) in (7.41), we have $\varphi_{,2}\nu_1 - \varphi_{,1}\nu_2 = 0$ or (Fig. 7.3)

$$\varphi_{,1}\, dx_1 + \varphi_{,2}\, dx_2 = 0. \tag{7.42}$$

It follows that $d\varphi = 0$ on C, the function φ thus being constant on the boundary of the cross section. Since an additive constant in φ is irrelevant for our problem (and since the cross section has been assumed to be simply connected), the boundary condition may be written

$$\varphi = 0 \quad \text{on} \quad C. \tag{7.43}$$

We are now faced with the problem of finding a function φ satisfying the differential equation (7.40) in the entire area A and the boundary condition (7.43). According to the theory of partial differential equations, this problem has a unique solution, provided the angle of twist ϑ is prescribed. Once the function $\varphi(x_1, x_2)$ is known, the stress components follow from (7.38); the strain components are

$$\varepsilon_{23} = -\frac{1}{2\mu}\varphi_{,1}, \qquad \varepsilon_{31} = \frac{1}{2\mu}\varphi_{,2}, \tag{7.44}$$

and from (7.39) we obtain the differential equations

$$w_{,1} = \frac{1}{\mu}\varphi_{,2} + \vartheta x_2, \qquad w_{,2} = -\frac{1}{\mu}\varphi_{,1} - \vartheta x_1 \tag{7.45}$$

for the warping function. On account of (7.40) the integrability condition $w_{,21} = w_{,12}$ is satisfied, and the function $w(x_1, x_2)$ can be obtained by integration.

The forces acting in an element $\mathrm{d}A$ of an arbitrary cross section are

$$\sigma_{31}\,\mathrm{d}A = \varphi_{,2}\,\mathrm{d}A, \qquad \sigma_{23}\,\mathrm{d}A = -\varphi_{,1}\,\mathrm{d}A. \tag{7.46}$$

Reduction of these forces to the point of intersection with the axis x_3 yields the resultant force

$$F_k = \int e_{3kl}\varphi_{,l}\,\mathrm{d}A \tag{7.47}$$

and the moment

$$M_i = \int e_{ijk}x_j e_{3kl}\varphi_{,l}\,\mathrm{d}A. \tag{7.48}$$

Applying the theorem of Gauss in the form (1.82) to (7.47), we obtain

$$F_k = e_{3kl}\int \varphi\nu_l\,\mathrm{d}s = 0 \tag{7.49}$$

on account of the boundary condition (7.43). Treating (7.48) in a similar manner and making use of $(1.29)_1$, we get

$$\begin{aligned} M_i &= -e_{ijk}e_{3lk}\int x_j\varphi_{,l}\,\mathrm{d}A = -(\delta_{i3}\delta_{jl}-\delta_{il}\delta_{j3})\int x_j\varphi_{,l}\,\mathrm{d}A \\ &= -\delta_{i3}\int x_j\varphi_{,j}\,\mathrm{d}A+\delta_{il}x_3\int \varphi_{,l}\,\mathrm{d}A \end{aligned} \tag{7.50}$$

since x_3 is a constant. On account of (7.43) and the theorem of Gauss, the second integral is again zero. Thus, $M_1 = M_2 = 0$ and

$$M_3 = -\int (x_j\varphi)_{,j}\,\mathrm{d}A+\int x_{j,j}\varphi\,\mathrm{d}A = -\int x_j\varphi\nu_j\,\mathrm{d}s+2\int \varphi\,\mathrm{d}A \tag{7.51}$$

or

$$M_3 = 2\int \varphi\,\mathrm{d}A. \tag{7.52}$$

It follows that the forces in any section reduce to a torque which is obtained, except for the factor 2, by integration of the stress function over the cross section.

As an example, let us consider the cross section of Figure 7.4 in the form of an equilateral triangle. Here the stress function can be shown to be

$$\varphi = \frac{\mu\vartheta}{6a}(a-x_1)(2a+x_1+\sqrt{3}x_2)(2a+x_1-\sqrt{3}x_2) \tag{7.53}$$

or

$$\varphi = \frac{\mu\vartheta}{6a}(a-x_1)[(2a+x_1)^2-3x_2^2]. \tag{7.54}$$

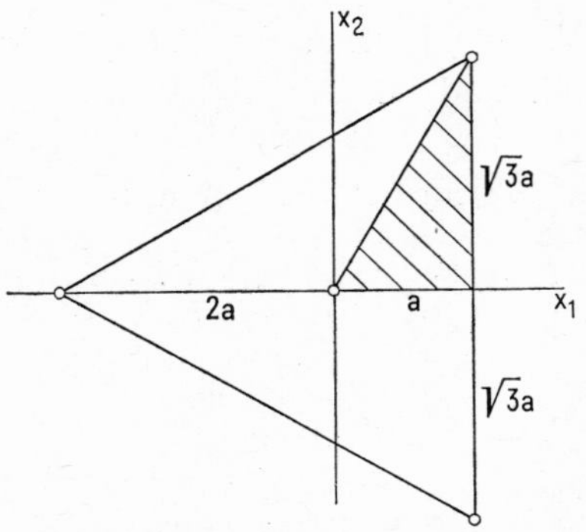

Fig. 7.4 Equilateral triangle

It satisfies the boundary condition (7.43). The stresses are

$$\begin{aligned}\sigma_{23} &= -\varphi_{,1} = \frac{\mu\vartheta}{2a}(x_1^2+2ax_1-x_2^2),\\ \sigma_{31} &= \varphi_{,2} = \frac{\mu\vartheta}{a}(x_1-a)x_2,\end{aligned} \tag{7.55}$$

and it is easy to verify that the differential equation (7.40) is satisfied. The torque (7.52) is twelve times the integral of φ over the shaded area in Figure 7.4:

$$M_3 = \frac{2\mu\vartheta}{a}\int_{x_1=0}^{a}\int_{x_2=0}^{\sqrt{3}x_1}(a-x_1)\left[(2a+x_1)^2-3x_2^2\right]\mathrm{d}x_1\,\mathrm{d}x_2, \tag{7.56}$$

and by carrying out the integration (P2) we obtain

$$M_3 = \frac{9\sqrt{3}}{5}\mu\vartheta a^4. \tag{7.57}$$

The differential equations (7.45) for the warping function become

$$w_{,1} = \frac{\vartheta}{a}x_1x_2, \qquad w_{,2} = \frac{\vartheta}{2a}(x_1^2-x_2^2), \tag{7.58}$$

and the function w itself, apart from an irrelevant additive constant, is found to be

$$w = \frac{\vartheta}{6a}(3x_1^2-x_2^2)x_2. \tag{7.59}$$

Returning to the general problem, we observe that the axis x_3 has been chosen at random, apart from its direction which is parallel to the generators of the cylinder. This seems to be inconsistent with the fact that the deformation leaves only the axis x_3 straight, whereas any material line parallel to it becomes a helix. The inconsistency, however, is only apparent: if (7.34) is to describe the deformation just discussed it is necessary for the displacements to remain small compared to the dimensions of the cross section. In this approximation the helices are still to be considered as (slightly oblique) straight lines, and the displacements (7.34) obtained for various choices of the axis x_3 are the same except for a rigid body displacement.

Problems

1. Show that two stress components $\sigma_{23}(x_1, x_2)$ and $\sigma_{31}(x_1, x_2)$ satisfying the equilibrium condition (7.37) can be expressed by (7.38) in terms of a stress function $\varphi(x_1, x_2)$.
2. Carry out the integration in (7.56).
3. Solve the torsion problem for the elliptic cross section

$$\frac{x_1^2}{a^2}+\frac{x_2^2}{b^2}=1,$$

using the stress function

$$\varphi=-\mu\vartheta\frac{a^2b^2}{a^2+b^2}\left(\frac{x_1^2}{a^2}+\frac{x_2^2}{b^2}-1\right).$$

Specialize the results for the circular cross section.

7.3. Crystals

Hooke's law (7.3) is based on the assumptions that the elastic solid is isotropic and that the deformations are small. The second assumption is responsible for the linearity of the connection between strains and stresses, the first one for the special form of (7.3). For an anisotropic solid subjected to small deformations, Hooke's law must be generalized in the form

$$\sigma_{ij}=C_{ijkl}\varepsilon_{kl}, \tag{7.60}$$

where C_{ijkl} is a fourth-order tensor characterizing the material. The $3^4=81$ components of C_{ijkl} are not independent. Since the stress tensor is symmetric, we have

$$C_{jikl}=C_{ijkl}. \tag{7.61}$$

On account of the symmetry of the strain tensor we are allowed to set

$$C_{ijlk} = C_{ijkl}. \tag{7.62}$$

This brings the number of independent components down to $6^2 = 36$. A further reduction results from the existence of a strain energy $W(\varepsilon_{ij})$. According to (5.33), equation (7.60) may be written

$$\frac{\partial W}{\partial \varepsilon_{ij}} = C_{ijkl}\varepsilon_{kl}. \tag{7.63}$$

It follows that

$$\frac{\partial^2 W}{\partial \varepsilon_{ij}\, \partial \varepsilon_{kl}} = C_{ijkl}, \tag{7.64}$$

and since the order of the two differentiations is irrelevant, we finally get

$$C_{klij} = C_{ijkl}. \tag{7.65}$$

Let us note that, contrary to the isotropic case, the existence of a strain energy is not a consequence of the stress-strain relationship. The additional symmetry condition (7.65) reduces the independent components of C_{ijkl} to $\frac{36}{2}+3 = 21$. From (7.63) we obtain the strain energy

$$W = \tfrac{1}{2} C_{ijkl}\varepsilon_{ij}\varepsilon_{kl}, \tag{7.66}$$

where an additive constant has been suppressed.

The symmetry relations (7.61), (7.62), and (7.65) are consequences of the definition of the elastic solid and the restriction to isothermal or adiabatic processes. These relations are valid even if the solid is completely anisotropic. Moreover, it is clear that the C_{ijkl} transform according to the rules laid down in Section 1.1.

The most important anisotropic solids are crystals. On account of symmetries in their molecular structure they allow for certain operations (rotations or reflections of the material with respect to a coordinate system fixed in space) such that their response is the same in the initial and in the final configurations. These *symmetry operations* are

(a) reflection with respect to a point,

(b) reflection with respect to a plane, called a *plane of symmetry*,

(c) rotation about an axis, called an *axis of symmetry*.

In case (c) the angle of rotation is always $2\pi/n$ ($n = 2, 3, 4, 6$). According to the value of n, the axis of rotation is called an n-fold symmetry axis.

Incidentally, it is obvious that any 4-fold axis of symmetry is also 2-fold and that any 6-fold axis is also a 3-fold axis of symmetry. The relative positions of symmetry elements are not arbitrary but restricted to a few simple possibilities. An example is a 2-fold axis and a plane of symmetry perpendicular to it; it is clear that in this case we have also symmetry with respect to their point of intersection.

For the following discussion it is convenient to keep the material element fixed and to subject the coordinate system to the symmetry operations. On account of (b), this implies that we admit left-handed coordinate systems. Since the present section is self-contained and restricted to the discussion of symmetries, this is perfectly legitimate in spite of the fact that from Section 1.2 on we have limited ourselves to right-handed coordinate systems.

The definition of a symmetry operation requires that for an arbitrary set of strain components the stresses following from (7.60) are the same in the various coordinate systems connected by symmetry relations. It follows that the components of C_{ijkl} are invariant with respect to symmetry operations, so that not only the six relations (7.60) but also the expression (7.66) for the strain energy are the same, term by term, in corresponding coordinate systems.

It is obvious that in general each symmetry operation implies a reduction in the number of independent components C_{ijkl}. From the point of view of crystallography there are 6 *classes* containing a total of 32 *subclasses*. Any subclass is characterized by a given set of symmetries and hence by a certain number of independent material constants C_{ijkl}, starting from the 21 constants already mentioned and reducing, with increasing number of symmetries, to 3, i.e. one more than the two constants λ and μ of the isotropic solid.

For the discussion of the crystal classes it is convenient to replace the form (7.60) of Hooke's law by

$$\begin{aligned} \sigma_{11} &= a_{11}\varepsilon_{11}+a_{12}\varepsilon_{22}+a_{13}\varepsilon_{33}+a_{14}\varepsilon_{23}+a_{15}\varepsilon_{31}+a_{16}\varepsilon_{12}, \\ \sigma_{22} &= a_{21}\varepsilon_{11}+a_{22}\varepsilon_{22}+ \dots \qquad\qquad\qquad\qquad +a_{26}\varepsilon_{12}, \\ &\vdots \\ \sigma_{12} &= a_{61}\varepsilon_{11}+a_{62}\varepsilon_{22}+ \dots \qquad\qquad\qquad\qquad +a_{66}\varepsilon_{12}, \end{aligned} \tag{7.67}$$

where the a_{ij} are new material constants replacing the C_{ijkl}. This may be less satisfying from an aesthetic viewpoint, but it has the advantage that now the material constants can be arranged as a matrix, which is symmetric on account of (7.65) and is therefore entirely specified by the table

	ε_{11}	ε_{22}	ε_{33}	ε_{23}	ε_{31}	ε_{12}
σ_{11}	a_{11}	a_{12}	a_{13}	a_{14}	a_{15}	a_{16}
σ_{22}		a_{22}	a_{23}	a_{24}	a_{25}	a_{26}
σ_{33}			a_{33}	a_{34}	a_{35}	a_{36}
σ_{23}				a_{44}	a_{45}	a_{46}
σ_{31}					a_{55}	a_{56}
σ_{12}						a_{66}

(7.68)

which might be completed by reflection with respect to the diagonal a_{11}, a_{22}, ..., a_{66}. The marginal elements on the left indicate in which one of the equations a given a_{ij} appears; the ones on top indicate the place in the equation. Apart from the factor $\frac{1}{2}$, the strain energy (7.66) is obtained by multiplying the right-hand sides of (7.67) in succession by ε_{11}, ε_{22}, ..., ε_{12} and by adding the results. Thus, the various terms of W are obtained by replacing the σ_{11}, ... in (7.68) by ε_{11}, ..., by multiplying each a_{ij} with the marginal elements at the left and on top, and by adding the factor $\frac{1}{2}$ in the terms corresponding to the diagonal elements. Like C_{ijkl}, the constants a_{ij} must be invariant with respect to any symmetry operation.

Let us now briefly review the various crystal classes. The *triclinic class* contains crystals belonging to two different subclasses. One of them exhibits no symmetry at all and hence is characterized by the 21 constants appearing

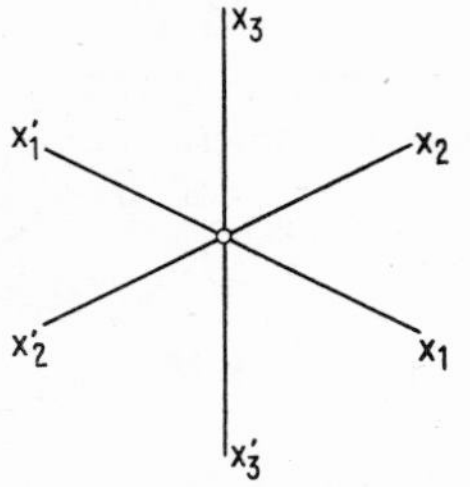

Fig. 7.5 Symmetry with respect to the origin

in (7.68). The second one admits symmetry with respect to a point. In Figure 7.5 the axes x_i are the original ones, whereas the reflected axes are indicated by x_i'. It is clear that the strain components ε_{ij}, interpreted as extensions and shear strains or calculated by means of (7.1), are the same in the two coordinate systems and that, therefore, the system (7.68) still contains 21 independent constants.

In the *monoclinic class* one of the subclasses contains a single 2-fold symmetry axis. Denoting it by x_3, we deduce from Figure 7.6, where again

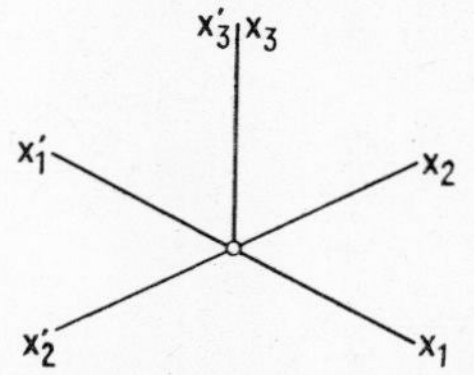

Fig. 7.6 2-fold symmetry axis x_3

the corresponding coordinate systems are denoted by x_i and x_i', that $\varepsilon_{23}' = -\varepsilon_{23}$ and $\varepsilon_{31}' = -\varepsilon_{31}$, whereas the remaining strain components are unaltered. Since Hooke's law (7.67) is to be the same in the two coordinate systems, the constants a_{14}, a_{15}, a_{24}, a_{25}, a_{34}, a_{35}, a_{46}, a_{56} must be zero, and (7.68) reduces to

$$\begin{matrix} a_{11} & a_{12} & a_{13} & 0 & 0 & a_{16} \\ & a_{22} & a_{23} & 0 & 0 & a_{26} \\ & & a_{33} & 0 & 0 & a_{36} \\ & & & a_{44} & a_{45} & 0 \\ & & & & a_{55} & 0 \\ & & & & & a_{66} \end{matrix} \tag{7.69}$$

The zero elements are (a) those appearing in the 4th and 5th line but not in the 4th or 5th column, (b) the ones appearing in the 4th and 5th column but not in the 4th or 5th line. The number of independent constants is 13. A second subclass of the monoclinic system contains a single symmetry plane. Denoting it by x_1, x_2, we conclude from Figure 7.7 that the ε_{ij} transform in the

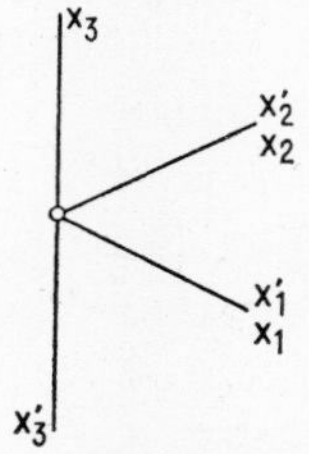

Fig. 7.7 Symmetry plane x_1, x_2

same manner as in the first subclass. Thus, the material constants are still given by (7.69), and the same is true for the third subclass, characterized by a 2-fold symmetry axis and a symmetry plane perpendicular to it.

In the *rhombic class* one of the subclasses contains two orthogonal 2-fold symmetry axes. Denoting them by x_3 and x_1, we conclude from Figure 7.8,

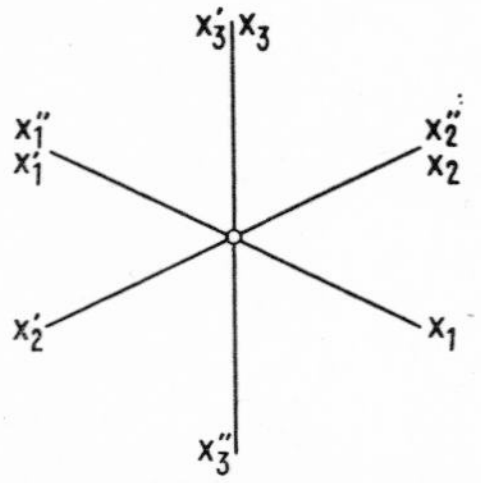

Fig. 7.8 2-fold symmetry axes x_3 and x_1

where the coordinate system x_i'' is the result of the two symmetry operations $x_i \to x_i'$ and $x_i' \to x_i''$, that also the axis x_2 is a 2-fold symmetry axis. The material constants are obtained by treating the 5th and 6th lines and columns in the same manner as the 4th and the 5th in (7.69). Thus, (7.69) reduces to

$$\begin{matrix} a_{11} & a_{12} & a_{13} & 0 & 0 & 0 \\ & a_{22} & a_{23} & 0 & 0 & 0 \\ & & a_{33} & 0 & 0 & 0 \\ & & & a_{44} & 0 & 0 \\ & & & & a_{55} & 0 \\ & & & & & a_{66} \end{matrix} \tag{7.70}$$

and the number of independent constants is 9. There are two more subclasses (P1) with the constants (7.70).

The *hexagonal class* contains 12 subclasses, characterized by 3- and 6-fold symmetry axes. We will not discuss them here but note that the various subclasses have 5–7 independent material constants (P2).

In the *tetragonal class* a first subclass has three orthogonal 2-fold symmetry axes, one of which is also 4-fold. Denoting it by x_3 and considering the rotation about $\pi/2$ of Figure 7.9, we obtain the strain transformations

$$\begin{matrix} \varepsilon_{11}' = \varepsilon_{22}, & \varepsilon_{22}' = \varepsilon_{11}, & \varepsilon_{33}' = \varepsilon_{33}, \\ \varepsilon_{23}' = -\varepsilon_{31}, & \varepsilon_{31}' = \varepsilon_{23}, & \varepsilon_{12}' = -\varepsilon_{12}, \end{matrix} \tag{7.71}$$

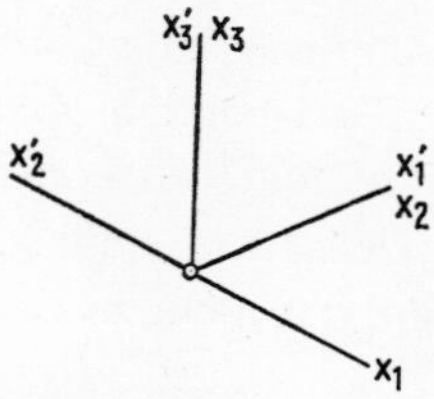

Fig. 7.9 2-fold symmetry axes x_1, x_2; 4-fold axis x_3

and we conclude that in (7.70) $a_{22} = a_{11}$, $a_{13} = a_{23}$, and $a_{55} = a_{44}$. Thus, the number of remaining constants is 6. There are six additional subclasses with 6–7 constants.

The *cubic class* contains a subclass with three orthogonal 2-fold symmetry axes, two of which are also 4-fold. Denoting them by x_3 and x_1, we obtain $a_{33} = a_{22} = a_{11}$, $a_{12} = a_{13} = a_{23}$, and $a_{66} = a_{55} = a_{44}$, the number of constants thus reducing to 3. There are four more subclasses with the same 3 constants.

From the cubic class an additional step leads to the isotropic solid with two constants (P3). Incidentally, some of the considered cases of anisotropy do not only occur in crystals but also in certain building materials. A timber board, e.g., may have three orthogonal symmetry planes. Such a material is called *orthotropic* and can be shown (P4) to have the material constants given by (7.70).

Incidentally, the isotropic material obeying (5.41) and the anisotropic solids considered here have a common property: a given stress component is in general not determined by the corresponding strain component alone, and the reverse is also true. If, e.g., ε_{11} is the only non-vanishing strain component, at least the stresses σ_{11}, σ_{22}, and σ_{33} are different from zero, irrespective of the choice of the coordinate system. If, on the other hand, $\varepsilon_{23} = \ldots = 0$ in a coordinate system of arbitrary orientation with respect to the material, (5.41) yields $\sigma_{23} = \ldots = 0$ and vice versa. It follows that in the isotropic case the principal axes of strain and stress are always the same. According to (7.67), this is only exceptionally true in the anisotropic solid.

Problems

1. Consider the two subclasses of the rhombic crystal class with (a) a 2-fold symmetry axis and a symmetry plane containing the axis, (b) the same symmetry elements and an additional symmetry plane perpendicular to the

symmetry axis. Show that each one of these subclasses has more symmetries and determine the number of independent material constants.

2. Determine the number of material constants for the subclass of the hexagonal system characterized by a single 3-fold symmetry axis. Hint: choose the symmetry axis so that it includes equal angles with the coordinate axes.

3. Formulate the expression for the strain energy of the cubic crystal class and show that the condition that the a_{ij} are invariant with respect to arbitrary rotations of the coordinate system yields (5.42) and hence Hooke's law (5.41) of the isotropic solid.

4. Show that the material constants of the orthotropic solid are given by (7.70).

7.4. Thermoelasticity

Thermoelasticity is concerned with elastic solids under conditions that are neither isothermal nor adiabatic. Returning to Section 5.1 we note that under these circumstances the response of the solid is determined by its specific free energy (5.27), which is dependent on the strain tensor and the temperature, whereas a strain energy cannot be defined.

Let us restrict ourselves again to small displacements and also to small deviations of the temperature from a reference value ϑ_0. Measuring the strains from the stress-free state at ϑ_0, we note that the values of ε_{ij} and of $\vartheta-\vartheta_0$ are small and that, therefore, an approximate theory can be obtained by expanding the function $\psi(\varepsilon_{ij}, \vartheta)$ in the vicinity of $\varepsilon_{ij} = 0$, $\vartheta = \vartheta_0$ into a power series and by truncating this series after the quadratic terms. The stresses are given by $(5.30)_1$. Since they are zero in the reference state $\varepsilon_{ij} = 0$, $\vartheta = \vartheta_0$, the expansion contains no linear terms in ε_{ij} alone. It is convenient to expand $\varrho\psi$, the free energy per unit volume, in place of ψ, and we obtain

$$\varrho\psi = \varrho\psi_0 - \varrho s_0(\vartheta-\vartheta_0) + \tfrac{1}{2}C_{ijkl}\varepsilon_{ij}\varepsilon_{kl} + c_{ij}\varepsilon_{ij}(\vartheta-\vartheta_0) - \frac{\varrho c}{2\vartheta_0}(\vartheta-\vartheta_0)^2, \quad (7.72)$$

where the coefficients ψ_0, s_0, C_{ijkl}, c_{ij}, and c are still free and the factors ϱ and ϱ/ϑ_0 have been added for convenience.

Applying (7.72) to $\varepsilon_{ij} = 0$ and $\vartheta = \vartheta_0$, we note that ψ_0 is the specific free energy in the reference state. We further observe that, on account of the forms of the third and fourth terms on the right, (7.72) is valid for anisotropic elastic solids. If we restrict ourselves from now on to the isotropic case, (7.72) can be simplified. In the first place, a comparison of (7.66) and

(5.42) shows that the third term may be written

$$\frac{\lambda}{2}\,\varepsilon_{ii}\varepsilon_{jj}+\mu\varepsilon_{ij}\varepsilon_{ij}\,, \tag{7.73}$$

where λ and μ are Lamé's constants. In the second place, the tensor c_{ij} must be isotropic and hence (P of Section 1.1) of the form $a\delta_{ij}$, where a is a scalar. Replacing a for convenience by $-(3\lambda+2\mu)\alpha$, where α is free, we write

$$-(3\lambda+2\mu)\,\alpha\varepsilon_{kk}(\vartheta-\vartheta_0) \tag{7.74}$$

for the fourth term on the right, obtaining

$$\begin{aligned}\varrho\psi = \varrho\psi_0-\varrho s_0(\vartheta-\vartheta_0)+\frac{\lambda}{2}\,\varepsilon_{ii}\varepsilon_{jj}+\mu\varepsilon_{ij}\varepsilon_{ij}\\ -(3\lambda+2\mu)\,\alpha\varepsilon_{kk}(\vartheta-\vartheta_0)-\frac{\varrho c}{2\vartheta_0}\,(\vartheta-\vartheta_0)^2\,.\end{aligned} \tag{7.75}$$

Applying $(5.30)_1$ to (7.75), we get the stresses

$$\sigma_{ij} = \varrho\,\frac{\partial\psi}{\partial\varepsilon_{ij}} = \lambda\varepsilon_{kk}\delta_{ij}+2\mu\varepsilon_{ij}-(3\lambda+2\mu)\,\alpha\delta_{ij}(\vartheta-\vartheta_0) \tag{7.76}$$

or

$$\sigma_{ij} = [\lambda\varepsilon_{kk}-(3\lambda+2\mu)\,\alpha(\vartheta-\vartheta_0)]\delta_{ij}+2\mu\varepsilon_{ij}\,. \tag{7.77}$$

Equation (7.77) is the thermoelastic generalization of Hooke's law (7.3). Its inversion can be calculated in the same way as (7.8). It reads (P1)

$$\varepsilon_{ij} = \frac{1}{2\mu}\left\{\sigma_{ij}+\left[2\mu\alpha(\vartheta-\vartheta_0)-\frac{\lambda}{3\lambda+2\mu}\,\sigma_{kk}\right]\delta_{ij}\right\}. \tag{7.78}$$

We note that the strains are the same as in (7.8) except for an additional term $\alpha(\vartheta-\vartheta_0)\delta_{ij}$ stemming from the increase in temperature. It corresponds to a uniform extension $\alpha(\vartheta-\vartheta_0)$ in all directions, i.e. to a dilatation $3\alpha(\vartheta-\vartheta_0)$, determined by the so-called *coefficient of thermal expansion*, α.

Applying $(5.30)_2$ to (7.75), we obtain the entropy per unit volume,

$$\varrho s = -\varrho\,\frac{\partial\psi}{\partial\vartheta} = \varrho s_0+(3\lambda+2\mu)\alpha\varepsilon_{kk}+\frac{\varrho c}{\vartheta_0}\,(\vartheta-\vartheta_0), \tag{7.79}$$

and we note that s_0 is the specific entropy in the reference state. The internal energy per unit volume follows from (4.34). It is given by

$$\varrho u = \varrho(\psi+\vartheta s) = \varrho(\psi_0+\vartheta_0 s_0)+\frac{\lambda}{2}\,\varepsilon_{ii}\varepsilon_{jj}+\mu\varepsilon_{ij}\varepsilon_{ij}$$

$$+(3\lambda+2\mu)\,\alpha\vartheta_0\varepsilon_{kk}+\frac{\varrho c}{2\vartheta_0}(\vartheta+\vartheta_0)(\vartheta-\vartheta_0) \qquad (7.80)$$

or

$$\varrho u = \varrho u_0+\frac{\lambda}{2}\,\varepsilon_{ii}\varepsilon_{jj}+\mu\varepsilon_{ij}\varepsilon_{ij}+(3\lambda+2\mu)\alpha\vartheta_0\varepsilon_{kk}+\frac{\varrho c}{2\vartheta_0}(\vartheta^2-\vartheta_0^2), \qquad (7.81)$$

where u_0 is the specific internal energy in the reference state. Equation (7.81) shows that the internal energy cannot be conceived as the sum of the strain energy and a "caloric energy"; its structure is more complicated. By partial differentiation with respect to the temperature, we finally obtain

$$\left.\frac{\partial u}{\partial \vartheta}\right|_{\vartheta_0} = c, \qquad (7.82)$$

and by comparing this to $(5.11)_2$ we see that c is the specific heat capacity at temperature $\vartheta = \vartheta_0$. With this statement, the last of the expansion coefficients in (7.75) has found its physical interpretation.

It has been mentioned that, in the set of differential equations available for the treatment of thermoelastic problems, (7.77) replaces (7.3). If we re-admit small material motions, noting that the acceleration may be written as $u_{i,00}$, we have, in place of (7.2),

$$\varrho u_{i,00} = \varrho f_i+\sigma_{ij,j}. \qquad (7.83)$$

The 15 scalar differential equations (7.1), (7.83), and (7.77) do not suffice, however, for the determination of the 16 unknowns u_i, ε_{ij}, σ_{ij}, and ϑ. The mechanical and the thermal problems are now coupled. To solve the whole complex, we resort to Fourier's law (5.9)

$$q_k = -\lambda'\vartheta_{,k}, \qquad (7.84)$$

where the thermal conductivity is now denoted by λ' to distinguish it from the first of Lamé's constants, and to the first fundamental law (4.63) in the form

$$\varrho u_{,0} = \sigma_{ij}\varepsilon_{ij,0}-q_{k,k}. \qquad (7.85)$$

With (7.84) and (7.85) we have introduced 4 additional scalar differential equations but only 3 additional unknowns q_i. We thus have obtained a complete set of 19 differential equations for the thermomechanical problem. They have to be supplemented, of course, by appropriate boundary and initial conditions.

Upon substitution of (7.81) and (7.77) in (7.85) we get

$$\lambda\varepsilon_{ii}\varepsilon_{jj,0}+2\mu\varepsilon_{ij}\varepsilon_{ij,0}+(3\lambda+2\mu)\alpha\vartheta_0\varepsilon_{kk,0}+\frac{\varrho c}{\vartheta_0}\vartheta\vartheta_{,0}$$
$$= [\lambda\varepsilon_{ii}-(3\lambda+2\mu)\,\alpha(\vartheta-\vartheta_0)]\varepsilon_{jj,0}+2\mu\varepsilon_{ij}\varepsilon_{ij,0}-q_{k,k} \tag{7.86}$$

or shorter

$$\varrho c\frac{\vartheta}{\vartheta_0}\vartheta_{,0} = -(3\lambda+2\mu)\alpha\vartheta\varepsilon_{jj,0}-q_{k,k}\,. \tag{7.87}$$

In the approximation considered here ϑ may be replaced by ϑ_0 on either side. Elimination of q_k by means of (7.84) finally yields

$$\varrho c\vartheta_{,0}+(3\lambda+2\mu)\alpha\vartheta_0\varepsilon_{kk,0} = \lambda'\vartheta_{,kk}\,. \tag{7.88}$$

This is *Duhamel's differential equation of heat conduction*, a generalization of (5.14). Whereas (5.14) has been derived subject to the assumption that the material is rigid, (7.88) takes the elastic dilatations into account. Incidentally, the temperature ϑ enters (7.88) only in the form of partial derivatives. It follows that ϑ might be interpreted here not as the absolute temperature but as the temperature difference with respect to ϑ_0.

By means of (7.88) the heat flow has been eliminated. We are therefore left with 16 scalar differential equations (7.1), (7.83), (7.77), and (7.88) for u_i, ε_{ij}, σ_{ij}, and ϑ. To simplify the problem, let us consider the stationary case, where the material is at rest and the heat flow is steady. Here the left-hand side of (7.83) is zero, and (7.88) reduces to the Laplace equation

$$\vartheta_{,kk} = \Delta\vartheta = 0. \tag{7.89}$$

The boundary conditions for the displacements and stresses have been discussed in Section 7.1. They must be supplemented now by thermal boundary conditions. In general, either the temperature ϑ or the normal component $q_i\nu_i$ of the heat flow are prescribed on a boundary, and on account of (7.84) this is equivalent to the statement that either

$$\vartheta \quad \text{or} \quad \vartheta_{,i}\nu_i = \frac{\partial\vartheta}{\partial\nu} \tag{7.90}$$

are given. By means of (7.89) and these boundary conditions, it is possible to determine the temperature field separately. Once this is done, the remaining equations can be used to solve the mechanical side of the problem. The original problem thus resolves into two simpler ones.

As an example, let us consider the cylindrical tube of Figure 7.10, and let us refer the heat flow to the unit length in the axial direction. We assume

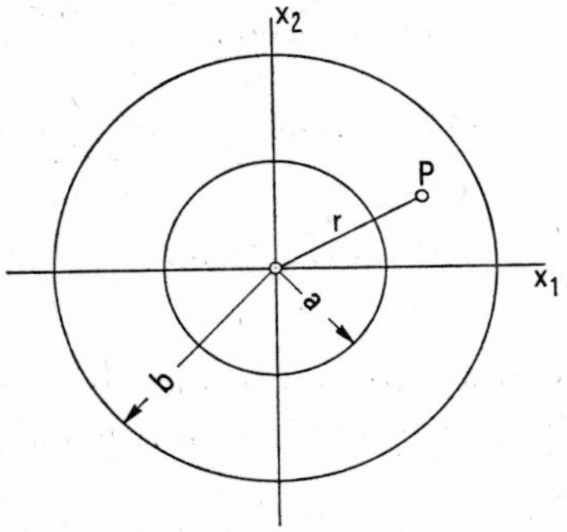

Fig. 7.10 Cross section of a cylindrical tube

that there is a uniform heat supply g per unit time along the interior boundary and that the exterior boundary is maintained at the constant temperature Θ. Instead of specifying the mechanical boundary conditions, we merely assume that they are symmetric with respect to any diameter and any cross section. Thus, the dependent variables are functions of r alone; their derivatives with respect to r will be denoted by a prime. Furthermore, the direction of the heat flow and the displacement is radial, and the principal axes of the strain and stress tensors have radial, circumferential, and axial directions.

In cylindrical coordinates and on account of the assumed symmetries, the differential equation (7.89) has (P2) the form

$$(r\vartheta')' = 0. \tag{7.91}$$

Its general solution is

$$\vartheta = A \ln r + B, \tag{7.92}$$

where A and B are integration constants. The heat flow, determined by (7.84), becomes (P2)

$$q = -\lambda'\vartheta' = -\lambda' \frac{A}{r}. \tag{7.93}$$

The boundary conditions

$$q = \frac{g}{2\pi a} \quad \text{for } (r = a), \qquad \vartheta = \Theta \quad \text{for } (r = b) \tag{7.94}$$

yield

$$A = -\frac{g}{2\pi\lambda'} \qquad B = \frac{g}{2\pi\lambda'} \ln b + \Theta. \tag{7.95}$$

Inserting this in (7.92), we obtain the final expression

$$\vartheta = \frac{g}{2\pi\lambda'} \ln \frac{b}{r} + \Theta \tag{7.96}$$

for the temperature distribution.

Denoting the displacement by u, it is easy to see (P3) that the only non-vanishing strain components are the radial and circumferential extension,

$$\varepsilon_{rr} = u', \qquad \varepsilon_{\alpha\alpha} = \frac{u}{r}. \tag{7.97}$$

According to (7.77) and (7.97), the non-vanishing stress components are

$$\begin{aligned} \sigma_{rr} &= \lambda\left(u' + \frac{u}{r}\right) - (3\lambda + 2\mu)\,\alpha(\vartheta - \Theta) + 2\mu u', \\ \sigma_{\alpha\alpha} &= \lambda\left(u' + \frac{u}{r}\right) - (3\lambda + 2\mu)\,\alpha(\vartheta - \Theta) + 2\mu\frac{u}{r}, \\ \sigma_{zz} &= \lambda\left(u' + \frac{u}{r}\right) - (3\lambda + 2\mu)\,\alpha(\vartheta - \Theta), \end{aligned} \tag{7.98}$$

where z denotes the axial direction and Θ is used as the reference temperature. In the absence of a body force the only nontrivial equilibrium condition (7.2) assumes (P4) the form

$$\sigma'_{rr} + \frac{1}{r}(\sigma_{rr} - \sigma_{\alpha\alpha}) = 0. \tag{7.99}$$

Inserting the stress components (7.98), we have

$$\lambda\left(u' + \frac{u}{r}\right)' - (3\lambda + 2\mu)\alpha\vartheta' + 2\mu u'' + \frac{2\mu}{r}\left(u' - \frac{u}{r}\right) = 0. \tag{7.100}$$

or shorter

$$(\lambda+2\mu)\left(u'+\frac{u}{r}\right)' = (3\lambda+2\mu)\alpha\vartheta' . \tag{7.101}$$

Integrating (7.101) and inserting (7.96), we obtain

$$(\lambda+2\mu)\left(u'+\frac{u}{r}\right) = -(3\lambda+2\mu)\alpha\frac{g}{2\pi\lambda'} \ln r + C, \tag{7.102}$$

where C is a constant. A final integration yields the displacement field

$$u = \frac{r}{2(\lambda+2\mu)}\left[(3\lambda+2\mu)\alpha\frac{g}{2\pi\lambda'}\left(\frac{1}{2}-\ln r\right)+C\right]+\frac{D}{r}, \tag{7.103}$$

where D is another constant (P5), to be determined, together with C, by means of the boundary conditions for the displacement and the stresses.

Problems

1. Verify (7.78).
2. Show that, in cylindrical coordinates and with the symmetries assumed in connection with Figure 7.10, (7.89) assumes the form (7.91) and that (7.84) becomes equivalent to the first equality (7.93).
3. Verify (7.97).
4. Show that in the special problem of Figure 7.10 the equilibrium conditions (7.2) reduce to (7.99).
5. Verify that (7.103) is the general integral of (7.102).

CHAPTER 8

INVISCID GASES

In this chapter the general theorems established in Chapters 2 through 5 will be applied to inviscid gases without internal parameters and in particular to the so-called ideal gas.

For the barotropic case much of the required material has already been prepared in Section 6.1. We do not want, however, to restrict ourselves from the beginning to this special case; therefore, an approach similar to the one used in thermoelasticity (Section 7.4) will be more appropriate. For a fluid this approach is somewhat simpler than for a solid; on the other hand, the fact that the dilatations are not small causes a certain complication.

8.1. Basic equations

The mechanical constitutive equations of an inviscid fluid without internal parameters have been discussed in Section 5.2; they are given by (5.15), $(5.18)_2$, and (5.19). The dissipative stress is identically zero; the quasiconservative stress reduces to the hydrostatic pressure, and the pressure is connected with the density and the temperature by the thermal equation of state, $p = p(\varrho, \vartheta)$. From a mechanical point of view the material is isotropic.

In the case of a gas the dilatations are not necessarily small. According to (5.2), however, the material is elastic, and its response is therefore determined by the caloric equation of state, $\psi = \psi(\varrho, \vartheta)$, i.e. by the specific free energy, expressed as a function of density and temperature. On account of (4.70) and (5.15), we have

$$\varrho\dot{\psi} = -pd_{jj} - \varrho s\dot{\vartheta}. \tag{8.1}$$

Using (2.14) and the continuity equation (6.1), we further obtain

$$d_{jj} = v_{j,j} = -\frac{\dot{\varrho}}{\varrho}. \tag{8.2}$$

From (8.1) and (8.2) we conclude that

$$\varrho\dot{\psi} = \frac{p}{\varrho}\dot{\varrho} - \varrho s\dot{\vartheta} \tag{8.3}$$

and that, therefore,

$$\frac{p}{\varrho^2} = \frac{\partial\psi}{\partial\varrho}, \qquad -s = \frac{\partial\psi}{\partial\vartheta}. \tag{8.4}$$

The first of these equations is the thermal equation of state; the second, already obtained in $(4.37)_2$, supplies the specific entropy, and the specific internal energy finally follows from (4.34).

The differential equations at our disposal for the investigation of moving gases are similar to the ones in Section 6.1. The motion is subject to the continuity equation (6.1), the theorem of linear momentum (6.2), and the constitutive equation $(6.3)_1$. As long as we do not limit ourselves to barotropic cases, $(6.3)_2$ must be replaced by the thermal equation of state. This equation, however, contains the temperature and makes it necessary to supplement the equations already mentioned by Fourier's law and by the first fundamental law (4.63) of thermodynamics. Assuming that the gas is isotropic with respect to heat conduction, we use Fourier's law in the form (5.9), obtaining in this manner the following set of equations:

$$\dot{\varrho} + \varrho v_{j,j} = \varrho_{,0} + (\varrho v_j)_{,j} = 0, \tag{8.5}$$

$$\varrho a_i = \varrho\dot{v}_i = \varrho(v_{i,0} + v_{i,j}v_j) = \varrho f_i - p_{,i}, \tag{8.6}$$

$$p = p(\varrho, \vartheta), \tag{8.7}$$

$$q_i = -\lambda\vartheta_{,i}, \tag{8.8}$$

$$\varrho\dot{u} = \frac{p}{\varrho}\dot{\varrho} - q_{j,j}. \tag{8.9}$$

In writing these equations down we have eliminated σ_{ij} by means of $(6.3)_1$ on the right-hand sides of (8.6) and (8.9); besides, we have used (8.2) in (8.9). The whole system represents 9 scalar differential equations for the unknowns ϱ, p, v_i, ϑ, and q_i.

It is in general not possible to separate the mechanical and the thermodynamic parts of the problem described by (8.5) through (8.9). This possibility exists, however, if the temperature ϑ can be eliminated from (8.7), i.e. if the process is *barotropic*. Before restricting ourselves to this case, let us add a few remarks concerning heat capacities.

In an arbitrary process the heat supply per unit mass and time is $-q_{j,j}/\varrho$. Let us define the *specific heat capacity* for the process considered as the quotient $c = -q_{j,j}/\varrho\dot{\vartheta}$ of this heat supply and the rate of temperature increase $\dot{\vartheta}$. It follows immediately that

$$\varrho c\dot{\vartheta} = -q_{j,j}, \tag{8.10}$$

where c depends on the independent state variables ϱ, ϑ and on the particular process. Provided we keep the density (and hence the volume per unit mass) constant, equation (8.10), now written as

$$\varrho c^{(v)}\dot{\vartheta} = -(q_{j,j})_{\varrho=\text{const}}, \tag{8.11}$$

defines the specific heat capacity $c^{(v)}$ for constant volume. Comparing (8.11) and (8.9), we obtain

$$c^{(v)}\dot{\vartheta} = (\dot{u})_{\varrho=\text{const}} \tag{8.12}$$

and hence

$$c^{(v)} = \frac{\partial u}{\partial \vartheta} \tag{8.13}$$

as in $(5.11)_2$. If, on the other hand, the pressure is maintained constant, (8.10) defines the corresponding specific heat capacity denoted by $c^{(p)}$.

The simplest case of the barotropic type is the *isothermal* process. Here (8.7) reduces immediately to $p = p(\varrho)$. It is easy to see, however, that also the *adiabatic* process is barotropic. Here (8.3) and (8.9) yield

$$\dot{\psi} = \dot{u} - s\dot{\vartheta}. \tag{8.14}$$

Comparing this to (4.68), we find $\dot{s} = 0$. In fact, since the gas is elastic and the heat conduction is zero, the material element experiences neither an entropy production nor an entropy supply, and it follows that $s(\varrho, \vartheta) =$ const. This equation permits to express ϑ as a function of ϱ, and the thermal equation of state reduces again to $p = p(\varrho)$.

As an example, let us consider the *ideal gas*, defined by the specific free energy

$$\psi = \int c^{(v)}\,\mathrm{d}\vartheta - \vartheta \int \frac{c^{(v)}}{\vartheta}\,\mathrm{d}\vartheta + R\vartheta \ln \varrho, \tag{8.15}$$

where R is a constant and the specific heat capacity $c^{(v)}$ is a function of the temperature alone. Applying $(8.4)_1$ to the function (8.15), we obtain the thermal equation of state

$$\frac{p}{\varrho} = R\vartheta \tag{8.16}$$

of the ideal gas; $(8.4)_2$ yields the specific entropy

$$s = \int \frac{c^{(v)}}{\vartheta}\,d\vartheta - R\ln\varrho, \tag{8.17}$$

and (4.34) supplies the specific internal energy

$$u = \int c^{(v)}\,d\vartheta. \tag{8.18}$$

Since $c^{(v)}$ is independent of ϱ, the same is true for u. It thus follows from (8.18) that $\dot{u} = c^{(v)}\dot{\vartheta}$, the first fundamental law therefore assuming the form

$$\varrho c^{(v)}\dot{\vartheta} = \frac{p}{\varrho}\dot{\varrho} - q_{j,j}. \tag{8.19}$$

Material differentiation of (8.16) yields

$$\frac{\dot{p}}{\varrho} - \frac{p}{\varrho^2}\dot{\varrho} = R\dot{\vartheta}. \tag{8.20}$$

It follows from (8.19) and (8.20) that

$$\varrho c^{(v)}\dot{\vartheta} = \dot{p} - \varrho R\dot{\vartheta} - q_{j,j} \tag{8.21}$$

and that, in consequence,

$$-(q_{j,j})_{p=\mathrm{const}} = \varrho(c^{(v)}+R)\dot{\vartheta}. \tag{8.22}$$

In view of (8.10), the specific heat capacity for constant pressure is

$$c^{(p)} = c^{(v)} + R. \tag{8.23}$$

From this equation, well known in thermodynamics, we conclude that R is the so-called *gas constant* referred to the unit mass. It is important to note, however, that in most texts on thermodynamics R and the specific heat capacities are referred to the mole with the result that R is independent of the particular gas considered. For our purposes it is more convenient to use the unit mass for reference, dividing the customary values by the molecular mass.

If the process is *isothermal*, (8.16) yields the barotropic equation

$$p = \alpha\varrho, \quad \text{where} \quad \alpha = R\vartheta \tag{8.24}$$

is a constant. Since, on the other hand, the *adiabatic* process is also isentropic, (8.17) yields

$$\dot{s} = c^{(v)}\frac{\dot{\vartheta}}{\vartheta} - R\frac{\dot{\varrho}}{\varrho} = 0. \tag{8.25}$$

Eliminating ϑ and $\dot{\vartheta}$ by means of (8.16) and (8.20) respectively, we obtain

$$c^{(v)}\frac{\dot{p}}{p} = (c^{(v)}+R)\frac{\dot{\varrho}}{\varrho} = c^{(p)}\frac{\dot{\varrho}}{\varrho}. \tag{8.26}$$

Provided $c^{(v)}$ and $c^{(p)}$ are constant, integration of (8.23) is simple, supplying

$$p = \beta\varrho^{\varkappa}, \quad \text{where} \quad \varkappa = \frac{c^{(p)}}{c^{(v)}} \tag{8.27}$$

and where β is an arbitrary constant.

In practice many processes are nearly isothermal or adiabatic. We will therefore limit ourselves for the remainder of this chapter to the barotropic case, dropping at the same time the restriction to ideal gases. Here the mechanical and thermodynamic parts of the problem can be separated. In fact, if the thermal equation of state (8.7) is replaced by the barotropic equation $\varrho = \varrho(p)$, the system (8.5) through (8.7) represents 5 differential equations for the mechanical unknowns ϱ, p, v_i alone. These equations are equivalent to (6.1) through (6.3), and this enables us to use all of the results obtained in Section 6.1:

Introducing the pressure function (6.5), we reduce Euler's differential equation (8.6) to the form (6.8). Assuming that the specific body force has a potential, we further obtain (6.13). On the strength of Thomson's vorticity theorem, we finally assume that the flow is irrotational. Thus, the velocity is the gradient of a potential, and (6.13) assumes the form (6.18).

It is sometimes possible to neglect the body force. Equation (6.18) then reduces to

$$\varphi_{,0}+P+\frac{v^2}{2} = 0. \tag{8.28}$$

Let us add the continuity equation (8.5) in the form

$$\varrho_{,0}+(\varrho v_j)_{,j} = 0 \tag{8.29}$$

and let us note that v_j might be expressed in terms of φ and that, once the barotropic equation is given, P may be considered as a function of ϱ. We are thus left with two differential equations (8.28) and (8.29) for ϱ and φ.

Problems

1. Consider a portion of the earth's surface, assuming that it is plane. In a first approximation, the atmosphere may be treated as an ideal gas the

temperature of which decreases according to

$$\vartheta = \vartheta_0\left(1-\frac{x_3}{h}\right), \tag{8.30}$$

where ϑ_0 is the temperature at sea level, x_3 the height above sea level and h the height in which the temperature would become zero if (8.30) were valid for arbitrary values of x_3. Investigate the equilibrium of the atmosphere under the influence of its weight, considering the gravity acceleration g as a constant. Denote the pressure at sea level by p_0 and determine the distributions $p(x_3)$ and $\varrho(x_3)$.

2. Solve P1 once more, replacing (8.30) by the assumption that the atmosphere is isothermal. Determine its height h.

3. Solve P1 once more, replacing (8.30) by the relation

$$p = \beta\varrho^{\varkappa} \tag{8.31}$$

between pressure and density. Determine also the temperature distribution $\vartheta(x_3)$ and the height h of the atmosphere.

4. The solutions of P1 and P3 imply a vertical heat flow in the atmosphere. In P3 this seems to be inconsistent with the fact that (8.31) corresponds to the adiabatic equation (8.27). Solve the dilemma.

8.2. Simple applications

The differential equations (8.28) and (8.29) are nonlinear. However, we can linearize them in the vicinity of an equilibrium state, and this possibility is important in the field of *acoustics*.

In the case of equilibrium, the velocity v_i is identically zero. On account of (8.28) the same is true for the pressure function, and it follows from (6.5) and the barotropic equation that the values p_0 and ϱ_0 of the pressure and density respectively are constant in space and time. In acoustics the perturbations of the equilibrium, i.e. the velocities and the deviations of the density and pressure from ϱ_0, p_0 respectively are small. Making use of this fact, we truncate the power series expansion of the function $p(\varrho)$,

$$p = p_0+\left(\frac{\mathrm{d}p}{\mathrm{d}\varrho}\right)_{\varrho_0}(\varrho-\varrho_0)+\ \ldots, \tag{8.32}$$

after the linear term, obtaining

$$p = p_0+c^2(\varrho-\varrho_0), \tag{8.33}$$

where

$$c^2 = \left(\frac{\mathrm{d}p}{\mathrm{d}\varrho}\right)_{\varrho_0}. \tag{8.34}$$

In the same approximation the pressure function becomes

$$P = \frac{p-p_0}{\varrho_0} = c^2\left(\frac{\varrho}{\varrho_0}-1\right), \tag{8.35}$$

where use has been made of (8.33), and it follows that

$$\varrho = \varrho_0\left(\frac{P}{c^2}+1\right). \tag{8.36}$$

Linearizing (8.28), we obtain

$$\varphi_{,0} = -P. \tag{8.37}$$

The linearized form of (8.29) is

$$\varrho_{,0}+\varrho_0 v_{j,j} = 0 \tag{8.38}$$

or, if we make use of (8.36) and (6.15),

$$\frac{\varrho_0}{c^2}P_{,0}+\varrho_0\varphi_{,jj} = 0. \tag{8.39}$$

It follows that

$$P_{,0} = -c^2\varphi_{,jj}. \tag{8.40}$$

Equations (8.37) and (8.40) yield

$$\varphi_{,00} = c^2\varphi_{,jj}, \qquad P_{,00} = c^2P_{,jj}. \tag{8.41}$$

In the approximation considered here the velocity potential and the pressure functions are thus subject to the co-called *wave equation.*

As an example, let us consider the case where φ and P depend on t and x_1 alone. The right-hand sides of (8.41) then reduce to $c^2\varphi_{,11}$ and $c^2P_{,11}$ respectively, and the general solution of, e.g. $(8.41)_1$ is

$$\varphi = f\left(t-\frac{x_1}{c}\right)+g\left(t+\frac{x_1}{c}\right), \tag{8.42}$$

where f and g are arbitrary functions subject only to the condition that they are twice differentiable. The two terms on the right represent plane waves perpendicular to the axis x_1 and moving in the directions of increasing or decreasing x_1 respectively with the velocity c. It thus turns out that the quantity c introduced by means of (8.34) is the velocity of propagation of these waves or, in acoustical terms, the *velocity of sound.*

Let us assume for a moment that an ideal gas is isothermal, i.e. that the propagation of an acoustic disturbancy is sufficiently slow for the temperature to remain practically constant. Equations (8.34) and (8.24) then yield

$$c = \sqrt{\frac{p_0}{\varrho_0}}. \tag{8.43}$$

If we assume, on the other hand, that the process is adiabatic, i.e. so fast that practically no heat exchange takes place, (8.34), applied to (8.27), yields

$$c = \sqrt{\varkappa \frac{p_0}{\varrho_0}}. \tag{8.44}$$

In the case of air $\varkappa$ is approximately 1,4, and if we evaluate (8.43) and (8.44) for a temperature of zero centigrades and for normal atmospheric pressure, we obtain 280 and 331 m/s respectively. Experiments confirm the higher value, and we conclude that acoustic phenomena are practically adiabatic processes.

Let us now return to the nonlinear problem, characterized by the differential equations (8.28) and (8.29), and let us apply them to a *steady potential flow.* Here (8.28) reduces to

$$P + \frac{v^2}{2} = 0. \tag{8.45}$$

The pressure function (6.5) contains an arbitrary constant p_0 representing, according to (8.45), the pressure at a stagnation point. In view of certain applications, e.g. uniform flow around an arbitrary solid, where the pressure at infinity is usually denoted by p_0, it is convenient to leave the significance of p_0 open, writing

$$P + \frac{v^2}{2} = \text{const} \tag{8.46}$$

in place of (8.45) and interpreting (8.46) as a *generalized Bernoulli equation.* Differentiation with respect to x_j yields

$$\frac{1}{2}(v_i v_i)_{,j} = -P_{,j} = -\frac{1}{\varrho} p_{,j} \tag{8.47}$$

or

$$v_i v_{i,j} = -\frac{1}{\varrho}\frac{\mathrm{d}p}{\mathrm{d}\varrho}\varrho_{,j} = -\frac{c^2}{\varrho}\varrho_{,j}, \tag{8.48}$$

where

$$c^2 = \frac{\mathrm{d}p}{\mathrm{d}\varrho}. \tag{8.49}$$

In analogy to (8.34), this quantity may be interpreted as the square of the so-called *local velocity of sound*. It depends on the value of ϱ and hence usually differs from point to point.

For a steady flow the continuity equation (8.29) may be written

$$\varrho v_{j,j} + \varrho_{,j} v_j = 0. \tag{8.50}$$

Multiplying (8.48) by v_j and using (8.50), we obtain

$$v_i v_j v_{i,j} = -\frac{c^2}{\varrho}\varrho_{,j} v_j = c^2 v_{j,j} \tag{8.51}$$

or

$$(c^2 \delta_{ij} - v_i v_j) v_{i,j} = 0. \tag{8.52}$$

This equation with the alternate form

$$(c^2 \delta_{ij} - \varphi_{,i}\varphi_{,j})\varphi_{,ij} = 0 \tag{8.53}$$

is the *fundamental equation of gas dynamics*. It is a nonlinear partial differential equation of the second order in φ. The nonlinearity is not only due to the presence of the second term between parentheses in (8.53) but also to the fact that c is a function of ϱ and thus depends, on account of (6.5) and (8.46), on v_i and hence on $\varphi_{,i}$.

If $v \ll c$, the second term between parentheses in (8.53) is small compared to the first one, and the fundamental equation reduces in a first approximation to the Laplace equation (6.20) governing the steady potential flow of an incompressible liquid. This observation justifies the approximate treatment of a gas as an incompressible fluid provided the velocity is small everywhere compared to the velocity of sound. If this is not the case, however, the nonlinear differential equation (8.53) must be used, and the problem is more difficult. In particular, the nonlinearity of (8.53) precludes the superposition of simple flows in the solution of more complicated problems.

There exists a practically important case where (8.53) can be linearized. If a uniform flow of velocity u in the direction x_1 is only slightly disturbed, e.g. by the presence of a slender body, the deviations of the velocity from u and the partial derivatives of the velocity components with respect to the coordinates x_i are small. Thus (8.52), in a first approximation, reduces to

$$c^2 v_{j,j} - v_1^2 v_{1,1} = 0. \tag{8.54}$$

Moreover, in the approximation considered here v_1 may be replaced by u and c by a constant, namely the velocity of sound at infinity. Thus (8.54) assumes the form

$$\left(1-\frac{u^2}{c^2}\right)v_{1,1}+v_{2,2}+v_{3,3} = 0, \tag{8.55}$$

and the corresponding linearized form of (8.53) becomes

$$\left(1-\frac{u^2}{c^2}\right)\varphi_{,11}+\varphi_{,22}+\varphi_{,33} = 0. \tag{8.56}$$

If $u < c$, the flow is referred to as *subsonic*; for $u > c$ we call it *supersonic*. In the first case the coefficients of $\varphi_{,11}, \ldots$ in (8.56) have the same sign, and the differential equation is referred to as *elliptic*. It follows from the theory of partial differential equations that the solutions are analytic. In the second case the signs of the terms in (8.56) are different. The differential equation is referred to as *hyperbolic*, and the solution may exhibit discontinuities. The distinction between the two types can also be made in the nonlinear case, but it is obvious that there the type depends on the local values of c and of the velocity of flow. Thus, the type of the differential equation may be different in different domains, and the fact that their boundaries are unknown presents an additional difficulty.

Problems

1. Write the wave equation in spherical coordinates. Specialize it for the case of a disturbancy depending only on r and t, where r is the distance from the origin. Find the general solution and discuss it.

2. Consider an infinite horizontal layer of thickness h (Fig. 8.1) of an ideal liquid under the influence of its weight, and investigate its surface waves, assuming that the displacements are small and that the motion is irrotational. Introduce the velocity potential φ and formulate the differential equation for φ and the boundary conditions for $x_3 = 0$ and $x_3 = h$. Hints: Since the free surface is always composed of the same material points, the material derivative of the pressure at the surface is zero. Moreover, for small

motions all nonlinear terms may be suppressed, and in the last boundary condition the free surface may be replaced by the plane $x_3 = h$.

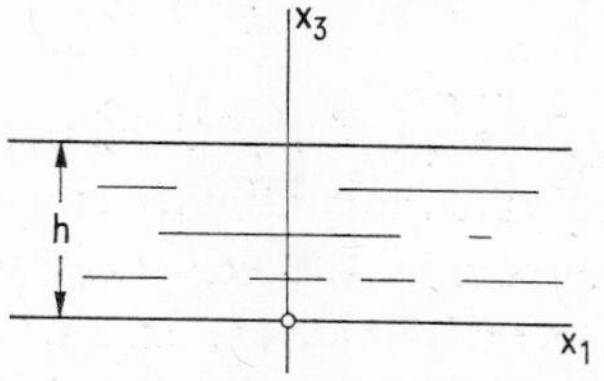

Fig. 8.1 Horizontal layer of an ideal liquid

3. Investigate and discuss those solutions of the last problem that have the form

$$\varphi = f(x_3) \sin \varkappa(x_1 - ct).$$

8.3. Subsonic and supersonic flow

For the study of the typical features of subsonic and supersonic flow, the velocity field of a source in a uniform stream is an appropriate example. It is convenient, however, to consider the problem in a coordinate system moving with the uniform flow, i.e. to assume that the fluid is at rest at infinity whereas the source moves along, e.g. the axis x_1.

Let us first consider a *source* at rest in an incompressible liquid which itself is at rest sufficiently far from the source. If the source is located at the origin O, the velocity potential is

$$\varphi = -\frac{g}{4\pi r}, \tag{8.57}$$

where g is a constant called the *strength* of the source and where $r = (x_j x_j)^{1/2}$ is the distance from the origin. The velocity is given by

$$v_k = \varphi_{,k} = \frac{g}{4\pi r^2} r_{,k} = \frac{g}{4\pi r^3} x_k; \tag{8.58}$$

it has the direction of the radius vector, and its magnitude decreases with the square of the distance from O. By means of (8.58) it is easy (P1) to verify that (8.57) satisfies the Laplace equation (6.20) everywhere outside the origin. The flow through a sphere of radius r with the center O is given by

$$4\pi r^2 (v_k v_k)^{1/2} = \frac{g}{r} (x_k x_k)^{1/2} = g \tag{8.59}$$

and hence is equal to the strength of the source, irrespective of the radius r.

If the strength of the source depends on the time, the potential

$$\varphi = -\frac{1}{4\pi r} g(t) \tag{8.60}$$

still satisfies Laplace's equation, which, for the potential flow of an incompressible liquid, is equivalent to the continuity equation (6.1). Except for the immediate vicinity of the source, the velocities and the deviations of the pressure from their value at infinity are small. Thus, the conditions for treatment with the methods of acoustics (Section 8.2) are satisfied if we now make the transition to a compressible fluid. Here any perturbation propagates with the velocity of sound. The potential (8.60) is therefore to be replaced by

$$\varphi = -\frac{1}{4\pi r} g\left(t - \frac{r}{c}\right). \tag{8.61}$$

It can in fact be shown (P2) that (8.61) is a solution of the wave equation $(8.41)_1$.

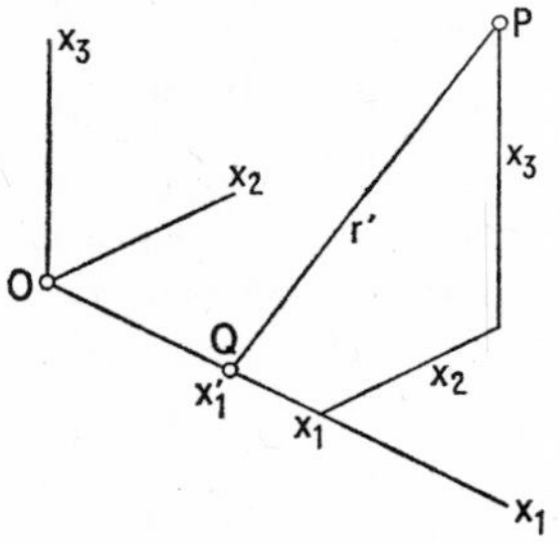

Fig. 8.2 Source on axis x_1

Let us next consider the case where the source is transferred to the point Q (Fig. 8.2) with coordinates $(x_1', 0, 0)$. The potential (8.61) in an arbitrary point P now becomes

$$\varphi = -\frac{1}{4\pi r'} g\left(t - \frac{r'}{c}\right), \tag{8.62}$$

where

$$r' = [(x_1 - x_1')^2 + x_2^2 + x_3^2]^{1/2} \tag{8.63}$$

is the distance of P from Q. If the source flows only instantaneously at time $t' = 0$, its strength can be represented by

$$g(t) = h\,\delta(t), \tag{8.64}$$

where h is a constant and $\delta(t)$ is *Dirac's delta function*, defined by

$$\delta(t) = \lim_{\tau \to 0} \begin{cases} \dfrac{1}{\tau} & \left(-\dfrac{\tau}{2} \leqslant t \leqslant \dfrac{\tau}{2}\right) \\ 0 & \left(|t| > \dfrac{\tau}{2}\right) \end{cases} \qquad (8.65)$$

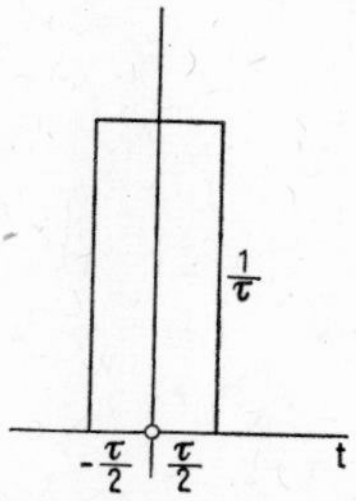

Fig. 8.3 Dirac's delta function

or, in geometrical terms, by the square of Figure 8.3, the base length of which tends to zero while its height increases in such a way that the area remains equal to unity. Since

$$\int_{-\infty}^{\infty} \delta(t)\,\mathrm{d}t = 1, \qquad (8.66)$$

the entire output of the source is h. Moreover, if $f(t)$ is an arbitrary continuous function, we obviously have

$$\int_{-\infty}^{\infty} \delta(t) f(t)\,\mathrm{d}t = f(0). \qquad (8.67)$$

If, finally, the source Q flows at time t', its strength is

$$g(t) = h\,\delta(t - t'), \qquad (8.68)$$

and the potential (8.62) becomes

$$\varphi = -\frac{h}{4\pi r'}\,\delta\left(t - t' - \frac{r'}{c}\right). \qquad (8.69)$$

Let us now consider a source Q of constant strength g moving with the velocity u along the axis x_1 in the direction of decreasing values of x_1. If the source passes the origin at time $t' = 0$, its equation of motion is

$$x_1' = -ut'. \qquad (8.70)$$

As an approximation, the moving source may be replaced by a sequence of stationary sources x_1' distributed along the axis x_1 and flowing instantaneously at the successive times $t' = -x_1'/u$. With increasing density of the distribution of sources, the approximation tends to the original situation provided we choose the output h per unit length such that

$$g\,\mathrm{d}t' = -h\,\mathrm{d}x_1' \tag{8.71}$$

or, on account of (8.70),

$$h = \frac{g}{u}. \tag{8.72}$$

The contribution of an element of the axis x_1 to the potential (8.69) now is

$$\mathrm{d}\varphi = -\frac{g}{4\pi u}\,\delta\left(t - t' - \frac{r'}{c}\right)\frac{1}{r'}\,\mathrm{d}x_1', \tag{8.73}$$

and the total potential is obtained by integration over the axis x_1. In particular, the potential at the point P at the moment $t = 0$ when the source passes the origin is given by

$$\varphi = -\frac{g}{4\pi u}\int_{-\infty}^{\infty}\delta\left(\frac{x_1'}{u} - \frac{r'}{c}\right)\frac{1}{r'}\,\mathrm{d}x_1', \tag{8.74}$$

where use has once more been made of (8.70).

On account of (8.63) the potential (8.74) is a function of x_1, x_2, x_3, i.e. of the location of P. The integrand also depends on x_1', while r' is a function of the variables already mentioned. For the evaluation of (8.74) it is necessary to decide on a single integration variable, and it is convenient to choose the time

$$\tau = \frac{x_1'}{u} - \frac{r'}{c}. \tag{8.75}$$

Substituting it in (8.74), we obtain

$$\varphi = -\frac{g}{4\pi u}\int \delta(\tau)\frac{1}{r'}\frac{\mathrm{d}x_1'}{\mathrm{d}\tau}\,\mathrm{d}\tau, \tag{8.76}$$

where x_1' and r' are to be considered as functions of τ and the limits of the integral are yet to be determined.

In the *subsonic* case ($u < c$) it follows immediately from (8.75) that for $x_1' \to \mp\infty$, also $\tau \to \mp\infty$. Besides, from (8.75) and (8.63) we obtain

$$\frac{d\tau}{dx_1'} = \frac{1}{u} + \frac{1}{c}\,\frac{x_1 - x_1'}{r'}\,. \tag{8.77}$$

Thus, $d\tau/dx_1' = 0$ implies

$$[(x_1' - x_1)^2 + x_2^2 + x_3^2]^{1/2} = \frac{u}{c}\,(x_1' - x_1), \tag{8.78}$$

where the square root has to be taken with the positive sign. For $u < c$ and an arbitrarily given point P, this equation has no solution since the right-hand side is smaller than the left-hand side. Thus, τ is monotonic in x_1', and integration of (8.76) between $-\infty$ and $+\infty$ yields, according to (8.67),

$$\varphi = -\frac{g}{4\pi u}\left(\frac{1}{r'}\,\frac{dx_1'}{d\tau}\right)_{\tau=0} \tag{8.79}$$

or, after some manipulation (P3),

$$\varphi = -\frac{g}{4\pi}\left[x_1^2 + \left(1 - \frac{u^2}{c^2}\right)(x_2^2 + x_3^2)\right]^{-1/2}. \tag{8.80}$$

For $c \to \infty$ (8.80) reduces to the potential of a source in an incompressible liquid. For finite values of c, on the other hand, Figure 8.4 supplies a geometric interpretation of the result. Equation (8.75) shows that for a given value of x_1' the condition $\tau = 0$ is the equation of the sphere on which the perturbations emanating from x_1' are located at the moment $t = 0$ when the source reaches the origin. Since the perturbations travel faster than the source, each of these spheres encloses the origin and also the spheres of smaller radius; besides, at $t = 0$ the influence of the source has already

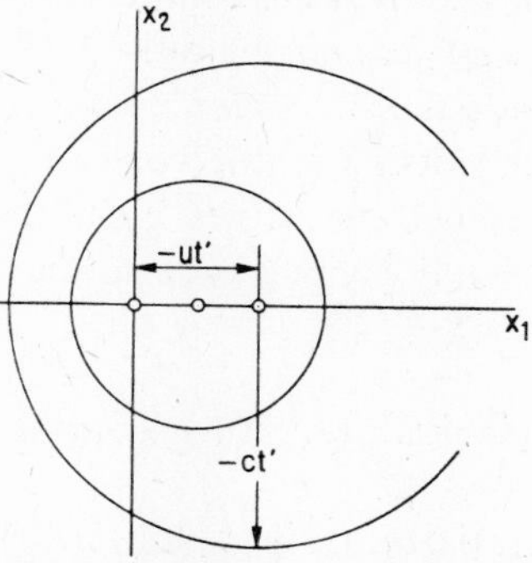

Fig. 8.4 Wave fronts for subsonic motion of a source

spread over the whole space. The fact that the distances between the spheres are smaller ahead of the moving source than behind it is equivalent to the so-called *Doppler effect*.

In the *supersonic* case ($u > c$) it follows from (8.75) that $\tau \to -\infty$ both for $x_1' \to \mp\infty$. Besides, (8.78) now has a solution for any given location of P. It follows that, while x_1' increases from $-\infty$ to $+\infty$, the variable τ increases from $-\infty$, reaching a maximum and decreasing again beyond limit. If, in this process, τ remains negative, (8.76) yields $\varphi = 0$. If τ becomes positive, it passes zero twice, and the potential becomes twice the value given by (8.80). The geometric interpretation of these results follows from Figure 8.5. Since the source travels faster than the perturbations, the spheres

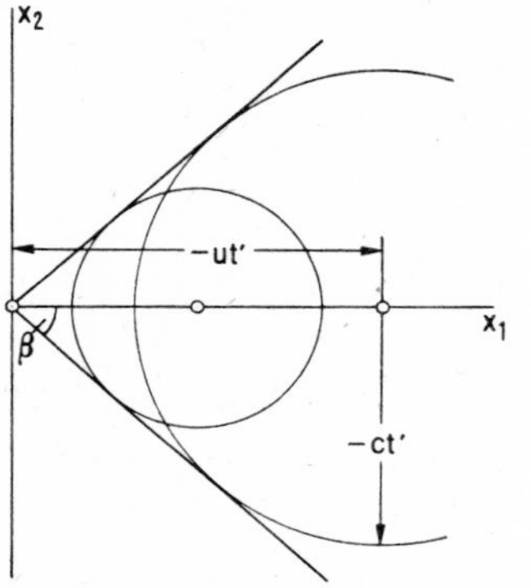

Fig. 8.5 Wave fronts for supersonic motion of a source

representing the locations of the disturbances at $t = 0$ do not intersect the plane x_2, x_3. They even lie inside an envelope: the semi-cone determined by the angle $\beta = \arcsin (c/u)$. The semi-cone is referred to as the *Mach cone*, the angle β as the *Mach angle*, and the ratio $M = u/c$ is called the *Mach number* of the flow. Inside the Mach cone any point P lies on two spheres corresponding to $\tau = 0$, outside of the cone on none of them. The cone enclosing the domain reached by the perturbations at $t = 0$ thus separates the regions where $\varphi = 0$ and where φ is twice the value given by (8.80).

Problems

1. Show that the potential (8.57) satisfies Laplace's equation (6.20) outside the origin.
2. Verify that the potential (8.61) is a solution of the wave equation $(8.41)_1$.
3. Calculate (8.80) from (8.79).

CHAPTER 9

VISCOUS FLUIDS

The fluids treated in Chapters 6 and 8 have been assumed to be inviscid. In the present chapter the viscosity will be taken into account, but the fluid will still be assumed to be free of internal parameters.

After a brief glance at viscous gases we will concentrate on incompressible viscous liquids. Section 9.3 will offer a brief outlook on the problem of turbulence, and in Section 9.4 the non-Newtonian liquid will be discussed.

9.1. Basic equations

The mechanical constitutive equations of a viscous fluid without internal parameters have been discussed in Section 5.2. The quasiconservative stress is determined by the specific free energy $\psi(\varrho, \vartheta)$. According to (5.15) and $(5.18)_2$ it consists of a hydrostatic pressure connected with the independent state variables ϱ and ϑ by the thermal equation of state. On account of the internal friction, $\sigma_{ij}^{(q)}$ is to be supplemented by a dissipative stress given by (5.20). Unlike the reversible stress, $\sigma_{ij}^{(d)}$ is not a state function but depends on the deformation rate and possibly on the temperature. If the fluid is isotropic, equation (5.20) assumes the special form (5.21). For the *Newtonian fluid* it reduces to the linear relationship (5.23), where λ and μ are the viscosity coefficients (possibly dependent on ϑ). We thus have

$$\sigma_{ij} = \sigma_{ij}^{(q)} + \sigma_{ij}^{(d)}, \tag{9.1}$$

where

$$\sigma_{ij}^{(q)} = -p\delta_{ij}, \qquad \sigma_{ij}^{(d)} = \lambda d_{kk}\delta_{ij} + 2\mu d_{ij}. \tag{9.2}$$

The specific power of dissipation is given by (4.38) and assumes the form

$$l^{(d)} = \frac{1}{\varrho}\sigma_{ij}^{(d)} d_{ij} = \frac{1}{\varrho}(\lambda d_{ii} d_{jj} + 2\mu d_{ij} d_{ij}). \tag{9.3}$$

Assuming that the second fundamental law of thermodynamics holds in the separated form (4.80), we obtain

$$\lambda d_{ii}d_{jj}+2\mu d_{ij}d_{ij} \geqslant 0, \qquad \vartheta_{,i}q_i = -\lambda'\vartheta_{,i}\vartheta_{,i} \leqslant 0, \tag{9.4}$$

where use has been made of Fourier's law (5.9) and where the thermal conductivity is denoted by λ' to distinguish it from the viscosity coefficient λ. From $(9.4)_2$ we obtain

$$\lambda' \geqslant 0 \tag{9.5}$$

in confirmation of the side condition in (5.9). Since

$$d_{ij}d_{ij} = (d'_{ij}+\tfrac{1}{3}d_{kk}\delta_{ij})\,(d'_{ij}+\tfrac{1}{3}d_{ll}\delta_{ij}) = d'_{ij}d'_{ij}+\tfrac{1}{3}d_{ii}d_{jj}\,, \tag{9.6}$$

$(9.4)_1$ may be written as

$$2\mu d'_{ij}d'_{ij}+(\lambda+\tfrac{2}{3}\mu)d_{ii}d_{jj} \geqslant 0. \tag{9.7}$$

In consequence,

$$\mu \geqslant 0, \qquad 3\lambda+2\mu \geqslant 0. \tag{9.8}$$

Incidentally, (9.7) confirms the observation already made in Section 5.2 that the resistance to distorsions is represented by the viscosity coefficient μ, whereas the so-called *bulk viscosity* $\lambda+\frac{2}{3}\mu$ represents resistance to dilatations.

Let us now consider an *ideal gas* of the Newtonian type and let us assume that λ and μ are constant. The differential equations governing the motion and the heat flow are generalizations of (8.5) through (8.9). The continuity equation may be used in either one of the two forms

$$\dot{\varrho}+\varrho v_{j,j} = \varrho_{,0}+(\varrho v_j)_{,j} = 0. \tag{9.9}$$

On account of (9.1), (9.2), and (2.14) we have

$$\sigma_{ij} = -p\delta_{ij}+\lambda v_{k,k}\delta_{ij}+\mu(v_{i,j}+v_{j,i}) \tag{9.10}$$

and hence

$$\sigma_{ij,j} = -p_{,i}+\mu v_{i,jj}+(\lambda+\mu)v_{j,ji}\,. \tag{9.11}$$

On substitution in the theorem of linear momentum (6.2) we obtain

$$\varrho(v_{i,0}+v_{i,j}v_j) = \varrho f_i-p_{,i}+\mu v_{i,jj}+(\lambda+\mu)v_{j,ji} \tag{9.12}$$

in place of (8.6). The thermal equation of state (8.7) assumes the special form (8.16), while Fourier's law remains unaltered except for the fact that the thermal conductivity is now denoted by λ'. We thus have

$$\frac{p}{\varrho} = R\vartheta, \qquad q_i = -\lambda'\vartheta_{,i} \tag{9.13}$$

in place of (8.7) and (8.8) respectively. Inserting (8.18), (9.10), (2.14), and $(9.13)_2$ in (4.63), we finally obtain the first fundamental law in the form

$$\varrho c^{(v)}(\vartheta_{,0}+\vartheta_{,j}v_j) = -pv_{j,j}+\lambda v_{i,i}v_{j,j}+\frac{\mu}{2}(v_{i,j}+v_{j,i})(v_{i,j}+v_{j,i})+\lambda'\vartheta_{,jj}. \tag{9.14}$$

In the special case where the bulk viscosity $\lambda+\frac{2}{3}\mu$ is zero, the Newtonian fluid becomes a *Navier–Stokes fluid* (Section 5.2). The dissipative stress $(9.2)_2$ is now

$$\sigma_{ij}^{(d)} = 2\mu(d_{ij}-\tfrac{1}{3}d_{kk}\delta_{ij}) = 2\mu d'_{ij}, \tag{9.15}$$

and the specific power of dissipation (9.3) becomes

$$l^{(d)} = \frac{2\mu}{\varrho}d'_{ij}d_{ij} = \frac{2\mu}{\varrho}d'_{ij}d'_{ij}. \tag{9.16}$$

The theorem of linear momentum (9.12) reads

$$\varrho(v_{i,0}+v_{i,j}v_j) = \varrho f_i-p_{,i}+\mu(v_{i,jj}+\tfrac{1}{3}v_{j,ji}), \tag{9.17}$$

and the first fundamental law (9.14) reduces to

$$\varrho c^{(v)}(\vartheta_{,0}+\vartheta_{,j}v_j) = -pv_{j,j}-\tfrac{2}{3}\mu v_{i,i}v_{j,j}+\frac{\mu}{2}(v_{i,j}+v_{j,i})(v_{i,j}+v_{j,i})+\lambda'\vartheta_{,jj}. \tag{9.18}$$

The differential equations governing the motion and the heat exchange in a Navier–Stokes fluid are (9.9), (9.17), $(9.13)_1$, and (9.18). They represent 6 differential equations for the unknowns ϱ, p, v_i, and ϑ. All of these relations are nonlinear.

In the *barotropic* case $(9.13)_1$ reduces to

$$p = p(\varrho). \tag{9.19}$$

Now (9.9), (9.17), and (9.19) represent 5 differential equations (of orders 0 through 2) for the unknowns ϱ, p, and v_i. The mechanical and the thermodynamic problems can thus be separated; once the first is solved, (9.18) supplies the heat exchange.

9.2. Incompressible Newtonian liquids

Let us specialize the differential equations established in Section 9.1 for an *incompressible* liquid of the Newtonian type. Here the continuity equation (9.9) assumes the same simple form

$$v_{j,j} = 0 \tag{9.20}$$

as in Section 6.2, and the theorem of linear momentum (9.12) therefore reduces to

$$v_{i,0}+v_{i,j}v_j = f_i-\frac{1}{\varrho}p_{,i}+\frac{\mu}{\varrho}v_{i,jj}\,. \tag{9.21}$$

This equation is also a consequence of (9.17) and (9.20). It follows that, in the incompressible case, the Newtonian fluid is also a Navier–Stokes fluid. Equation (9.21) is referred to as the *Navier–Stokes differential equation* of the incompressible fluid, and the quotient μ/ϱ is sometimes denoted as the *kinematic viscosity*.

The two relations (9.20) and (9.21) represent 4 scalar differential equations for the unknowns v_i and p; they suffice for the determination of the mechanical flow. Since the pressure is now a reaction, the thermal equation of state $(9.13)_1$ must be dropped, and the heat exchange is subject to the equation

$$\varrho c(\vartheta_{,0}+\vartheta_{,j}v_j) = \frac{\mu}{2}(v_{i,j}+v_{j,i})(v_{i,j}+v_{j,i})+\lambda'\vartheta_{,jj} \tag{9.22}$$

obtained from (9.14) or (9.18) if we take (9.20) into account and note that, since ϱ is constant, there exists only one specific heat capacity c. On account of (2.14) and $(1.52)_2$, a more compact form of (9.22) is

$$\varrho c\vartheta_{,0} = -\varrho c\vartheta_{,j}v_j+\lambda'\vartheta_{,jj}+4\mu d_{(2)}\,, \tag{9.23}$$

where $d_{(2)}$ is the second basic invariant of the rate of deformation. In this form it becomes apparent that the first fundamental law establishes the connection between the temperature increase in a given spatial point, the heat convection, the heat conduction, and the heat production by internal friction.

Let us assume now for a moment that $\mu = 0$. The fluid considered here then becomes the ideal liquid of Chapter 6. In fact, the continuity equation (9.20) is identical with (6.19), and for $\mu = 0$ the differential equation of motion (9.21) reduces to (6.4). Whereas the differential equations of the ideal liquid are of the first order, (9.21) is of the second order in v_i, and this implies that we need more boundary conditions in a Newtonian velocity field than in an ideal liquid. In Section 6.2 we have assumed that along a solid boundary at rest the flow is tangential. In the presence of friction, however, the assumption that the material points in contact with a boundary do not move with respect to it seems more realistic, and it provides the needed additional condition. Along a solid boundary at rest we thus have

$$v_i = 0 \tag{9.24}$$

in place of (6.23).

In an irrotational flow the velocity v_i is the gradient of a potential φ, and the continuity equation (9.20) assumes the form (6.20). Thus the last term in (9.21), written as $(\mu/\varrho)\varphi_{,ijj}$, is zero. It follows that any irrotational flow of an ideal liquid satisfies the differential equations (9.20) and (9.21) and hence qualifies as a solution also for the incompressible Newtonian liquid. However, the value of such solutions, as the ones discussed in Sections 6.2 and 6.3, is restricted by the fact that in the great majority of cases solid boundaries are present, where the boundary conditions for the Newtonian liquid are not satisfied.

The exact treatment of flows subject to (9.20) and (9.21) is more difficult than the corresponding problem for the ideal liquid, not so much for the presence of the additional term in (9.21), but for the fact that Thomson's vortex theorem (Section 6.1) becomes invalid and that, therefore, the concept of a velocity potential must be dropped. In certain simple cases, however, exact solutions can be obtained. One of them is the *parallel flow*

$$v_1 = v(x_2, x_3, t), \qquad v_2 = v_3 = 0, \tag{9.25}$$

where the absence of x_1 as an argument of v is required by the continuity condition (9.20). The differential equation of motion, (9.21), reduces to

$$v_{,0} = f_1 - \frac{1}{\varrho} p_{,1} + \frac{\mu}{\varrho} v_{,jj} \tag{9.26}$$

and

$$p_{,2} = \varrho f_2, \qquad p_{,3} = \varrho f_3, \tag{9.27}$$

where the summation in the last term of (9.26) extends over the subscripts 2 and 3 alone.

As a first application, let us consider the case where the specific body force is zero. According to (9.27) p is a function of x_1 and t alone and, on account of (9.26) and the fact that v is independent of x_1, the pressure gradient $p_{,1}$ is a function of t alone:

$$p_{,1} = f(t). \tag{9.28}$$

If the pressure gradient vanishes, (9.26) reduces to

$$v_{,0} = \frac{\mu}{\varrho} v_{,jj}. \tag{9.29}$$

This equation is of the same type as the differential equation of heat conduction, (5.14); it is the two-dimensional version of the so-called *diffusion equation.*

If, on the other hand, the flow is assumed to be *steady*, the pressure gradient (9.28) is different from zero but constant:

$$p_{,1} = -k'. \tag{9.30}$$

Provided k' is positive, the pressure is a linearly decreasing function of x_1. Equation (9.26) now reduces to

$$v_{,jj} = \Delta v = -\frac{k'}{\mu}, \tag{9.31}$$

i.e. to the two-dimensional version of Poisson's differential equation already encountered in the form (7.40). If the flow takes place inside a cylindrical tube with generators parallel to x_1, the boundary condition (9.24) has to be added, and the problem becomes analogous to the torsion problem of Section 7.2.

As a particular case, let us consider the flow between two infinite planes at rest (Fig. 9.1). Here v is independent of x_3, and the solution of the differential equation

$$v_{,22} = -\frac{k'}{\mu} \tag{9.32}$$

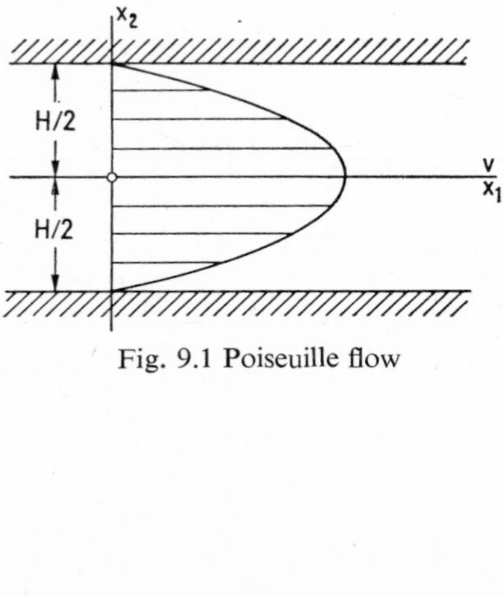

Fig. 9.1 Poiseuille flow

and the boundary condition $v(\pm H/2) = 0$ is

$$v = \frac{k'}{2\mu}\left(\frac{H^2}{4} - x_2^2\right). \tag{9.33}$$

It corresponds to the parabolic velocity profile of Figure 9.1.

Another particular case is the parallel flow in a tube of circular cross section (Fig. 9.1. where $H/2$ is to be replaced by the radius R), referred to as a *Poiseuille flow*. On account of the rotational symmetry, the velocity v depends only on the distance r from the axis, and Poisson's differential equation (9.31) becomes (P1)

$$v'' + \frac{1}{r}v' = -\frac{k'}{\mu}, \tag{9.34}$$

where the prime denotes differentiation with respect to r. This differential equation, supplemented by the boundary condition $v(R) = 0$ and the symmetry condition $v'(0) = 0$, yields

$$v = \frac{k'}{4\mu}(R^2 - r^2). \tag{9.35}$$

The velocity profile is again parabolic.

As another application of (9.26) and (9.27) let us consider the steady flow, independent of x_3, of a heavy liquid in a channel (Fig. 9.2) of infinite depth,

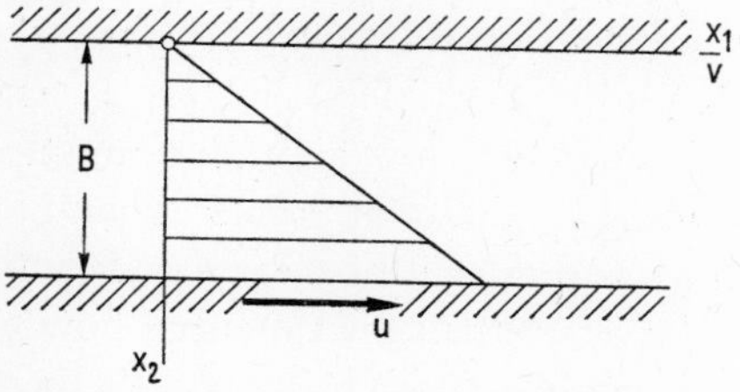

Fig. 9.2 Infinite channel with a moving wall

assuming that one of the two vertical walls is at rest while the other moves with the velocity u. Letting the axis x_3 point downwards, we have $f_1 = f_2 = 0$ and $f_3 = g$, where g is the acceleration of gravity. Provided the pressure gradient has no component in the direction x_1, equations (9.26) and (9.27) reduce to

$$v_{,22} = 0, \qquad p_{,2} = 0, \qquad p_{,3} = \varrho g. \tag{9.36}$$

The pressure becomes

$$p = p_0 + \varrho g x_3 , \tag{9.37}$$

where p_0 is a constant. Since the surfaces p = const are horizontal planes, the free surface is horizontal too, and p_0 may be interpreted as the pressure acting on the free surface $x_3 = 0$. The differential equation $(9.36)_1$, together with the boundary conditions $v(0) = 0$ and $v(B) = u$, yields the straight velocity profile

$$v = \frac{u}{B} x_2 . \tag{9.38}$$

The Navier–Stokes differential equation plays an important role in the theory of flight. It has been mentioned in Section 8.2 that, for velocities that are small compared to the velocity of sound, the air flowing past an airfoil may be treated as an incompressible liquid. In Section 6.3 the uniform flow past a circular cylinder has been studied, and it was pointed out that the flow past an airfoil with a more general cross section may be obtained from it by conformal mapping. An essential feature of this theory was the indeterminacy, in (6.42), of the circulation c determining the location of the stagnation points in Figure 6.5 and on the corresponding non-circular profile.

Experiments show that for low viscosities, e.g. in the case of air, the plane uniform flow past a solid obstacle looks similar to the one discussed in Section 6.3, except for the motion in the wake of the obstacle. Based on this observation, Prandtl developed his *boundary layer theory*, assuming that the viscosity of the liquid is negligible except for a narrow layer along the profile, where the deformation rate is high on account of the considerable velocity gradient between the material points adhering to the profile and the ones in the free flow, and except for the wake where the material elements leaving the boundary layer disturb the potential flow otherwise prevailing outside the boundary layer.

In Prandtl's theory the motion outside the boundary layer (and the wake) is treated as a potential flow determined by (9.20) and (9.21), the last equation formulated with $\mu = 0$ and with $f_i = 0$ provided the influence of the body force is negligible. Within the boundary layer (9.21) is applied with $\mu \neq 0$, and the fact that v_i is practically tangential to the profile permits certain simplifications resulting in the so-called *boundary layer equations*. Considering a smooth and only slightly curved element of the profile, and the component of (9.21) perpendicular to it, we note that the contributions of the velocity and the acceleration may be neglected in a first approximation and that, therefore, the pressure gradient has practically no component perpendicular to the profile. The pressure within the boundary layer is thus practically the same as in the adjacent potential flow. If the profile (Fig. 9.3) has a sharp trailing edge which originally is not a stagnation point, the

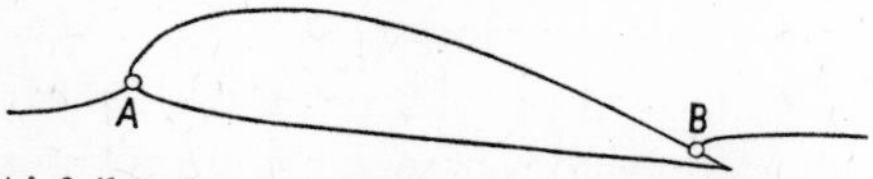

Fig. 9.3 Airfoil before stagnation point B moves to trailing edge

velocity is large in its vicinity, and it follows from Bernoulli's equation (6.22) that the pressure is small. There results a considerable pressure decrease in the boundary layer towards the edge, and under normal flight conditions this decrease moves the stagnation pont to the trailing edge (cf. Fig. 6.6.) The location of the corresponding stagnation point B on the circle of Figure 6.5 determines the constant c and hence the circulation responsible for the lift experienced by the profile. This mechanism explains the importance of a sharp trailing edge, and the whole theory shows that the potential flows studied in Section 6.3 keep their significance within the boundary layer theory.

It is obvious that the friction in the boundary layer also contributes to the drag. The main part of the drag, however, is due to two vortex systems, one of them caused by the finite length of the wing and the other by a peculiar unsteady manner in which the boundary layer leaves the trailing edge.

Problems

1. Show that in cylindrical coordinates the Poisson differential equation for $v(r)$ has the form (9.34).

2. Verify (9.35).

3. Provided a motion is sufficiently slow, the quadratic term $v_{i,j}v_j$ in (9.21) may be neglected. Under this condition the steady parallel flow of an incompressible viscous liquid past a sphere of radius R is given by

$$v_i = u\left[\frac{3}{4}\,\frac{Rx_ix_3}{r^3}\left(\frac{R^2}{r^2}-1\right)-\frac{1}{4}\,\delta_{i3}\,\frac{R}{r}\left(3+\frac{R^2}{r^2}\right)+\delta_{i3}\right],$$

where u is the velocity at infinity and where $r^2 = x_jx_j$. Verify this velocity field and determine the pressure distribution, the resultant force and the reusltant moment acting on the sphere.

9.3. Turbulence

The exact solutions discussed in Section 9.2 are well confirmed by experiments in cases where the viscosity coefficient is sufficiently large. For small values of μ, however, experience shows that on the so-called *laminar flow* considered so far, there are superposed irregular fluctuations referred to as *turbulence.* Although this phenomenon has not yet found a complete solution, a few aspects of it are fairly obvious and will be briefly mentioned on account of their similarity to the irregular molecular motion typical for thermodynamic processes.

Let us consider an incompressible liquid without body forces. The local form of the continuity equation is given by (9.20):

$$v_{j,j} = 0. \tag{9.39}$$

According to (3.8) the linear momentum theorem, formulated for a finite spatial volume becomes

$$\int \varrho v_{i,0} \, \mathrm{d}V = \int (\sigma_{ij} - \varrho v_i v_j) \nu_j \, \mathrm{d}A, \tag{9.40}$$

and the corresponding local form is given by (3.9):

$$\varrho \dot{v}_i = \sigma_{ij,j}. \tag{9.41}$$

By means of (2.38) the first fundamental law (4.60) may be reformulated for a finite spatial volume. Making use of (9.39) and of the theorem of Gauss (1.79), we obtain (P1) the global equation

$$\int \varrho(u+\tfrac{1}{2}v_i v_i)_{,0} \, \mathrm{d}V = \int (\sigma_{ij} v_i - q_j)\nu_j \, \mathrm{d}A - \int \varrho(u+\tfrac{1}{2}v_i v_i) v_j \nu_j \, \mathrm{d}A \tag{9.42}$$

with the corresponding local form (4.63):

$$\varrho \dot{u} = \sigma_{ij} d_{ij} - q_{j,j}. \tag{9.43}$$

We now consider the case where the velocity v_i appears as the sum of a slowly varying (secular) term $\bar{v}_i$ representing the mean velocity evaluated for phenomenologically short but microscopically long time intervals, and a rapid statistical fluctuation v_i' about this mean value. We thus have

$$v_i = \bar{v}_i + v_i'; \tag{9.44}$$

similar decompositions hold for the deformation rate d_{ij}, the stress σ_{ij}, and the internal energy u. The concept of a mean value implies that for any function of the form

$$f = \bar{f} + f' \tag{9.45}$$

the mean fluctuation is zero:

$$\bar{f}' = 0. \tag{9.46}$$

We further assume that the processes of differentiation and averaging are interchangeable, i.e. that

$$(f_{,0})^- = \bar{f}_{,0}, \qquad (f_{,i})^- = \bar{f}_{,i}, \tag{9.47}$$

and if we finally postulate that

$$(\bar{f}g)^- = \bar{f}\bar{g}, \tag{9.48}$$

this equation, together with (9.45) and (9.46), yields

$$(fg)^- = \bar{f}\bar{g} + (f'g')^- . \tag{9.49}$$

Applying $(9.47)_2$ and (9.46) to the continuity equation (9.39), we obtain

$$\bar{v}_{j,j} = 0. \tag{9.50}$$

Thus, not only the velocity v_i, but also its secular part $\bar{v}_i$ is subject to the continuity equation. Taking the mean values in the theorem of linear momentum (9.40) and making use of $(9.47)_1$ and (9.49), we also get

$$\int \varrho \bar{v}_{i,0} \, dV = \int (\bar{\sigma}_{ij} - \varrho \bar{v}_i \bar{v}_j - \varrho (v_i' v_j')^-) \nu_j \, dA. \tag{9.51}$$

With the notation

$$\sigma_{ij}^* = \bar{\sigma}_{ij} - \varrho (v_i' v_j')^- \tag{9.52}$$

(9.51) reduces to

$$\int \varrho \bar{v}_{i,0} \, dV = \int (\sigma_{ij}^* - \varrho \bar{v}_i \bar{v}_j) \nu_j \, dA. \tag{9.53}$$

This means that the linear momentum theorem remains valid for the mean flow provided the mean stress is supplemented by the so-called *Reynolds stress* $-\varrho(v_i' v_j')^-$ representing an additional dissipative effect. The local form of (9.53) reads

$$\varrho \dot{\bar{v}}_i = \sigma_{ij,j}^*, \tag{9.54}$$

where the material derivative on the left-hand side is to be formed with the mean velocity.

If we treat the expression (9.42) for the first fundamental law in an analogous manner, we obtain (P2),

$$\begin{aligned} \int \varrho[\bar{u} + \tfrac{1}{2}\bar{v}_i\bar{v}_i + \tfrac{1}{2}(v_i'v_i')^-]_{,0} \, dV &= \int [\bar{\sigma}_{ij}\bar{v}_i + (\sigma_{ij}'v_i')^- - q_j]\nu_j \, dA \\ - \int \varrho[\bar{u}\bar{v}_j + (u'v_j')^- + \tfrac{1}{2}\bar{v}_i\bar{v}_i\bar{v}_j &+ \bar{v}_i(v_i'v_j')^- + \tfrac{1}{2}(v_i'v_i')^- \bar{v}_j + \tfrac{1}{2}(v_i'v_i'v_j')^-]\nu_j \, dA. \end{aligned} \tag{9.55}$$

With the help of the notations

$$u^* = \bar{u}+\tfrac{1}{2}(v_i'v_i')^- \tag{9.56}$$

and

$$q_j^* = q_j-(\sigma_{ij}'v_i')^-+\varrho(u'v_j')^-+\frac{\varrho}{2}(v_i'v_i'v_j')^- = q_j-(\sigma_{ij}'v_i')^-+\varrho[(u'+\tfrac{1}{2}v_i'v_i')v_j']^-, \tag{9.57}$$

9.55) may be written (P3) in the more compact form

$$\int \varrho(u^*+\tfrac{1}{2}\bar{v}_i\bar{v}_i)_{,0}\,\mathrm{d}V = \int (\sigma_{ij}^*\bar{v}_i-q_j^*)\nu_j\,\mathrm{d}A-\int \varrho(u^*+\tfrac{1}{2}\bar{v}_i\bar{v}_i)\bar{v}_j\nu_j\,\mathrm{d}A, \tag{9.58}$$

which is equivalent (P4) to the local relation

$$\varrho\dot{u}^* = \sigma_{ij}^*\bar{d}_{ij}-q_{j,j}^*. \tag{9.59}$$

Comparing (9.58) and (9.42), we see that the first fundamental law remains valid for the mean flow provided the mean stress is supplemented by the Reynolds stress, the mean specific internal energy by the second term on the right-hand side of (9.56), and the heat flow by the additional terms appearing in (9.57). The interpretation of these extra terms is straightforward. According to (9.56), u^* is the sum of the mean internal energy and the kinetic energy of the turbulent fluctuations. Similarly, q_j^* is composed of the heat flow, the energy flow caused by the rate of work of the statistical surface stresses, and the statistical energy convection through the boundary.

It is obvious that there exists a close connection between the phenomenon considered here and the mechanical explanation of the thermodynamic laws in terms of molecular motion. In fact, if we set $q_j = 0$ and interpret u as the potential energy of the internal forces, (9.42) becomes equivalent to (3.29), written for $f_k = 0$. The only difference between the equations reflects the point of view of the observer: (3.29) is the material version of the mechanical energy theorem, while (9.42) is the corresponding statement for a spatial volume, obtained from (3.29) with the help of (2.38) and the theorem of Gauss (1.79). To take care of the micro-motion, three modifications become necessary: First, the quantities in (9.42) are to be replaced by their mean (or phenomenological) values. Secondly the internal energy u^* now is the sum (9.56) of the average potential of the internal forces and the kinetic energy of the molecular motion. Finally there appears an energy flow q_j^* composed, according to (9.57), of the rate of work of the micro-stresses on the surface and of the statistical energy convection through the boundary.

Problems

1. Verify (9.42).
2. Verify (9.55).
3. Show that, on account of (9.56) and (9.57), equation (9.55) reduces to (9.58).
4. Obtain (9.59) from (9.58).

9.4. Non-Newtonian liquids

The stress in a viscous liquid is composed, according to (9.1), of a quasi-conservative and a dissipative part. On account of our definition of a fluid (Section 5.1) $\sigma_{ij}^{(q)}$ is still given by $(9.2)_1$ even if the fluid is non-Newtonian. However, $(9.2)_2$ has to be replaced by a nonlinear constitutive equation for $\sigma_{ij}^{(d)}$. It has been shown in Section 5.2 that in the case of an isotropic fluid this equation is of the form (5.21) and hence reads

$$\sigma_{ij}^{(d)} = f\delta_{ij}+gd_{ij}+hd_{ik}d_{kj}, \tag{9.60}$$

where f, g, and h are functions, dependent on the material and possibly on the temperature, of the basic invariants $d_{(1)}, \ldots$. The specific power of dissipation (4.38) is

$$l^{(d)} = \frac{1}{\varrho}\sigma_{ij}^{(d)}d_{ij} = \frac{1}{\varrho}(fd_{ii}+gd_{ij}d_{ji}+hd_{ij}d_{jk}d_{ki}). \tag{9.61}$$

In terms of the basic invariants (1.52) of the deformation rate, equation (9.61) assumes the form

$$l^{(d)} = \frac{1}{\varrho}\{fd_{(1)}+g(2d_{(2)}+d_{(1)}^2)+h[3d_{(3)}+\tfrac{3}{2}(2d_{(2)}+d_{(1)}^2)d_{(1)}-\tfrac{1}{2}d_{(1)}^3]\}. \tag{9.62}$$

According to the second fundamental law (in the separated form) the rate of dissipation work is positive or zero. The functions f, g, h are therefore not arbitrary, but subject to the condition that (9.62) is non-negative for arbitrary d_{ij}.

Let us restrict ourselves from now on to *incompressible liquids*. Here d_{ij} is a deviator, and according to (1.69) we have

$$d_{(1)} = d_{ii} = 0, \qquad d_{(2)} = \tfrac{1}{2}d_{ij}d_{ji}, \qquad d_{(3)} = \tfrac{1}{3}d_{ij}d_{jk}d_{ki}. \tag{9.63}$$

The isotropic part of the dissipative stress (9.60) is

$$\tfrac{1}{3}\sigma_{kk}^{(d)}\delta_{ij} = f+\tfrac{2}{3}hd_{(2)}. \tag{9.64}$$

It must be zero since in an incompressible liquid it does not contribute to the power of dissipation. It follows that $f = -\frac{2}{3}hd_{(2)}$ and that $\sigma_{ij}^{(d)}$ is a deviator of the form

$$\sigma_{ij}^{(d)} = g(d_{(2)}, d_{(3)})d_{ij} + h(d_{(2)}, d_{(3)})\,(d_{ik}d_{kj} - \tfrac{2}{3}d_{(2)}\delta_{ij}). \tag{9.65}$$

The liquid obeying the constitutive equations $(9.2)_1$ and (9.65) has first been studied by Reiner [9] and Rivlin [10] and is therefore called a *Reiner–Rivlin liquid.* Its power of dissipation, obtained from (9.62) by setting $d_{(1)} = 0$, is

$$l^{(d)} = \frac{1}{\varrho}(2gd_{(2)} + 3hd_{(3)}). \tag{9.66}$$

The functions g and h are again subject to the condition that (9.66) is non-negative for arbitrary deviators d_{ij}.

According to (9.63) the basic invariants are of the second and third degrees respectively in d_{ij}. Considering d_{ij} as small, let us develop the functions g and h into power series

$$g = 2(g^{(0)} + g^{(2)}d_{(2)} + g^{(3)}d_{(3)} + g^{(4)}d_{(2)}^2 + \ldots) \tag{9.67}$$

and

$$h = 4(h^{(0)} + h^{(2)}d_{(2)} + \ldots), \tag{9.68}$$

where the superscripts of the expansion coefficients $g^{(0)}, g^{(2)}, \ldots, h^{(0)}, \ldots$ indicate the orders of the corresponding expressions in terms of d_{ij}. Truncating the series in an appropriate manner, we obtain approximations of various orders for (9.65). The *linear* approximation is obtained by reducing the series (9.67) to its constant term and by dropping all terms in (9.68). Equation (9.65) thus becomes

$$\sigma_{ij}^{(d)} = 2g^{(0)}d_{ij}. \tag{9.69}$$

This is the constitutive equation of an incompressible Newtonian liquid of viscosity $\mu = g^{(0)}$. For the *second-order* approximation, we have to admit the constant term in (9.68), and we obtain

$$\sigma_{ij}^{(d)} = 2g^{(0)}d_{ij} + 4h^{(0)}(d_{ik}d_{kj} - \tfrac{2}{3}d_{(2)}\delta_{ij}). \tag{9.70}$$

The *third-order* approximation becomes

$$\sigma_{ij}^{(d)} = 2(g^{(0)} + g^{(2)}d_{(2)})d_{ij} + 4h^{(0)}(d_{ik}d_{kj} - \tfrac{2}{3}d_{(2)}\delta_{ij}), \tag{9.71}$$

and by extending this process, approximations of higher orders may be obtained.

The constitutive equation $\sigma_{ij}^{(d)} = 2\mu d_{ij}$ of the incompressible Newtonian liquid is sometimes generalized in a different manner, namely by assuming that μ is not a constant (possibly dependent on the temperature), but a function of the basic invariants of the deformation rate. This generalization reads

$$\sigma_{ij}^{(d)} = 2\mu(d_{(2)}, d_{(3)})d_{ij}. \tag{9.72}$$

Since only the viscosity coefficient is nonlinear, the liquid obeying (9.72) is referred to as *quasilinear*. It represents a special case of the non-Newtonian liquid, obtained from (9.65) by equating the function h to zero.

There is an essential difference between the response of a Newtonian or a quasilinear liquid on one hand and the behavior of the nonlinear liquid subject to the general form of (9.65) on the other. In liquids of the first two types the absence of an arbitrary component d_{ij} of the deformation rate implies that the corresponding dissipative stress $\sigma_{ij}^{(d)}$ vanishes too. In the truly nonlinear liquid subject to (9.65) this need not be the case. However, if $d_{ij} = 0$ $(i \neq j)$, i.e. if all of the shear rates are zero, the same is true, also according to (9.65), for the shear stresses $\sigma_{ij}^{(d)}$ $(i \neq j)$. It follows (and this is a consequence of the isotropy of the fluid) that the principal directions of the strain rate and the stress coincide.

As an example, let us consider the case of *simple shear*, defined in an appropriate coordinate system by the plane velocity field

$$v_1 = cx_2, \qquad v_2 = v_3 = 0, \tag{9.73}$$

where c is a constant. Figure 9.4 shows the deformation of an element which at time $t = 0$ is a cuboid. The only non-zero component of the velocity gradient is $v_{1,2} = c$. It follows that $d_{12} = d_{21} = c/2$ are the only non-vanishing components of the deformation rate. If the liquid is Newtonian or quasilinear, (9.69) or (9.72) respectively yield a single dissipative stress

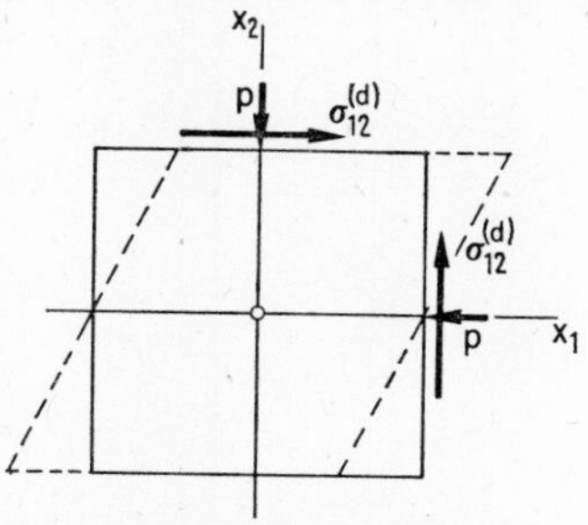

Fig. 9.4 Stresses in simple shear

component $\sigma_{12}^{(d)} = \sigma_{21}^{(d)}$. The corresponding stress state, supplemented by the hydrostatic pressure, is indicated on two sides of the cuboid in Figure 9.4. In the truly nonlinear case, however, the stress state is more complicated. According to (9.63) the basic invariants of the rate of deformation are $d_{(2)} = d_{12}^2$ and $d_{(3)} = 0$. In matrix notation the square of d_{ij} is

$$d_{ik}d_{kj} = \begin{pmatrix} d_{12}^2 & 0 & 0 \\ 0 & d_{12}^2 & 0 \\ 0 & 0 & 0 \end{pmatrix}. \tag{9.74}$$

The dissipative stress (9.65) therefore becomes

$$\sigma_{ij}^{(d)} = g\begin{pmatrix} 0 & d_{12} & 0 \\ d_{12} & 0 & 0 \\ 0 & 0 & 0 \end{pmatrix} + \tfrac{1}{3}h\begin{pmatrix} d_{12}^2 & 0 & 0 \\ 0 & d_{12}^2 & 0 \\ 0 & 0 & -2d_{12}^2 \end{pmatrix}. \tag{9.75}$$

Now the dissipative stress tensor does not consist of shear stresses alone but also of the normal stresses represented by the second matrix. The appearance of these normal stresses in spite of the absence of the corresponding extensions is sometimes denoted as a *cross effect*. However, this term is not well-defined; it is not used, e.g. in the case of an elastic solid, where, according to (7.10), the uniaxial stress is not only accompanied by an axial extension, but also by lateral contractions.

In a nonlinear liquid effects are possible which do not occur in the linear or the quasilinear case and hence appear quite unexpected. As an example, let us consider the so-called *Couette flow* (Fig. 9.5). Let the space between two semi-infinite circular cylinders with axis z be filled with an incompressible liquid. The cylinder of radius r_2 is at rest, the other one, of radius r_1, rotates with the constant angular velocity ω assumed to be so small that the inertia

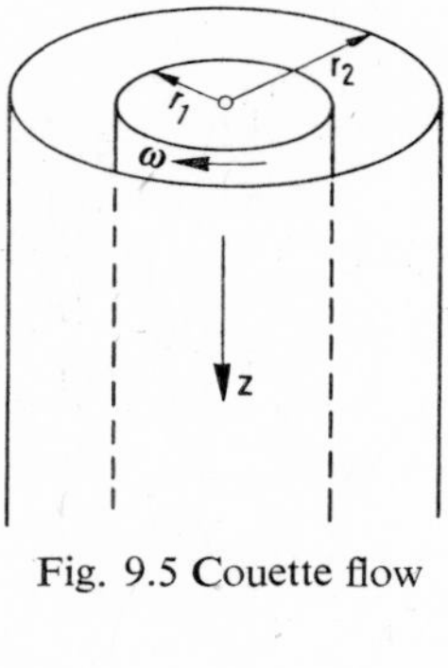

Fig. 9.5 Couette flow

forces in the moving liquid may be neglected. Using cylindrical coordinates r, α, z, we assume that of the three velocity components the radial and the axial ones are zero and that the circumferential component v depends on r alone. Thus, the deformation of an arbitrary element of the liquid is a simple shear, and it is easy to see (P2) that the only non-vanishing component of the deformation rate is

$$d_{r\alpha} = \tfrac{1}{2}\left(v' - \frac{v}{r}\right), \tag{9.76}$$

where the prime denotes differentiation with respect to r. It is independent of α and z. Since the element retains its volume, the continuity condition is satisfied. The stress tensor is given by $(9.2)_1$ and (9.75), the last equation written with $d_{12} = d_{r\alpha}$. On account of the weight, the hydrostatic pressure may depend on r and z; the dissipative stress, however, is a function of r alone. The only total stress components possibly different from zero are σ_{rr}, $\sigma_{\alpha\alpha}$, σ_{zz}, and $\sigma_{r\alpha}$; they are independent of α. Considering an element bounded by two adjacent coaxial cylinders, two planes containing the axis z and two planes perpendicular to it, one easily obtains (P3) the equilibrium conditions

$$\frac{\partial\sigma_{rr}}{\partial r} + \frac{\sigma_{rr}-\sigma_{\alpha\alpha}}{r} = 0, \qquad \frac{\partial\sigma_{r\alpha}}{\partial r} + 2\frac{\sigma_{r\alpha}}{r} = 0, \qquad \frac{\partial\sigma_{zz}}{\partial z} + \gamma = 0, \tag{9.77}$$

where γ is the specific weight of the liquid. The boundary conditions for the velocity are

$$v(r_1) = \omega r_1, \qquad v(r_2) = 0, \tag{9.78}$$

and the boundary condition for the stress tensor requires that on the free surface the shear stresses are zero and the negative normal stress is equal to the atmospheric pressure p_0.

To solve the problem, it is convenient to consider various approximations of the constitutive equation (9.65). The *first-order* approximation corresponds to the Newtonian liquid. Here $g = 2g^{(0)}$ and $h = 0$. From $(9.2)_1$ and (9.75) we obtain

$$\sigma_{rr} = \sigma_{\alpha\alpha} = \sigma_{zz} = -p, \qquad \sigma_{r\alpha} = 2g^{(0)}d_{r\alpha}. \tag{9.79}$$

Substituting this in the equilibrium conditions (9.77), we have

$$p = \gamma z + c, \tag{9.80}$$

where c is a constant, and

$$d'_{r\alpha}+2\frac{d_{r\alpha}}{r}=0, \tag{9.81}$$

the prime denoting differentiation with respect to r. According to (9.80) the pressure is constant on horizontal planes. Thus, the free surface also is horizontal, and if it contains the origin, the boundary conditions for the stresses are satisfied provided we set $c = p_0$. Integration of (9.81) yields

$$d_{r\alpha}=\frac{A}{r^2}, \tag{9.82}$$

where A is a constant. The differential equation (9.76) for $v(r)$ thus becomes

$$v'-\frac{v}{r}=\frac{2A}{r^2}. \tag{9.83}$$

Its general solution is

$$v=-\frac{A}{r}+Br, \tag{9.84}$$

where B is another constant. The boundary conditions (9.78) require that

$$A=Br_2^2=-\frac{r_1^2r_2^2\omega}{r_2^2-r_1^2}, \tag{9.85}$$

and the velocity field finally becomes

$$v=\frac{r_1^2\omega}{r_2^2-r_1^2}\left(\frac{r_2^2}{r}-r\right). \tag{9.86}$$

In the case of the *quasilinear* liquid the constitutive equations are $(9.2)_1$ and (9.72). The stress components are still given by (9.79), except that the constant $g^{(0)}$ is to be replaced by the function $\mu(d_{r\alpha}^2)$. The equilibrium conditions $(9.77)_1$ and $(9.77)_3$ still supply (9.80); the field of the hydrostatic stress is thus the same as in the linear case, and the free surface is horizontal. However, from the equilibrium condition $(9.77)_2$ we obtain

$$(\mu d_{r\alpha})'+2\frac{\mu d_{r\alpha}}{r}=0 \tag{9.87}$$

in place of (9.81) and hence a velocity field different from (9.86).

Let us finally consider the simplest truly *nonlinear* liquid with the constitutive equation (9.70). Here $g = 2g^{(0)}$ and $h = 4h^{(0)}$. From $(9.2)_1$ and (9.75) we obtain

$$\sigma_{rr} = \sigma_{\alpha\alpha} = -p + \tfrac{4}{3}h^{(0)}d_{r\alpha}^2, \qquad \sigma_{zz} = -p - \tfrac{8}{3}h^{(0)}d_{r\alpha}^2 \tag{9.88}$$

and

$$\sigma_{r\alpha} = 2g^{(0)}d_{r\alpha}. \tag{9.89}$$

Since the shear stress (9.89) is the same as in the linear case, the equilibrium condition $(9.77)_2$ yields (9.81) and hence the velocity field (9.86) of the Newtonian liquid. On the other hand, substitution of (9.88) in $(9.77)_1$ and $(9.77)_3$ supplies

$$\frac{\partial}{\partial r}(p - \tfrac{4}{3}h^{(0)}d_{r\alpha}^2) = 0, \qquad \frac{\partial}{\partial z}(p - \tfrac{4}{3}h^{(0)}d_{r\alpha}^2) = \gamma, \tag{9.90}$$

where use has been made of the fact that $d_{r\alpha}$ is independent of z. Integration yields

$$p = \tfrac{4}{3}h^{(0)}d_{r\alpha}^2 + \gamma z + C. \tag{9.91}$$

Making use of (9.82) we have

$$p = \tfrac{4}{3}h^{(0)}\frac{A^2}{r^4} + \gamma z + C, \tag{9.92}$$

where A is given by (9.85) and C is an arbitrary constant. The normal stresses (9.88) finally are

$$\sigma_{rr} = \sigma_{\alpha\alpha} = -\gamma z - C, \qquad \sigma_{zz} = -4h^{(0)}\frac{A^2}{r^4} - \gamma z - C. \tag{9.93}$$

According to (9.92), the surfaces on which p is constant are no longer horizontal planes. It follows that the free surface is not horizontal. In fact, on account of $(9.93)_2$ the normal stress σ_{zz} is a function of r on any horizontal plane, and this is inconsistent with the boundary condition $\sigma_{zz} = -p_0$ on a horizontal free surface.

It follows from (9.88) and (9.89) that $\sigma_{zz} \neq \sigma_{rr}$ and that z is a principal direction of the stress at any point. This is in contradiction to the fact that the elements of the inclined free surface are principal elements of stress. We conclude that the boundary conditions for the stress cannot be satisfied and that in the nonlinear case the motion is more complicated than the one considered here, based on the assumption that $v_r = v_z = 0$ and $v_\alpha = v(r)$. All the same, (9.92) and (9.93) may be interpreted as indications of the so-called *Weissenberg* effect [11], i.e. of the fact that certain fluids climb towards the inner or outer cylinder.

There are other cases where the nonlinear liquid shows an unexpected behavior. One of them is the Poiseuille flow treated in Section 9.2 in connection with Figure 9.1. Ericksen has shown [12] that in a tube of noncircular cross section body forces are necessary to maintain a parallel flow of the type (9.25). Green and Rivlin [13] have treated the flow without body forces and have shown for a simple nonlinear constitutive equation that the parallel flow in a tube of elliptic cross section is accompanied by a secondary motion of circulatory character as indicated in Figure 9.6. The problem will be treated in more detail in Section 16.3.

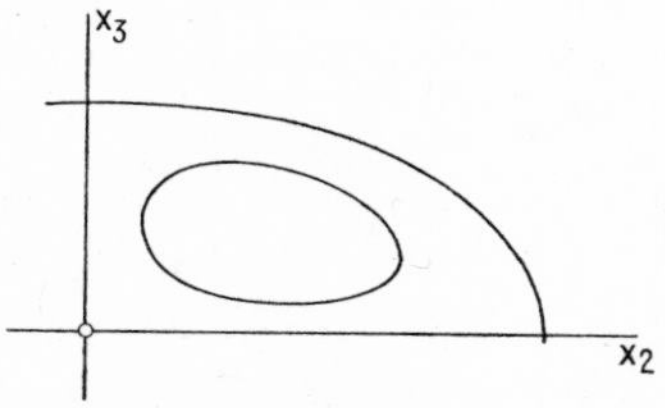

Fig. 9.6 Green–Rivlin effect

The constitutive equations (9.72) for the quasilinear liquid and (9.65) for the Reiner–Rivlin liquid have been presented here in the manner they usually appear in literature. We will reconsider them in Chapter 16, and we will show there that, on account of the orthogonality condition, only certain special forms of these equations are admissible.

Problems

1. Treat the steady parallel flow between two infinite plates at rest (Fig. 9.1) for a nonlinear incompressible liquid with the constitutive equations $(9.2)_1$ and (9.70).
2. Verify (9.76).
3. Verify the equilibrium conditions (9.77).

CHAPTER 10

PLASTIC BODIES

The elastic body has been defined in Section 5.1 by the absence of internal parameters and of a dissipative stress tensor. The purely viscous body, on the other hand, was defined by the absence of internal parameters and of quasiconservative stresses. The incompressible liquid discussed in Chapter 9 is an example of a purely viscous material. In its simplest form the *plastic body* to be treated here is a limiting case of a viscous liquid.

Motions of plastic bodies are usually so slow that they can be treated by means of the continuity equation and the equilibrium condition. We will concentrate here on the constitutive relations, assuming that the bdy is isotropic, and we will consider viscoplastic and perfectly plastic materials, strain-hardening and elastic-plastic bodies. In the last case internal parameters will appear.

10.1. Viscoplastic bodies

Let us consider an *incompressible* viscous liquid. The stress tensor (9.1) consists of a quasiconservative and a dissipative part. The first is given by $(9.2)_1$, where the hydrostatic pressure is a reaction, and for the dissipative stress we adopt the *quasilinear* constitutive equation (9.72).

In the special case of simple shear (9.73), $d_{12} = d_{21}$ is the only non-vanishing component of the deformation rate, and $d_{(2)} = d_{12}^2$ is the only basic invariant which is different from zero. The only non-vanishing dissipative stress is $\sigma_{12}^{(d)} = \sigma_{21}^{(d)}$, and the connection between d_{12} and $\sigma_{12}^{(d)}$ is given by the function $\mu(d_{(2)}, d_{(3)})$, which, in this simple case, reduces to $\mu(d_{12}^2)$.

Figure 10.1 illustrates the type of relationship $\sigma_{12}^{(d)}(d_{12})$ in which we are interested at present. The smooth curve might be represented by

$$\sigma_{12}^{(d)} = 2ad_{12} + k \operatorname{arc\,tan} \frac{d_{12}}{b}, \tag{10.1}$$

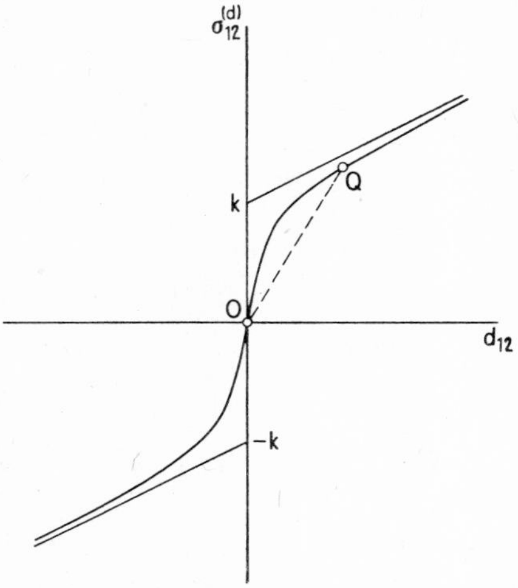

Fig. 10.1 Simple shear of a viscoplastic body

where a, b, and k are arbitrary positive constants. We thus have

$$2\mu(d_{12}^2) = 2a + k\,\frac{\text{arc}\tan(d_{12}/b)}{d_{12}}, \tag{10.2}$$

and we note that 2μ is the inclination of the straight segment connecting the origin O with a point Q on the curve. For small shear rates $|d_{12}|$ the viscosity μ is high; for large ones it decreases towards the value a. For obvious reasons these properties are particularly welcome in materials like paint.

For $k = 0$ the liquid would be Newtonian, and (10.1) would be the equation of a straight line passing through the origin. For $b \to 0$, on the other hand, the smooth curve in Figure 10.1 tends towards the broken line consisting of the segment

$$\sigma_{12}^{(d)2} \leqslant k^2 \quad (d_{12} = 0) \tag{10.3}$$

of the axis $\sigma_{12}^{(d)}$, and the two rays

$$\sigma_{12}^{(d)} = \left(2a + \frac{k}{|d_{12}|}\right) d_{12} = \left(2a + \frac{k}{\sqrt{d_{12}^2}}\right) d_{12} \quad (d_{12} \neq 0). \tag{10.4}$$

The corresponding viscosity coefficient is given by

$$2\mu(d_{12}^2) = 2a + \frac{k}{\sqrt{d_{12}^2}}, \tag{10.5}$$

and the specific power of dissipation is

$$l^{(d)} = \frac{1}{\varrho}(\sigma_{12}^{(d)}d_{12}+\sigma_{21}^{(d)}d_{21}) = \frac{2}{\varrho}\left(2a+\frac{k}{\sqrt{d_{12}^2}}\right)d_{12}^2. \tag{10.6}$$

It is plotted against d_{12} in Figure 10.2. In the Newtonian case ($k = 0$) the curve would be a parabola; in the present case it has a corner at O.

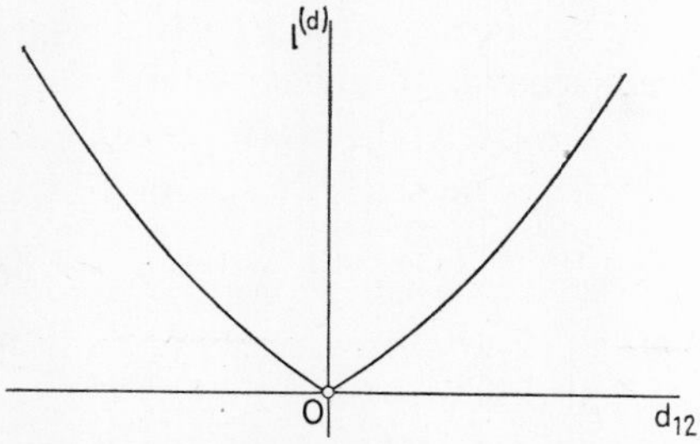

Fig. 10.2 Specific rate of dissipation work

The material with the constitutive equations (10.3) and (10.4) in simple shear has first been studied by Bingham [14]. The dissipative shear stress $\sigma_{12}^{(d)}$ is identical with the deviatoric (or the total) shear stress and may be denoted by σ'_{12}. Figure 10.1 shows that for $|\sigma'_{12}| \leqslant k$ the material remains rigid (although it must be classified, according to its definition, as a liquid). We call k its *yield stress* and observe that for $|\sigma'_{12}| \geqslant k$ the shear rate has the sign of σ'_{12} and is proportional to $|\sigma'_{12}|-k$.

The body just considered is referred to as *viscoplastic*. Its response to an arbitrary deformation is described by $(9.2)_1$ and (9.72), where the function $2\mu(d_{(2)}, d_{(3)})$ is to be chosen in such a way that it reduces to (10.5) in the case of simple shear. The simplest possible assumption is that μ depends on $d_{(2)}$ alone; in fact, it will be shown in Section 16.1 that, on account of the orthogonality condition, this is the only choice consistent with the condition that the material is quasilinear. Since in simple shear $d_{(2)} = d_{12}^2$, the generalization of (10.5) reads

$$2\mu(d_{(2)}) = 2a+\frac{k}{\sqrt{d_{(2)}}}. \tag{10.7}$$

Equation (10.4) thus becomes

$$\sigma'_{ij} = \left(2a+\frac{k}{\sqrt{d_{(2)}}}\right)d_{ij} \quad (d_{ij} \neq 0), \tag{10.8}$$

if use is made of the identity $\sigma_{ij}^{(d)} = \sigma'_{ij}$. For $d_{ij} \to 0$ it follows from (10.8) that $\sigma'_{(2)} = \sigma'_{ij}\sigma'_{ij}/2 \to k^2$; in consequence, the generalization of (10.3) is

$$\sigma'_{(2)} \leqslant k^2 \quad (d_{ij} = 0). \tag{10.9}$$

The constitutive relations (10.8) and (10.9) have been established by Hohenemser and Prager [15]. According to (10.9), the stress deviator is limited, but indetermined as long as $d_{ij} = 0$. Once $d_{ij} \neq 0$, the stress deviator (10.8) consists of two parts: the first corresponds to a Newtonian liquid of viscosity a; the second is characterized by the yield stress k and represents what is called a plastic stress. Conversely, the material remains rigid as long as the inequality (10.9) is satisfied. The stress states for which

$$\sigma'_{(2)} = k^2, \tag{10.10}$$

where k is the yield stress in simple shear, define the so-called *yield limit*. For $\sigma'_{(2)} > k^2$ the material deforms according to (10.8). The specific power of dissipation is

$$l^{(d)} = \frac{1}{\varrho}\sigma'_{ij}d_{ij} = \frac{2}{\varrho}\left(2a+\frac{k}{\sqrt{d_{(2)}}}\right)d_{(2)}. \tag{10.11}$$

Equation (10.10) is the so-called *yield condition of v. Mises*. If the stresses are interpreted as coordinates in a cartesian system of 9 (or, on account of the symmetry of σ_{ij}, of 6) dimensions, (10.10) defines a hypersurface called *yield surface*, which encloses the origin. The representation becomes particularly simple if we refer it to the system of principal axes in physical space. The state of stress in an arbitrary point is then represented by the principal stresses $\sigma_{\mathrm{I}}, \ldots$ and hence by a point Q with coordinates $\sigma_{\mathrm{I}}, \ldots$ in the 3-dimensional stress space of Figure 10.3. The points representing hydrostatic

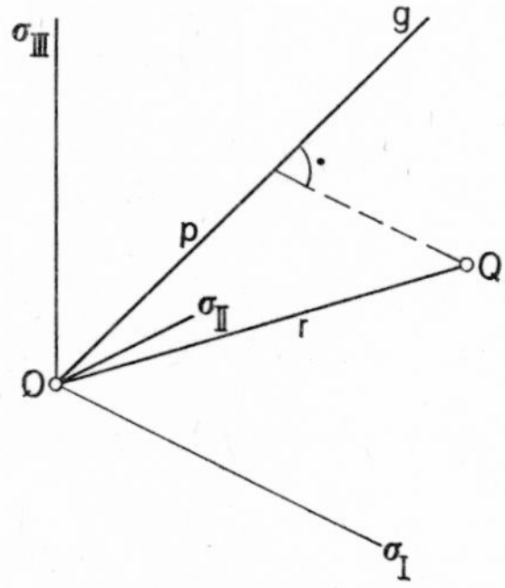

Fig. 10.3 Space of the principal stresses

stress states are situated on the straight line g at equal angles with the positive coordinate axes. On the other hand, deviatoric stress states correspond to points on the normal plane of g passing through O. Since the yield condition (10.10) is independent of a hydrostatic stress, the yield surface is a cylinder with axis g. On account of (1.69) and (1.70), the invariant appearing in (10.10) may be written as

$$\begin{aligned} \sigma'_{(2)} &= \tfrac{1}{2}\sigma'_{ij}\sigma'_{ij} = \tfrac{1}{2}(\sigma_{ij}-\tfrac{1}{3}\sigma_{kk}\delta_{ij})\,(\sigma_{ij}-\tfrac{1}{3}\sigma_{ll}\delta_{ij}) \\ &= \tfrac{1}{2}(\sigma_{ij}\sigma_{ij}-\tfrac{1}{3}\sigma_{ii}\sigma_{jj}) = \sigma_{(2)}+\tfrac{1}{3}\sigma_{(1)}^{2}. \end{aligned} \tag{10.12}$$

Making use of (1.59), we further obtain

$$\begin{aligned} \sigma'_{(2)} &= -\sigma_{\mathrm{II}}\sigma_{\mathrm{III}}- \ldots +\tfrac{1}{3}(\sigma_{\mathrm{I}}+ \ldots)^2 = \tfrac{1}{3}\sigma_{\mathrm{I}}^2+ \ldots -\tfrac{1}{3}\sigma_{\mathrm{II}}\sigma_{\mathrm{III}}- \ldots \\ &= \tfrac{1}{2}[\sigma_{\mathrm{I}}^2+ \ldots -\tfrac{1}{3}(\sigma_{\mathrm{I}}+ \ldots)^2]; \end{aligned} \tag{10.13}$$

the yield condition thus assumes the form

$$\sigma_{\mathrm{I}}^2+ \ldots -\tfrac{1}{3}(\sigma_{\mathrm{I}}+ \ldots)^2 = 2k^2. \tag{10.14}$$

In linear elasticity it can be shown (see for instance [4], p. 275) that, except for a constant factor, the left-hand side of (10.14) is the part of the strain energy that is due to distortion. Maxwell considered the distortion energy responsible for the transition from the elastic response to flow (in ductile materials) or fracture (in brittle ones). Therefore (10.10) is also referred to as *Maxwell's yield condition.* With the notations of Figure 10.3 equation (10.14) finally becomes

$$r^2-p^2 = 2k^2. \tag{10.15}$$

It follows that the yield surface is a circular cylinder of radius $k\sqrt{2}$ and axis g.

It is convenient to represent the state of stress by means of a vector $\boldsymbol{\sigma} = (\sigma_{\mathrm{I}}, \ldots)$, the radius vector of Q in Figure 10.3. Since the principal axes of the deformation rate and of the stress coincide, the deformation rate may be represented by a similar vector $\boldsymbol{d}(d_{\mathrm{I}}, \ldots)$ in the same coordinate system. Figure 10.4 shows an axonometric representation of the system $\sigma_{\mathrm{I}}, \ldots,$ obtained by projecting the space of Figure 10.3 in the direction gO onto the deviatoric plane $\sigma_{\mathrm{I}}+ \ldots = 0$. The yield cylinder appears as a circle of radius $k\sqrt{2}$. The projection of the stress point Q is Q′, and the projection $\boldsymbol{\sigma}'$ of the vector $\boldsymbol{\sigma}$ represents the deviatoric stress. For stress points Q inside the cylinder the vector $\boldsymbol{d}$ is zero. For points on or outside the cylinder we obtain the components of $\boldsymbol{d}$ by rewriting (10.8) in terms of principal values and solving for $d_{\mathrm{I}}, \ldots$. In any case, the vector $\boldsymbol{d}$ has the direction of $\boldsymbol{\sigma}'$.

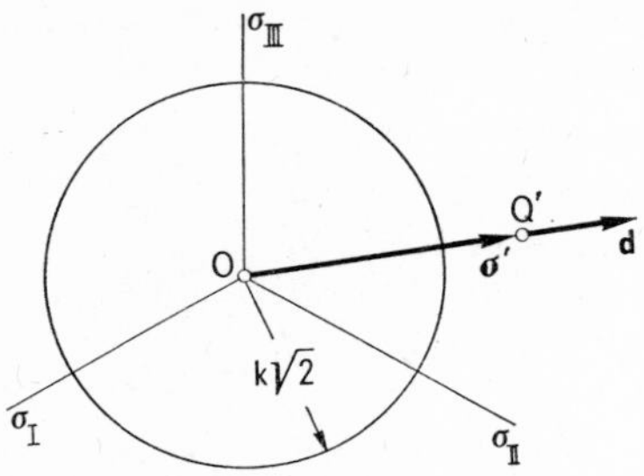

Fig. 10.4 Vector representation of stress and deformation rate

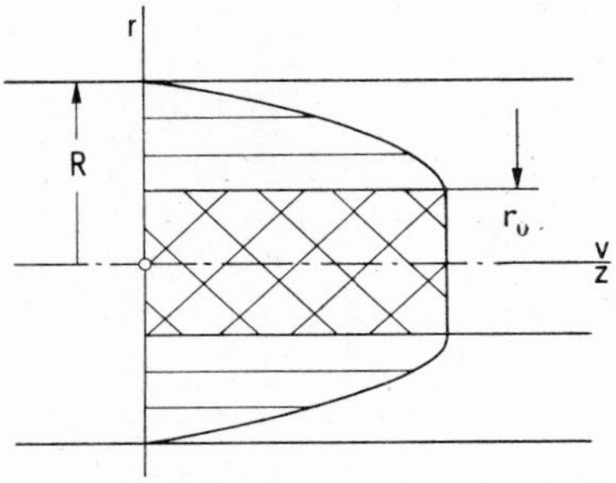

Fig. 10.5 Poiseuille flow

As a simple example, let us consider the *Poiseuille flow* (Section 9.2) of a viscoplastic material in a tube of circular cross section (Fig. 10.5). Using cylindrical coordinates r, α, z, and assuming that

$$v_z = v(r) \tag{10.16}$$

is the only velocity component, we note that the continuity condition is satisfied. The deformation is a simple shear, and

$$d_{rz} = \tfrac{1}{2}v', \tag{10.17}$$

where the prime denotes differentiation with respect to r, is the only non-vanishing deformation rate. The only deviatoric stress (10.8) is

$$\sigma_{rz} = av' - k \quad (v' \neq 0). \tag{10.18}$$

The remaining stress components are

$$\sigma_{rr} = \sigma_{\alpha\alpha} = \sigma_{zz} = -p, \tag{10.19}$$

where $p(r, z)$ is the hydrostatic pressure. The equilibrium condition $\sigma_{ij,j} = 0$, formulated in cylindrical coordinates and adapted to our problem (P1), yields

$$\frac{\partial \sigma_{rr}}{\partial r} + \frac{1}{r}(\sigma_{rr} - \sigma_{\alpha\alpha}) + \frac{\partial \sigma_{rz}}{\partial z} = 0, \qquad \frac{\partial \sigma_{rz}}{\partial r} + \frac{1}{r}\sigma_{rz} + \frac{\partial \sigma_{zz}}{\partial z} = 0 \tag{10.20}$$

or, on account of (10.18) and (10.19),

$$\frac{\partial p}{\partial r} = 0, \qquad av'' + \frac{1}{r}(av' - k) - \frac{\partial p}{\partial z} = 0. \tag{10.21}$$

It follows from $(10.21)_1$ that p is independent of r. Denoting the constant pressure gradient $\partial p/\partial z$ by $-k'$ as in Section 9.2, we obtain

$$v'' + \frac{1}{r}\left(v' - \frac{k}{a}\right) + \frac{k'}{a} = 0 \tag{10.22}$$

in place of $(10.21)_2$. The general solution of (10.22) is

$$v = -\frac{1}{2a}\left(\frac{k'}{2}r^2 - 2kr + A \ln r + B\right), \tag{10.23}$$

where A and B are constants. The boundary condition $v(R) = 0$ determines B and yields

$$v = \frac{1}{2a}\left[\frac{k'}{2}(R^2 - r^2) - 2k(R - r) + A \ln \frac{R}{r}\right]. \tag{10.24}$$

Since for $k \to 0$ (10.24) must tend towards the solution (9.35) of the Newtonian liquid, the constant A is zero, and we have

$$v = \frac{1}{4a}[k'(R^2 - r^2) - 4k(R - r)]. \tag{10.25}$$

On account of (10.25) the shear rate (10.17) is

$$d_{rz} = -\frac{1}{4a}(k'r - 2k). \tag{10.26}$$

Its magnitude decreases with the distance from the wall and becomes zero for $r = r_0 = 2k/k'$. In consequence, the parabolic solution (10.25) is valid only in the region $r_0 \leqslant r \leqslant R$ adjacent to the wall, whereas for $0 \leqslant r \leqslant r_0$ the material moves as a rigid core (Fig. 10.5).

The incompressible viscoplastic body considered in this section is not the most general material of this type, for the constitutive relations (10.8) and (10.9) have been derived on the basis of (9.72). The use of a function μ dependent on $d_{2)}$ alone does not imply any restriction, since it has been pointed out that the depencence on $d_{(3)}$ would violate the orthogonality condition. However, (9.72) is only a special case of the most general constitutive equation of an incompressible viscous liquid, and the definition of a general viscoplastic material ought to be based on (9.65). If this is done, one obtains yield cylinders of noncircular cross section in the space $\sigma_{\mathrm{I}}, \ldots,$ and it further turns out that the directions of corresponding vectors $\boldsymbol{\sigma}'$ and $\boldsymbol{d}$ may be different. These aspects will be considered in more detail in the next section.

Problems

1. Derive the equilibrium conditions (10.20) for Poiseuille flow.

2. Discuss the plane flow of Figure 9.1 for an incompressible viscoplastic material obeying the constitutive relations $(9.2)_1$, (10.8), and (10.9).

10.2. Perfectly plastic bodies

Let us modify Figure 10.1 in such a manner that the rays emanating from the points $\pm k$ on the axis $\sigma_{12}^{(d)}$ become parallel to the axis d_{12} (Fig. 10.6).

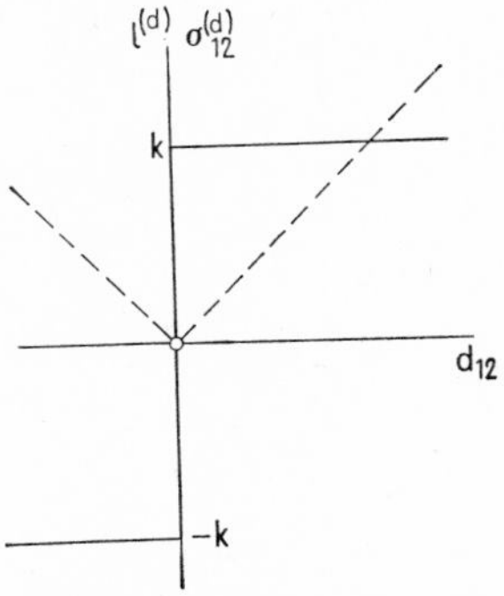

Fig. 10.6 Simple shear of a perfectly plastic body

The corresponding material is referred to as *perfectly plastic* or, more precisely, as *rigid-perfectly-plastic* since it remains rigid up to the yield limit. Its response in simple shear is obtained from (10.3) and (10.4) by setting

$a = 0$. We thus have

$$\sigma_{12}^{(d)2} \leqslant k^2 \qquad (d_{12} = 0), \tag{10.27}$$

$$\sigma_{12}^{(d)} = \frac{k}{\sqrt{d_{12}^2}} d_{12} \qquad (d_{12} \neq 0). \tag{10.28}$$

Equation (10.28) is homogeneous of order 0 in d_{12}. The shear stress $\sigma_{12}^{(d)}$ cannot exceed the yield stress k; it has the constant value $\pm k$ as long as the material deforms, and the shear rate remains indetermined except for its sign. The specific power of dissipation, indicated in Figure 10.6 by the dashed lines, is

$$l^{(d)} = \frac{2}{\varrho} \sigma_{12}^{(d)} d_{12} = \frac{2k}{\varrho} \sqrt{d_{12}^2}\,. \tag{10.29}$$

The generalizations of (10.27) and (10.28) corresponding to (10.9) and (10.8) are

$$\sigma'_{(2)} \leqslant k^2 \qquad (d_{ij} = 0), \tag{10.30}$$

$$\sigma'_{ij} = \frac{k}{\sqrt{d_{(2)}}} d_{ij} \qquad (d_{ij} \neq 0), \tag{10.31}$$

if use is made of the identity $\sigma_{ij}^{(d)} = \sigma'_{ij}$. As long as $d_{ij} = 0$, the stress deviator is limited but indetermined. For $d_{ij} \neq 0$ we obtain from (10.31)

$$\sigma'_{(2)} = \tfrac{1}{2}\sigma'_{ij}\sigma'_{ij} = \tfrac{1}{2}\frac{k^2}{d_{(2)}} d_{ij}d_{ij} = k^2; \tag{10.32}$$

it follows that the stress remains at the yield limit. Conversely, the material remains rigid as long as the inequality (10.30) is satisfied; for stresses at the yield limit the ratios between the deformation rates (but not the deformation rates themselves) are determined by (10.31). The specific power of dissipation is

$$l^{(d)} = \frac{1}{\varrho} \sigma'_{ij} d_{ij} = \frac{2k}{\varrho} \sqrt{d_{(2)}}\,. \tag{10.33}$$

In the 3-dimensional space of the principal stresses (Fig. 10.7), the yield surface

$$\sigma'_{(2)} = k^2 \tag{10.34}$$

is a circular cylinder of radius $k\sqrt{2}$ as in Figure 10.4. For stress states represented by vectors $\boldsymbol{\sigma} = (\sigma_{\mathrm{I}}, \ldots)$ with end points Q inside the cylinder, the vector $\boldsymbol{d} = (d_{\mathrm{I}}, \ldots)$ representing the deformation rate is zero. During plastic

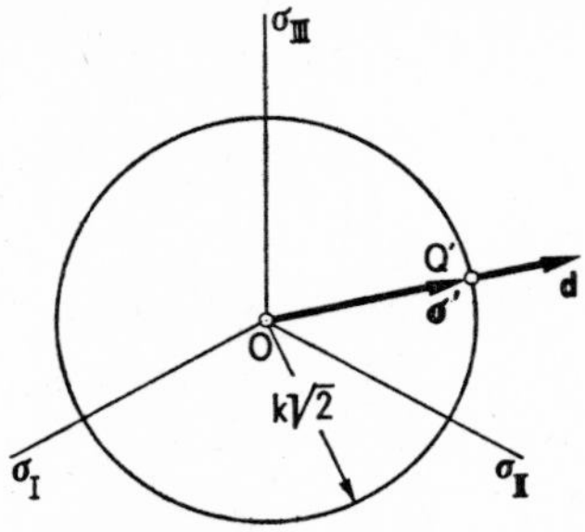

Fig. 10.7 Maxwell's yield surface

flow the point Q remains on the yield surface. The corresponding vector $\boldsymbol{d}$ has the direction of $\boldsymbol{\sigma}'$, but its magnitude remains indetermined.

The response of the plastic body considered here is described by *Maxwell's yield condition*

$$X(\sigma'_{(2)}) = \sigma'_{(2)} - k^2 = 0, \tag{10.35}$$

where $X\chi$ is denoted as the *yield function*, and the so-called *flow rule of v. Mises* (10.31), which may be written as

$$d_{ij} = \nu \sigma'_{ij}, \tag{10.36}$$

where ν is a non-negative proportionality factor of indetermined magnitude. The material obeying (10.35) and (10.36) is still a special case of the perfectly plastic body. Since (9.72) is a particular form of (9.65), it is possible to define more general perfectly plastic bodies the response of which is described by (10.27) and (10.28) in simple shear.

In the first place, the yield function of such a more general material will not have the simple form (10.35). On account of the assumed isotropy, however, it depends on stress invariants alone, and since in an incompressible liquid the dissipative stress is the deviatoric part of the stress tensor, the yield function has the form $X(\sigma'_{(2)}, \sigma'_{(3)})$. It can be defined in such a way that it is negative at the origin. The yield condition reads

$$X(\sigma'_{(2)}, \sigma'_{(3)}) = 0, \tag{10.37}$$

and it is obvious that the yield surface in the stress space of Figure 10.3 is still a (generally noncircular) cylinder with axis g.

As an example, Figure 10.8 shows the cross section of a regular hexagonal prism representing the *yield condition of Tresca*. It is based on the assumption

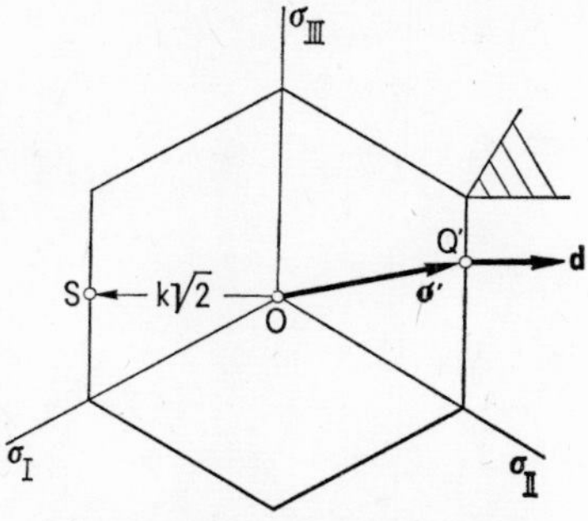

Fig. 10.8 Tresca's yield surface

that the maximal shear stress is responsible for the onset of yield. According to (3.26), the maximal shear stress is given by one or the other of the expressions $\frac{1}{2}|\sigma_{II}-\sigma_{III}|$, ... The yield condition thus reads

$$\sigma_{II}-\sigma_{III}, \ldots = \pm 2k, \tag{10.38}$$

and the six equations (10.38) represent the faces of the prism. The yield function may be written as

$$X = [(\sigma_{II}-\sigma_{III})^2-4k^2]\,[(\sigma_{III}-\sigma_I)^2-4k^2]\,[(\sigma_I-\sigma_{II})^2-4k^2], \tag{10.39}$$

and it is not difficult to verify (P1) that an alternate form of (10.39) with the structure of (10.37) is

$$X = 4\sigma'^3_{(2)}-27\sigma'^2_{(3)}-36k^2\sigma'^2_{(2)}+96k^4\sigma'_{(2)}-64k^6. \tag{10.40}$$

The points where the axes σ_I, ... intersect the prism of Figure 10.8 correspond to uniaxial states of stress; the projections of these points onto the deviatoric plane $\sigma_I + \ldots = 0$ are the corners of the hexagon. If, on the other hand, $\sigma_{12} = \sigma_{21} = 0$ is the only non-vanishing stress component, the principal stresses are $\sigma_I = -\sigma_{II} = \pm k$, $\sigma_{III} = 0$. One of the corresponding points in Figure 10.8 is denoted by S; it lies in the deviatoric plane, and its distance from O is

$$(\sigma_I'^2 + \ldots)^{1/2} = k\sqrt{2}. \tag{10.41}$$

Let us return to the general case of a perfectly plastic material. To describe its response, it is necessary to generalize not only Maxwell's yield condition, but also the flow rule (10.36) of v. Mises. In particular, it is to be expected that the components of the two tensors σ'_{ij} and d_{ij} are not proportional any more. According to v. Mises [16], there is a close connection between yield condition and flow rule, known as the *theory of the plastic potential.* It postulates that (a) the yield surface is (at least weakly) *convex* and that (b) the deformation rate is given by

$$d_{ij} = \nu \frac{\partial X}{\partial \sigma_{ij}}, \quad \text{where} \quad \nu \geqslant 0 \tag{10.42}$$

and where X is the yield function, also denoted as the *plastic potential.* Equation (10.42) implies that (e.g. in the space of Figure 10.3) the vector $\boldsymbol{d}$ representing the deformation rate is orthogonal to the yield surface in the end point of the vector $\boldsymbol{\sigma}$, that it points outwards (i.e. away from O) and that its magnitude is indetermined. Originally, the statement (a) was a mere assumption, corroborated to a certain extent by the particular cases of Figures 10.7 and 10.8, and the postulate (b) was merely supported by an argument of Drucker [17] concerning the energy to be extracted from the material. It will be shown, however, in Section 17.1 that the theory of the plastic potential is essentially a consequence of the orthogonality condition.

It is clear that a theory based on (10.42) is consistent only if the partial derivatives $\partial X/\partial \sigma_{ij}$ are the components of a symmetric deviator. In order to verify this, we note that on account of (10.12)

$$\sigma'_{(2)} = \sigma_{(2)} + \tfrac{1}{3}\sigma_{(1)}^2 \tag{10.43}$$

and that for $\sigma'_{(3)}$ (P2 of Section 3.2) a similar relation

$$\sigma'_{(3)} = \sigma_{(3)} + \tfrac{1}{3}\sigma_{(2)}\sigma_{(1)} + \tfrac{2}{27}\sigma_{(1)}^3 \tag{10.44}$$

holds. It follows (P2) that

$$\frac{\partial \sigma'_{(2)}}{\partial \sigma_{ij}} = \sigma'_{ij}, \qquad \frac{\partial \sigma'_{(3)}}{\partial \sigma_{ij}} = \sigma'_{ik}\sigma'_{kj} - \tfrac{2}{3}\sigma'_{(2)}\delta_{ij} \tag{10.45}$$

and that, therefore,

$$\frac{\partial X}{\partial \sigma_{ij}} = \frac{\partial X}{\partial \sigma'_{(2)}}\sigma'_{ij} + \frac{\partial X}{\partial \sigma'_{(3)}}(\sigma'_{ik}\sigma'_{kj} - \tfrac{2}{3}\sigma'_{(2)}\delta_{ij}). \tag{10.46}$$

The right-hand side is in fact a symmetric deviator.

On account of $(10.45)_1$, the theory of the plastic potential, applied to Maxwell's yield function (10.35), immediately supplies the flow rule (10.36) of v. Mises. In the case of Tresca's yield condition (10.38), the vector $\boldsymbol{d}$ corresponding to a stress point Q on one of the faces of the prism in Figure

10.8 is orthogonal to this face. At the edges of the prism, the direction of $\boldsymbol{d}$ is indetermined. However, if we consider the prism as the limiting case of a smooth cylinder, it becomes obvious that $\boldsymbol{d}$ lies within the angular area confined by the normals of the adjacent faces and shaded in one of the corners in Figure 10.8.

In general, the theory of the plastic potential establishes the connection between a vector $\boldsymbol{\sigma}$ and the direction of the corresponding $\boldsymbol{d}$. Even this connection, however, is not always single-valued. If the yield surface has a corner or an edge, there is an infinity of directions $\boldsymbol{d}$ corresponding to the same $\boldsymbol{\sigma}$. If, on the other hand, the yield surface contains plane faces or at least straight lines, there is an infinity of vectors $\boldsymbol{\sigma}$ corresponding to the same direction $\boldsymbol{d}$. The last case always occurs in the cylindrical yield surfaces considered above; in fact, the plastic response remains unaffected by an additional hydrostatic stress. In spite of these ambiguities, the specific power of dissipation

$$l^{(d)} = \frac{1}{\varrho} \sigma'_{ij} d_{ij} = \frac{1}{\varrho} \sigma_{ij} d_{ij} \tag{10.47}$$

is uniquely determined by the deformation rate. In fact, (10.47) may be written in the form

$$l^{(d)} = \frac{1}{\varrho} \boldsymbol{\sigma}' \cdot \boldsymbol{d} = \frac{1}{\varrho} \boldsymbol{\sigma} \cdot \boldsymbol{d}, \tag{10.48}$$

and it is easy to see that the scalar product is completely determined by $\boldsymbol{d}$ irrespective of the presence of singularities or linear subspaces in the yield surface. This property is essential in connection with the orthogonality principle and will be further discussed in Chapter 14.

Figure 10.9 shows part of a yield surface, a vector $\boldsymbol{d}$ representing plastic flow, and the corresponding vector $\boldsymbol{\sigma}$. If $\boldsymbol{\sigma}^*$ represents an arbitrary state of

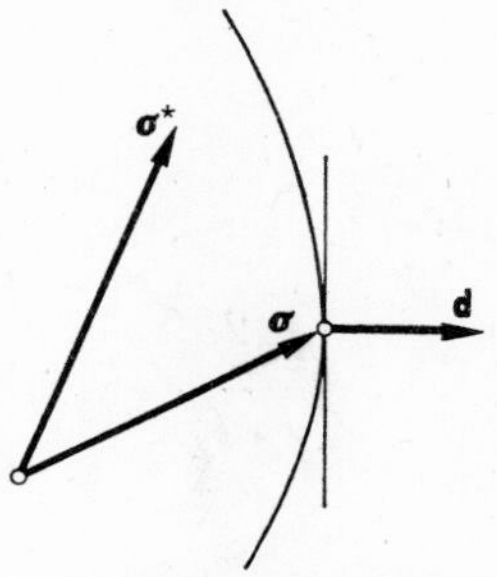

Fig. 10.9 Maximal rate of dissipation work

stress at or below the yield limit, it follows from the theory of the plastic potential that

$$\frac{1}{\varrho}\,\boldsymbol{\sigma}^*\boldsymbol{\cdot d} \leqslant \frac{1}{\varrho}\,\boldsymbol{\sigma}\boldsymbol{\cdot d} = l^{(d)}. \tag{10.49}$$

Thus, the actual power of dissipation corresponding to a given deformation rate d_{ij} is never less than the fictitious power of dissipation calculated from the actual deformation rate and a stress state σ^*_{ij} at or below the yield limit. This result, independently obtained by v. Mises [16], Taylor [18], Hill [19], and generalized by Koiter [20], is called the *principle of maximal dissipation rate.* Its inversion is also true: the theory of the plastic potential is a consequence of the principle of maximal power of dissipation.

Let us close this section with a remark concerning (10.42). Since $\nu = 0$ is admitted, a given element of a plastic body may remain rigid even though its state of stress is at the yield limit. In this case, we refer to the element as *plastified*, and we note that actual flow may be prevented by the rigid behavior of the vicinity until plastification has spread over a sufficiently large portion of the body.

As a simple example which can be treated by means of the constitutive relations (10.27), (10.28), and hence is independent of the particular form of the yield surface and does not require the application of the theory of the plastic potential, let us consider a cylindrical shaft of radius R, subjected to a torque $M > 0$. We have assumed so far that the material behaves as a rigid body as long as there is no plastic flow. In practice, however, a trace of elastic response is always present, and this allows us to adopt the elastic stress distribution for the nonplastified domain. In cylindrical coordinates r, α, z the only nonvanishing stress is thus the shear stress

$$\sigma_{\alpha z} = \frac{2M}{\pi R^4}\,r \quad (\sigma_{\alpha z} \leqslant k), \tag{10.50}$$

and the corresponding shear rate $d_{\alpha z}$ is (nearly) zero. The elements adjacent to the surface are the first to become plastified when

$$\sigma_{\alpha z}(R) = \frac{2M}{\pi R^3} = k. \tag{10.51}$$

The corresponding torque is

$$M_0 = \frac{\pi}{2}\,kR^3. \tag{10.52}$$

With M increasing beyond M_0, a plastified region enclosing a rigid cylindrical core spreads towards the axis. Plastic flow and, in fact, collapse sets in when the rigid core disappears, i.e. when $\sigma_{\alpha z} = k$ everywhere. The corresponding torque is

$$M_p = 2\pi k \int_0^R r^2 \, \mathrm{d}r = \frac{2\pi}{3} kR^3 = \tfrac{4}{3} M_0 . \qquad (10.53)$$

Problems

1. Derive (10.40) from (10.39).
2. Prove the two equations (10.45).

10.3. Plane problems

Many practical applications of the theory of plasticity are concerned either with plane stress as defined at the end of Section 3.2 or with a plane deformation rate (Section 2.1). If the plastic flow is plane, the deformation rate is plane as well.

In any case, we will have to discuss the intersection of the yield surface in the space $\sigma_{\mathrm{I}}, \ldots$ (i.e. of the cylinder or prism discussed in Section 10.2) with a plane passing through the origin, and since the yield surface is convex, the curve or polygon representing this intersection is convex, too.

In order to treat *plane stress*, we choose the coordinate system so that $\sigma_{\mathrm{III}} = 0$. The intersection of the yield surface with the plane $\sigma_{\mathrm{I}}, \sigma_{\mathrm{II}}$ is closed and encloses the origin; it is referred to as the *yield locus.*

In the case of Maxwell's yield condition, the yield locus obtained from the circular cylinder of Figure 10.7 is an ellipse (Fig. 10.10) with axes bisect-

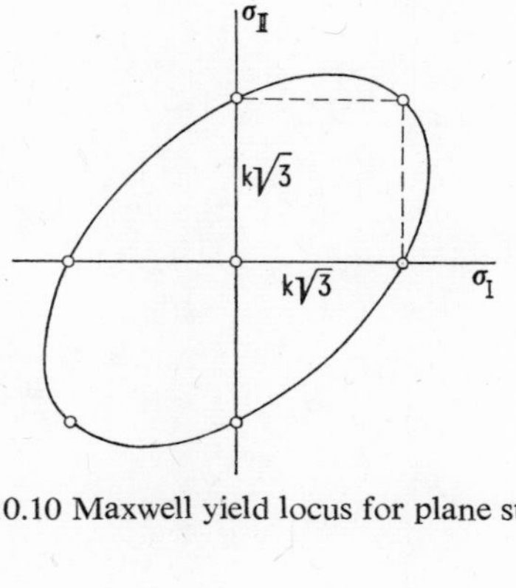

Fig. 10.10 Maxwell yield locus for plane stress

ing the angles between the coordinate axes. On account of (10.13) the yield condition (10.35) may be written as

$$\sigma_{\mathrm{I}}^2 + \ \ldots \ - \sigma_{\mathrm{II}}\sigma_{\mathrm{III}} - \ \ldots \ = 3k^2 . \tag{10.54}$$

Putting $\sigma_{\mathrm{III}} = 0$, we obtain the equation

$$\sigma_{\mathrm{I}}^2 + \sigma_{\mathrm{II}}^2 - \sigma_{\mathrm{I}}\sigma_{\mathrm{II}} = 3k^2 \tag{10.55}$$

of the yield locus. It passes through the six points marked in Figure 10.10. The yield stress in uniaxial tension or compression is $k\sqrt{3}$.

In the case of Tresca's yield condition, the yield locus obtained from the regular hexagonal prism of Figure 10.8 is an irregular hexagon. In fact, the yield condition (10.38) reduces to

$$\sigma_{\mathrm{I}} = \pm 2k, \qquad \sigma_{\mathrm{II}} = \pm 2k, \qquad \sigma_{\mathrm{I}} - \sigma_{\mathrm{II}} = \pm 2k, \tag{10.56}$$

and these equations represent the six sides of the yield locus of Figure 10.11. The yield stress in uniaxial tension or compression is $2k$.

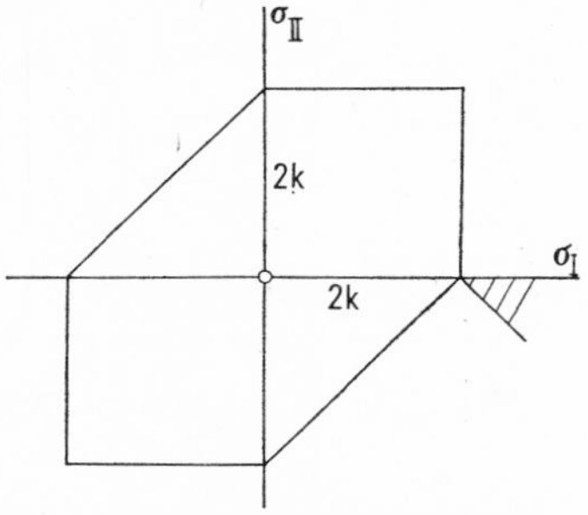

Fig. 10.11 Tresca yield locus for plane stress

It is important to note that the vector $\boldsymbol{d}$ corresponding to a stress state $\boldsymbol{\sigma} = (\sigma_{\mathrm{I}}, \sigma_{\mathrm{II}}, 0)$ at the yield limit does not lie in the plane $\sigma_{\mathrm{I}}, \sigma_{\mathrm{II}}$. It is orthogonal to the yield surface in the end point of $\boldsymbol{\sigma}$, and it follows that its projection $\boldsymbol{d}'$ onto the stress plane is orthogonal to the yield locus. The third component of $\boldsymbol{d}$, however, is generally different from zero; it is determined by $d_{\mathrm{I}} + d_{\mathrm{II}} + d_{\mathrm{III}} = 0$, since the material is incompressible.

On the basis of Maxwell's yield function (10.35), equations $(10.45)_1$ and (3.27) yield

$$\frac{\partial X}{\partial \sigma_{\mathrm{I}}} = \sigma_{\mathrm{I}}' = \tfrac{1}{3}(2\sigma_{\mathrm{I}} - \sigma_{\mathrm{II}} - \sigma_{\mathrm{III}}), \ \ldots \ . \tag{10.57}$$

Making use of $(10.42)_1$ and the condition $\sigma_{III} = 0$ of plane stress, we obtain

$$d_I = \nu(2\sigma_I - \sigma_{II}), \qquad d_{II} = \nu(2\sigma_{II} - \sigma_I), \qquad d_{III} = -\nu(\sigma_I + \sigma_{II}), \quad (10.58)$$

where the factor $\frac{1}{3}$ has been included in ν. The component d_{III} is not zero except in the case $\sigma_{II} = -\sigma_I$ of simple shear, but the projection $\boldsymbol{d}' = (d_I, d_{II}, 0)$ of $\boldsymbol{d}$ is clearly orthogonal to the yield locus (10.55) at the end point of $\boldsymbol{\sigma} = (\sigma_I, \sigma_{II}, 0)$. In uniaxial tension σ_I (10.58) reduces to $\boldsymbol{d} = \nu(2\sigma_I, -\sigma_I, -\sigma_I)$, where the last two components represent the lateral contraction corresponding to the extension in the direction of σ_I.

For Tresca's yield function (10.39), the vector $\boldsymbol{d}$ can be discussed in a similar manner (P1). At the corners of the yield locus its direction is indetermined. For uniaxial tension, e.g., $\boldsymbol{d}'$ lies anywhere in the shaded area of Figure 10.11. It follows that d_{II} and d_{III} are confined to the interval $-d_I \ldots 0$. We note here a remarkable lack of symmetry: under Tresca's yield condition the lateral contractions in uniaxial tension need not be equal; they are merely subject to the condition $d_{II} + d_{III} = -d_I$.

Let us now consider the case of a *plane deformation rate*, defined by $d_{III} = 0$. Since it is deviatoric, the vector $\boldsymbol{d}$ has the form $\boldsymbol{d} = (d_I, -d_I, 0)$, and the principal axes may be numbered so that $d_I > 0$. It is not necessary here to specify the yield condition [21]. Figure 10.12 shows part of an arbitrary yield locus in the deviatoric plane, which also contains the vector $\boldsymbol{d}$. Since the invariants of the stress deviator are symmetric with respect to $\sigma'_I, \ldots$, the yield locus is symmetric with respect to the projections of the axes $\sigma_I, \ldots$ onto the deviatoric plane. The unit vector in the direction of $\boldsymbol{d}$ is given by $\boldsymbol{e} = \left(1/\sqrt{2}, -1/\sqrt{2}, 0\right)$.

A straight line containing at least one point of the yield locus, but none of the points of its interior is called a *supporting line*. Examples are all tangents since the yield locus is convex, but also certain straight lines passing through a possible corner. Let us consider a point traveling clockwise along the yield locus. On account of the convexity of this curve, the supporting line of the point as well as its exterior normal rotate continuously in the clockwise sense. For reasons of symmetry there is a supporting line (Fig. 10.12) perpendicular to the projection of the axis σ_I in the point A_1, and an analogous statement holds for A_2. It follows that the point A the exterior normal of which has the direction of $\boldsymbol{e}$ lies on the arc A_1A_2 of the yield locus.

Let us consider a stress state $\sigma_I, \ldots$ the deviatoric part of which is represented by the radius vector $\boldsymbol{\sigma}'$ of A. The distance AA_{III} in Figure 10.12

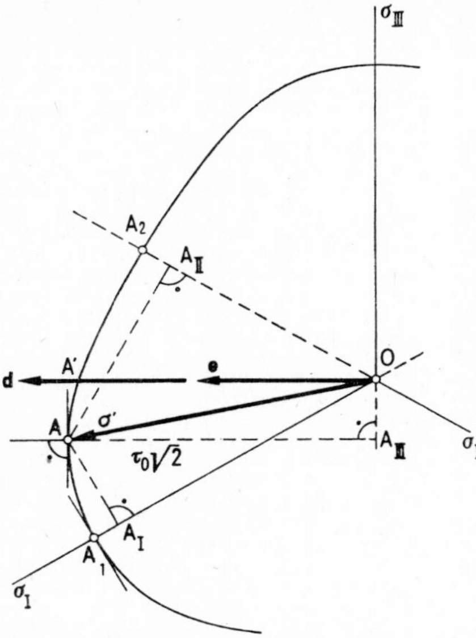

Fig. 10.12 Yield locus in the deviatoric plane

is the salar product

$$\boldsymbol{\sigma}'\cdot\boldsymbol{e} = \boldsymbol{\sigma}\cdot\boldsymbol{e} = \frac{1}{\sqrt{2}}(\sigma_{\mathrm{I}}-\sigma_{\mathrm{II}}); \tag{10.59}$$

except for a factor $1/\sqrt{2}$ it is one of the principal shear stresses (Section 3.2). In a similar manner, the distances AA_{I} and AA_{II} are found to be

$$\frac{1}{\sqrt{2}}(\sigma_{\mathrm{III}}-\sigma_{\mathrm{II}}), \qquad \frac{1}{\sqrt{2}}(\sigma_{\mathrm{I}}-\sigma_{\mathrm{III}}) \tag{10.60}$$

respectively. The location of A on the arc A_1A_2 implies that AA_{III} is the largest of the three distances. It follows that $AA_{\mathrm{III}} = \tau_0\sqrt{2}$, where

$$\tau_0 = \tfrac{1}{2}(\sigma_{\mathrm{I}}-\sigma_{\mathrm{II}}) \tag{10.61}$$

is the maximal shear stress at the physical point considered. Incidentally, since the distances (10.60) are positive, the principal stress σ_{III} perpendicular to the plane of the deformation rate is always the intermediate principal stress.

According to the theory of the plastic potential, the deviatoric stress corresponding to the deformation rate $\boldsymbol{d}$ is represented by the radius vector $\boldsymbol{\sigma}'$ of A.

The scalar product (10.59) is thus equal to $\tau_0\sqrt{2}$. We therefore obtain the simple yield condition $\sigma_{\text{I}}-\sigma_{\text{II}} = 2\tau_0$ or

$$(\sigma_{\text{I}}-\sigma_{\text{II}})^2 = 4\tau_0^2 \tag{10.62}$$

if we eventually drop the restriction $d_{\text{I}} > 0$ in the choice of the coordinate system. The corresponding yield locus in the plane σ_{I}, σ_{II} consists of the two straight lines in Figure 10.13; the intermediate principal stress σ_{III} is irrelevant.

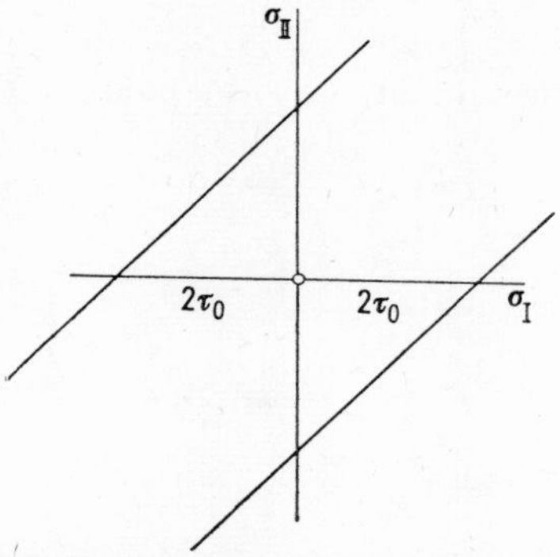

Fig. 10.13 Yield locus for plane deformation rate

It is remarkable that (10.62) holds irrespective of the particular form of the general yield condition (10.37). It should further be noted that in general the maximal shear stress τ_0 is not reached in simple shear. A possibile case of simple shear, corresponding to $\sigma_{\text{I}} = -\sigma_{\text{II}} = k$, $\sigma_{\text{III}} = 0$, is represented by the point A' in Figure 10.12. For yield loci that are not only symmetric with respect to the projections of the axes $\sigma_{\text{I}}, \ldots$ onto the deviatoric plane, but also with respect to their normals (as the yield loci of Figures 10.7 and 10.8), τ_0 is equal to k, and the yield condition becomes

$$(\sigma_{\text{I}}-\sigma_{\text{II}})^2 = 4k^2\,. \tag{10.63}$$

Problems

1. Discuss the direction of the vector $\boldsymbol{d}$ analytically for Tresca's yield function (10.39).

2. Show that the distances AA_{I} and AA_{II} in Figure 10.12 are given by (10.60).

3. Formulate the yield condition (10.62) for an arbitrary coordinate system with axis x_3 perpendicular to the plane of the deformation rate.

10.4. Generalizations

The concept of a perfectly plastic body has been developed in Section 10.2 from the diagram of Figure 10.6 corresponding to simple shear. Since the material is isotropic, it is obvious that the yield limit has the absolute value k irrespective of the sign of σ_{12}. In a uniaxial state of stress σ_I, however, the yield limit of certain materials depends on the sign of σ_I, and it is usually larger in compression than in tension. The inclusion of $\sigma'_{(3)}$ in the general yield function (10.37) takes care of this effect, which occurs, e.g. in the case of the yield locus of Figure 10.12. The ratio of the two yield limits is restricted by the condition that the yield locus is convex. The equilateral triangle of Figure 10.14 represents the case for which the ratio is a maximum, amounting to 2.

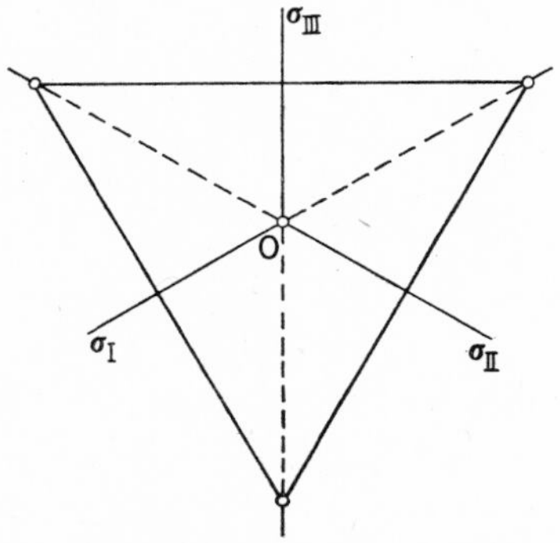

Fig. 10.14 Limiting case of yield locus in the deviatoric plane

In a perfectly plastic body the yield condition is always the same, no matter how the material is deformed. In certain practically important materials, however, it depends on the strain history. If this is the case, it is convenient to consider the dissipative stress $\sigma_{ij}^{(d)}$ as a function of the strain ε_{ij} rather than the deformation rate, despite the lack of uniqueness connected with this representation. Figure 10.15 shows the idealized stress-strain diagram for simple shear of a so-called *hardening material.* It consists of the segment $-k \; \ldots \; +k$ on the axis σ_{12} and of two parallel oblique lines. For $\sigma_{12} > 0$ plastic flow sets in at k. In order to maintain it, the shear stress must be increased. If the sample is unloaded from, say, point A in the diagram, the point illustrating the response drops to the axis ε_{12}, where its abscissa represents the remaining plastic shear strain. In reloading, the point moves up again to A and follows the oblique line, as if no unloading had

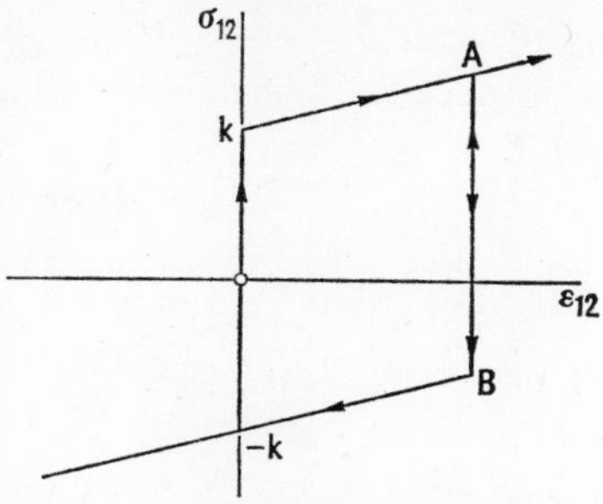

Fig. 10.15 Stress-strain diagram of a hardening material

taken place. Thus, the amount of plastic flow up to the point A increases the yield limit, and this is referred to as *hardening*. If, however, the sample, after unloading from A, is reloaded in the opposite sense, the representative point reaches the yield limit at B. Hardening is thus accompanied by a weakening effect for reverse loading. This phenomenon is known as the *Bauschinger effect*.

The response in simple tension and compression is similar. Representing it in the manner of Figure 10.15, one has to bear in mind, though, that part of the total tension is due to the hydrostatic pressure $(9.2)_1$ and that the plastic constitutive relations connect the time derivative of ε_{ij} with the dissipative part $\sigma_{ij}^{(d)} = \sigma'_{ij}$ of the stress tensor. To generalize these considerations for arbitrary strains and stresses, we note that the original yield limits in Figure 10.15 are the end points of the segment $-k \ldots +k$ on the axis $\sigma_{12} = \sigma'_{12}$. During plastic flow, this segment is displaced along its axis, the point representing the stress moving with one of its ends. Something similar must happen to the yield surface in Figure 10.12 if the stress deviator σ'_{ij} is arbitrary. In a 6- or 9-dimensional space σ'_{ij} (Fig. 10.16) the original yield surface

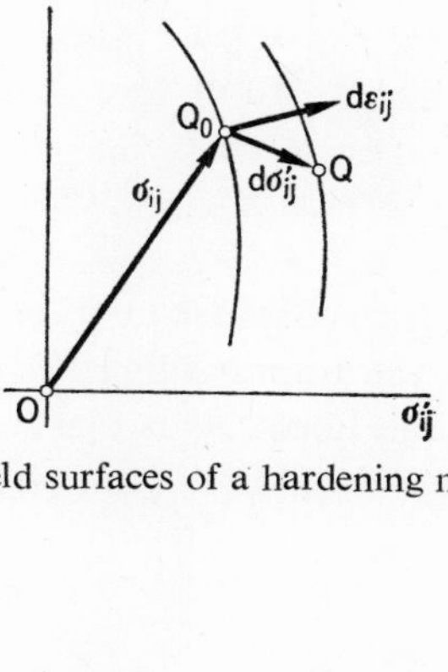

Fig. 10.16 Yield surfaces of a hardening material

is given by (10.37). Any increase $d\sigma'_{ij}$ of the stress deviator beyond original yield displaces the yield surface in a certain manner, but so that it always contains the stress point Q.

Various theories for this displacement have been proposed. The simplest of them assumes that the yield surface expands, retaining its shape and its situation relative to the coordinate axes. Since in this case the isotropy of the material is obviously preserved during plastic flow, this theory is referred to as *isotropic hardening*. It is clear, however, that it is in contradiction to the Bauschinger effect. The so-called theory of *kinematic hardening* avoids this drawback by assuming that the yield surface retains its shape but moves in a translation. This theory implies that the material becomes anisotropic during plastic flow. It requires a hardening law, i.e. a statement concerning the direction of the translation. Prager, who proposed kinematic hardening [22], assumed that the yield surface moves in the direction of the vector $d\varepsilon_{ij}$ (or d_{ij}, i.e. in the direction of the exterior normal). It can be shown [23], however, that this simple rule, provided it holds in a 9-dimensional space, is in general not valid in subspaces as, e.g., the 6-dimensional stress space or the ones used in plane stress or strain. The assumption [23] that the yield surface moves in the direction of $d\sigma_{ij}$ removes this disadvantage. However, none of the theories discussed here seems realistic since experiments indicate that the yield surface also suffers a deformation. In short, the problem is complicated and still unsolved.

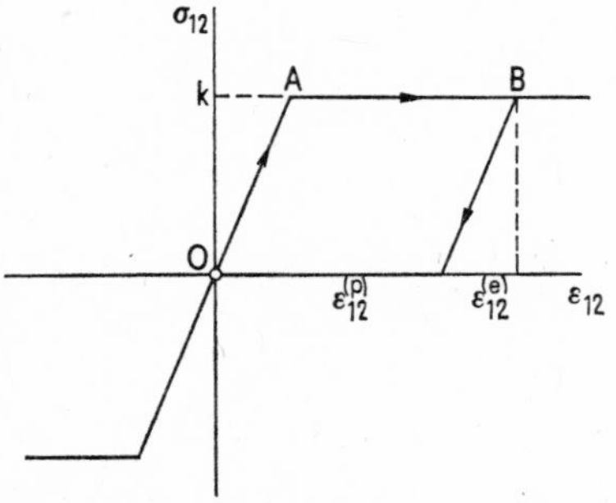

Fig. 10.17 Stress-strain diagram of an elastic-perfectly-plastic material

Another deviation from the behavior illustrated by Figure 10.15 for simple shear is given by the stress-strain diagram of Figure 10.17. Here the initial response, represented by the line OA, is elastic. At A plastic flow sets in, and upon unloading from B the elastic shear strain is recuperated. A material of this type is called *elastic-plastic* or, more precisely, *elastic-perfectly-plastic* if the absence of hardening is to be stressed.

It is customary to denote the elastic shear by $\varepsilon_{12}^{(e)}$ to distinguish it from the plastic shear $\varepsilon_{12}^{(p)}$ still present after complete unloading. The total shear in an arbitrary phase of the process can always be represented in the form $\varepsilon_{12} = \varepsilon_{12}^{(e)} + \varepsilon_{12}^{(p)}$. The obvious generalization for arbitrary strains is

$$\varepsilon_{ij} = \varepsilon_{ij}^{(e)} + \varepsilon_{ij}^{(p)}. \tag{10.64}$$

The strain tensor thus appears as the sum of an elastic and a plastic contribution. The first one, $\varepsilon_{ij}^{(e)}$, is assumed to be connected with the stress σ_{ij} by Hooke's law (5.41), provided the process is either isothermal or adiabatic. The second contribution, $\varepsilon_{ij}^{(p)}$, is assumed to be a deviator the rate of which, $d_{ij}^{(p)} = \dot{\varepsilon}_{ij}^{(p)}$, is connected with the stress deviator σ'_{ij} by the theory of the plastic potential (10.42). Replacing (5.41) by (7.4) and (7.5), we thus have the constitutive equation (10.64), supplemented by

$$\varepsilon_{kk} = \frac{1}{3\lambda + 2\mu}\,\sigma_{kk}, \qquad \varepsilon_{ij}^{(e)\prime} = \frac{1}{2\mu}\,\sigma'_{ij}, \tag{10.65}$$

$$\dot{\varepsilon}_{ij}^{(p)} = d_{ij}^{(p)} = \nu\,\frac{\partial X}{\partial \sigma_{ij}}, \tag{10.66}$$

where λ and μ are Lamé's constants, $X(\sigma'_{(2)}, \sigma'_{(3)})$ the plastic potential, and where $\nu \geqslant 0$. The specific power of σ_{ij} is

$$\frac{1}{\varrho}\,\sigma_{ij}\dot{\varepsilon}_{ij} = \frac{1}{\varrho}\,(\sigma_{ij}\dot{\varepsilon}_{ij}^{(e)} + \sigma_{ij}d_{ij}^{(p)}), \tag{10.67}$$

and the specific power of dissipation is given by

$$l^{(d)} = \frac{1}{\varrho}\,\sigma_{ij}d_{ij}^{(p)} = \frac{1}{\varrho}\,\sigma'_{ij}d_{ij}^{(p)}. \tag{10.68}$$

Let us note that the elastic-plastic body is our first example of a material with internal parameters. In fact, its local state is not determined by the total strains ε_{ij} and the temperature ϑ alone. We also need either the elastic strains or the plastic ones; on account of (10.64) one of these two sets is sufficient. The set we choose represents the internal parameters α_{ij} in the sense of Section 4.2. The corresponding forces β_{ij} are zero; their quasiconservative and dissipative parts are connected by (4.42).

Problems

1. Consider the yield locus of Figure 10.14. Denote its yield limit in simple shear by k and determine the yield limits in simple tension and compression, as well as the maximal shear stress.

CHAPTER 11

VISCOELASTICITY

In Chapters 7 and 8 we have treated various types of elastic materials (solids, gases). Chapter 9 has been concerned with purely viscous bodies (liquids), and in Chapter 10 we have dealt with some limiting cases of the viscous liquid: the viscoplastic and the rigid-plastic body.

In the present chapter, we will be concerned with materials the response of which is partly elastic and partly viscous. They are referred to as *visco-elastic*, and we will find that some of them are to be classified as solids, others as liquids. Most of these materials have internal parameters. The elastic-plastic body encountered in Section 10.4 may be considered as a limiting case of a viscoelastic material.

11.1. One-dimensional models

In order to study the response of a viscoelastic body, let us consider a material element in the form of a cuboid subjected to uniaxial stress, and let us replace the element by a model consisting of springs and dashpots. A simple example has been given in Figure 4.1; it consists of a spring and a dashpot connected in series.

If the dashpot is eliminated, Figure 4.1 reduces to the model of an *elastic* body. The force σ and the elongation ε represent in turn the normal stress in the element and the corresponding extension. Provided the spring is linear and the spring constant is E, the connection between ε and σ is given by

$$\sigma = E\varepsilon. \tag{11.1}$$

If, on the other hand, the spring is dropped, the model represents a *purely viscous* body. Provided the dashpot is linear and the viscosity coefficient is

F, we have

$$\sigma = F\dot{\varepsilon} \tag{11.2}$$

in place of (11.1).

It is obvious that models of this type have to be regarded as massless. Since they represent material elements, the temperature must be the same all over the model; therefore, the deformation has to be considered as an isothermal process. In the case of the spring (11.1), the stress is quasiconservative, and the corresponding free energy

$$\Psi = \tfrac{1}{2}E\varepsilon^2 \tag{11.3}$$

may be considered, as in Section 5.3, as the deformation energy. In the case of the dashpot (11.2), the stress is dissipative, and the corresponding dissipation function $\sigma\dot{\varepsilon}$ (Section 4.1) is

$$\Phi = F\dot{\varepsilon}^2 . \tag{11.4}$$

We will presently discuss more complicated spring-dashpot models. Before doing so, however, let us return to Section 10.4 and let us ask for the models representing some of the plastic bodies treated there. The stress-strain diagram of the *rigid-perfectly-plastic* body is the one of Figure 10.6 except for the notations; in the present context the diagram shows σ as a function of ε. The corresponding model might be obtained as a limiting case of a nonlinear dashpot. A simpler model, however, is given by a block subject to dry friction (Fig. 11.1). In a similar manner, the model of Figure 11.2 represents the *elastic-perfectly-plastic* material with the stress-strain diagram of Figure 10.17. Like the model of Figure 4.1, it has an internal parameter; as such, either the elongation of the spring or the displacement of the block may be used.

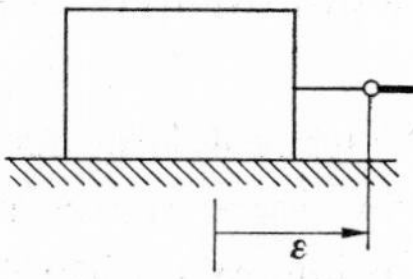

Fig. 11.1 Model of a rigid-perfectly-plastic material

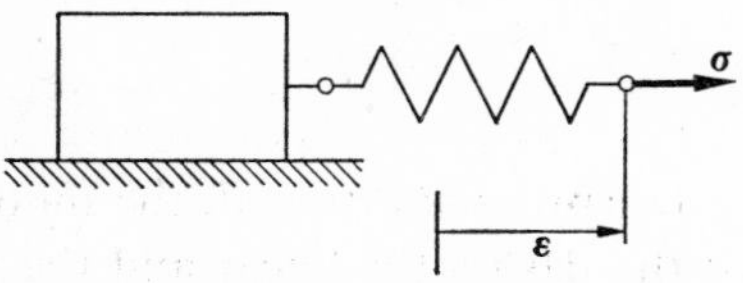

Fig. 11.2 Model of an elastic-perfectly-plastic material

It has been mentioned that the models for viscoelastic bodies consist of springs and dashpots, and we assume at present that all of these elements are linear. The simplest example is represented by Figure 11.3; it is referred

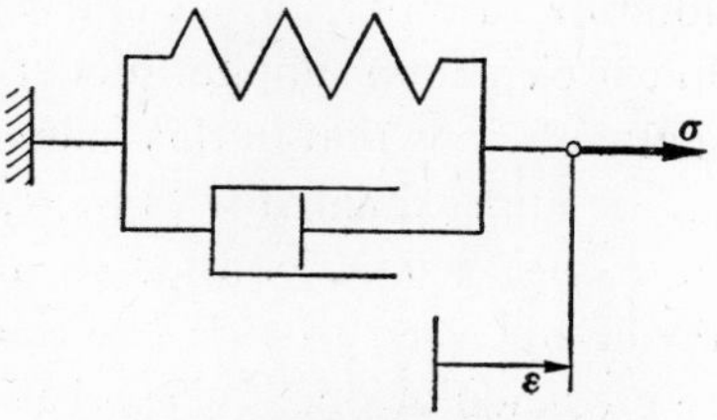

Fig. 11.3 Kelvin model

to as the *Kelvin model* and consists of a spring and a dashpot in parallel connection. Since the extension ε determines the configuration, the model is free of internal parameters. If the constants E and F characterize the two elements, the corresponding stresses are

$$\sigma' = E\varepsilon, \qquad \sigma'' = F\dot{\varepsilon}, \tag{11.5}$$

and the total stress σ is given by

$$\sigma = \sigma' + \sigma'' = E\varepsilon + F\dot{\varepsilon}. \tag{11.6}$$

The response of the model is governed by this differential equation. With the notations $q_0 = E$ and $q_1 = F$, (11.6) assumes the form

$$\sigma = q_0\varepsilon + q_1\dot{\varepsilon}. \tag{11.7}$$

Another example, referred to as the *Maxwell model*, has already been given in Figure 4.1. Its configuration is determined by the total extension ε and an internal parameter, e.g. the extension α of the spring. The two elements are subject to the same stress, and the extensions are determined by

$$\alpha = \frac{\sigma}{E} \quad \text{and} \quad (\varepsilon - \alpha)^{\cdot} = \frac{\sigma}{F}. \tag{11.8}$$

Eliminating α from $(11.8)_1$ and $(11.8)_2$, we obtain the differential equation

$$\frac{\sigma}{F} + \frac{\dot{\sigma}}{E} = \dot{\varepsilon} \tag{11.9}$$

or

$$\sigma + p_1\dot{\sigma} = q_1\dot{\varepsilon}, \tag{11.10}$$

where $p_1 = F/E$ and $q_1 = F$.

It is clear that the addition of more springs and dashpots (together with the admission of nonlinear elements) supplies models of more complicated structures, and it is equally obvious that in this manner the actual behavior of a given material may be approximated with increasing accuracy. A few of these more elaborate models will be considered in the problem section. Let us discuss, instead, the test used to compare the actual behavior of a material to the response predicted by one or the other of the simpler models.

The *standard test* (cf. [24]) consists of two phases. During the first of them the originally unstrained model is subjected to an external stress σ_0, instantaneously applied at time $t = 0$ and maintained constant during a time interval $0 < t < t_1$. In the second phase $t > t_1$ the stress is not controlled directly, but the extension is kept at the level ε_1 attained at the end of the first phase. The stress acting up to the end of the first phase may be represented by

$$\sigma(t) = \sigma_0\,\Delta(t), \tag{11.11}$$

where $\Delta(t)$ is the *Heaviside* function, defined by

$$\Delta(t) = \lim_{\tau \to 0} \begin{cases} 0 & (t \leqslant -\tau/2) \\ \frac{1}{2} + \dfrac{t}{\tau} & (-\tau/2 \leqslant t \leqslant \tau/2) \\ 1 & (t \geqslant \tau/2) \end{cases} \tag{11.12}$$

or, in geometrical terms, by the step function obtained from Figure 11.4 by letting $\tau \to 0$. Incidentally, a comparison of Figures 11.4 and 8.3 shows that the Heaviside function $\Delta(t)$ may be interpreted as the integral of the Dirac function $\delta(t)$ defined in Section 8.3.

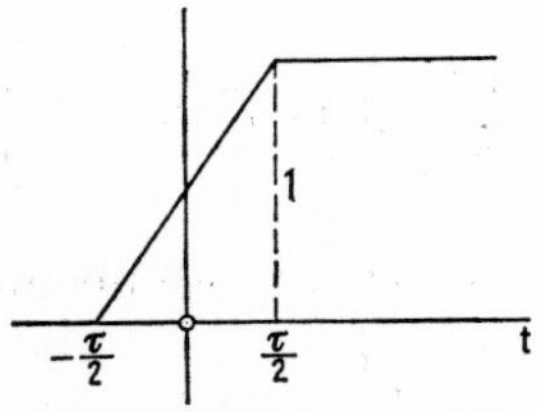

Fig. 11.4 Heaviside function

Let us subject the Kelvin model (Fig. 11.3) to the standard test. During the first phase, the differential equation (11.7) applies with $\sigma = \sigma_0$. Its integral is

$$\varepsilon = c \exp\left(-\frac{q_0}{q_1} t\right) + \frac{\sigma_0}{q_0}, \tag{11.13}$$

where c is a constant. At $t = 0$ the discontinuity of σ implies that $\dot{\varepsilon}$ is discontinuous. However, ε is continuous, and it follows that the initial condition is $\varepsilon(t = 0) = 0$. In consequence, $c = -\sigma_0/q_0$ and hence

$$\varepsilon = \frac{\sigma_0}{q_0}\left[1 - \exp\left(-\frac{q_0}{q_1} t\right)\right] \quad (0 \leqslant t \leqslant t_1). \tag{11.14}$$

In Figure 11.5 σ and ε are plotted against t. For $t_1 \to \infty$ ε would tend asymptotically towards σ_0/q_0. The increase of ε while σ is constant is referred

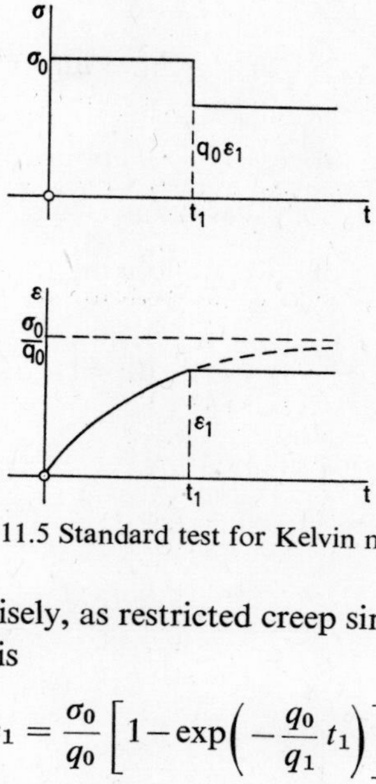

Fig. 11.5 Standard test for Kelvin model

to as *creep* or, more precisely, as restricted creep since ε never reaches σ_0/q_0. The value of ε at $t = t_1$ is

$$\varepsilon_1 = \frac{\sigma_0}{q_0}\left[1 - \exp\left(-\frac{q_0}{q_1} t_1\right)\right]. \tag{11.15}$$

During the second phase ε is maintained at the value ε_1. The differential equation (11.7) thus yields

$$\sigma = q_0 \varepsilon_1 = \sigma_0\left[1 - \exp\left(-\frac{q_0}{q_1} t_1\right)\right] < \sigma_0 \quad (t > t_1). \tag{11.16}$$

The value of σ remains constant but different from σ_0, corresponding to the discontinuity of $\dot{\varepsilon}$ at $t = t_1$. The drop of σ from σ_0 to the value (11.16) is referred to as *relaxation* or, more precisely, as instant relaxation. Qualitatively, the behavior illustrated by the two curves of Figure 11.5 is easily explained in mechanical terms by the response of the two elements of the Kelvin model.

In the case of the Maxwell model (Fig. 4.1), the first phase of the standard test is governed by the differential equation (11.10) with $\sigma = \sigma_0$. Integration yields

$$\varepsilon = \frac{\sigma_0}{q_1} t + c, \tag{11.17}$$

where c is a constant. At $t = 0$, $\dot{\varepsilon}$ becomes infinite with $\dot{\sigma}$. It follows that c cannot be determined by means of the initial condition $\varepsilon(t = 0) = 0$. However, by integration of (11.10) over the interval $-\tau/2 \leqslant t \leqslant \tau/2$ we obtain

$$\int_{-\tau/2}^{\tau/2} \sigma \, dt + p_1[\sigma(\tau/2) - \sigma(-\tau/2)] = q_1[\varepsilon(\tau/2) - \varepsilon(-\tau/2)]. \tag{11.18}$$

Since σ remains finite, the process $\tau \to 0$, applied to (11.18), supplies

$$p_1 \sigma_0 = q_1 \varepsilon_0, \tag{11.19}$$

where ε_0 is the initial value of ε we are looking for. It follows that $c = \varepsilon_0 = (p_1/q_1)\sigma_0$ and hence

$$\varepsilon = \frac{\sigma_0}{q_1}(t + p_1) \quad (0 < t \leqslant t_1). \tag{11.20}$$

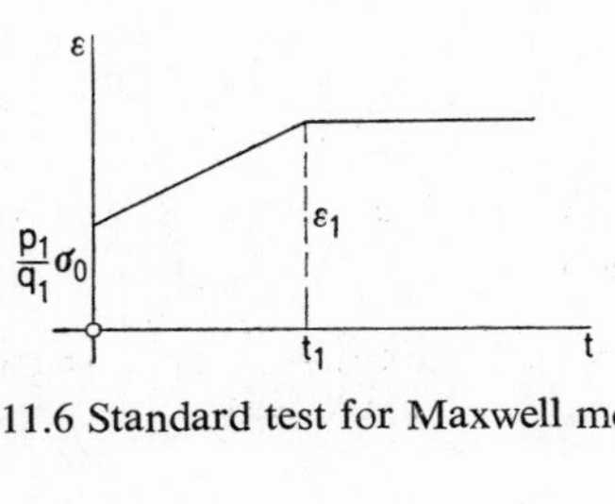

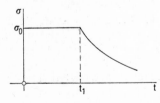

Fig. 11.6 Standard test for Maxwell model

In Figure 11.6 σ and ε are plotted against t. The first phase is again characterized by creep or, more precisely, by unrestricted creep since $\varepsilon \to \infty$ for $t_1 \to \infty$. The value of ε at $t = t_1$ is

$$\varepsilon_1 = \frac{\sigma_0}{q_1}(t_1+p_1). \tag{11.21}$$

During the second phase ε is kept at the value ε_1. From the differential equation (11.10) we obtain

$$\sigma = c \exp\left(-\frac{t}{p_1}\right) \quad (t > t_1). \tag{11.22}$$

The discontinuity of $\dot{\varepsilon}$ at $t = t_1$ implies that $\dot{\sigma}$ is discontinuous. However, σ remains continuous, and it follows that $c = \sigma_0 \exp(t_1/p_1)$. Thus,

$$\sigma = \sigma_0 \exp\left(-\frac{t-t_1}{p_1}\right) \quad (t \geqslant t_1). \tag{11.23}$$

The stress decreases exponentially while ε remains constant: the second phase is characterized by gradual relaxation. The curves of Figure 11.6 are again easily explained in mechanical terms by the response of the two elements of the Maxwell model.

If we add more springs, dashpots, Kelvin and Maxwell models to the models already considered, we obtain systems (like the ones of Figures 11.7 and 11.8) with a more refined response. Provided all of the elements are linear, the response is always described by a differential equation of the form

$$\sigma + p_1\dot{\sigma} + p_2\ddot{\sigma} + \ldots = q_0\varepsilon + q_1\dot{\varepsilon} + q_2\ddot{\varepsilon} + \ldots, \tag{11.24}$$

which is an obvious generalization of (11.1), (11.2), (11.7), and (11.10). For the treatment of more complicated models it is convenient to use Laplace transforms. They allow us to replace the differential equations by algebraic relations.

Problems

1. Establish the special form of the differential equation (11.24) for the model of Figure 11.7 and show that the coefficients are subject to the inequality $q_1 > p_1 q_0$. Find $\varepsilon(t)$ for the first and $\sigma(t)$ for the second phase of the standard test.

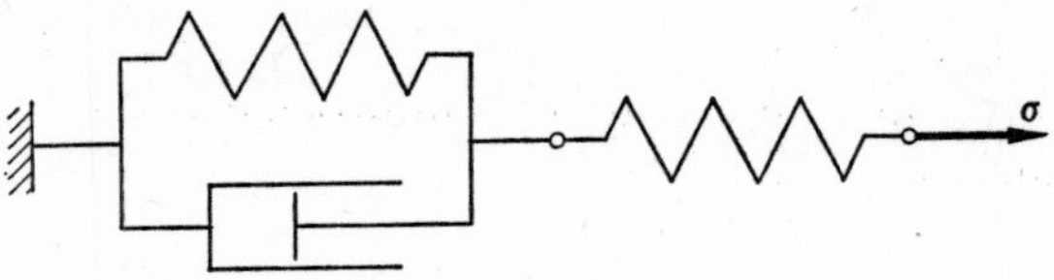

Fig. 11.7 Kelvin model plus spring

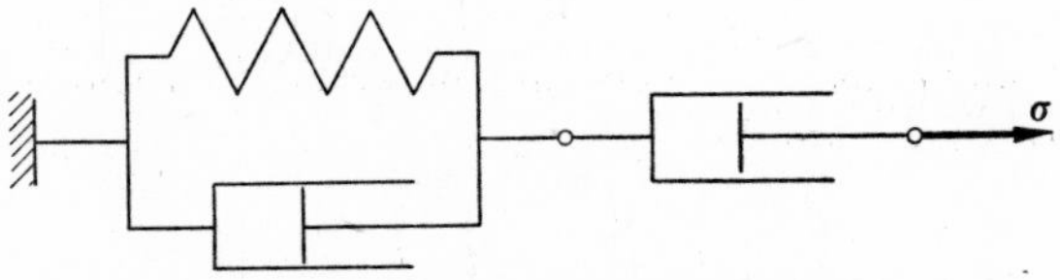

Fig, 11.8 Kelvin model plus dashpot

2. Solve the same problem for the model of Figure 11.8 and prove the inequality $p_1 q_1 > q_2$.

11.2. Hereditary integrals

The response of linear models has been described in Section 11.1 by particular forms of the differential equation (11.24). For certain purposes it is more convenient to describe this response by means of integral equations.

In the first phase of the standard test, also referred to as the *creep phase*, the stress was given by $\sigma(t) = \sigma_0 \Delta(t)$. The corresponding extension depends on the model. However, on account of the linearity of the models considered, $\varepsilon(t)$ is proportional to σ_0 [as in (11.14) or (11.20)] and may hence be written as

$$\varepsilon(t) = \sigma_0 I(t), \tag{11.25}$$

where the function $I(t)$ is characteristic for the particular model and is called its *creep compliance*. It is clear that $I(t)$ is zero for $t < 0$ and increases monotonically (i.e. at least as a weakly monotonic function) for $t > 0$.

Let us now modify the creep phase of the standard test by the assumption that the unstressed model is subjected to an extension pulse described by $\varepsilon(t) = \varepsilon_0 \Delta(t)$. The corresponding stress depends on the model and is proportional to ε_0. It may be written as

$$\sigma(t) = \varepsilon_0 J(t), \tag{11.26}$$

where the so-called *relaxation modulus* $J(t)$ is again characteristic for the particular model. It is zero for $t < 0$, possibly infinite for $t = 0$ (as in the case of a single dashpot) and decreases monotonically (i.e. at least as a weakly monotonic function) for $t > 0$.

In Table 11.1 four simple models are listed. The second column contains the differential equations established in Section 11.1, the third the creep

model	diff. equation	$I(t)$ $(t \geqq 0)$	$J(t)$ $(t \geqq 0)$
spring	$\sigma = q_0 \varepsilon$	$\frac{1}{q_0}$	q_0
dashpot	$\sigma = q_1 \dot{\varepsilon}$	$\frac{1}{q_1} t$	$q_1 \delta(t)$
Kelvin	$\sigma = q_0 \varepsilon + q_1 \dot{\varepsilon}$	$\frac{1}{q_0}\left[1 - \exp\left(-\frac{q_0}{q_1} t\right)\right]$	$q_0 + q_1 \delta(t)$
Maxwell	$\sigma + p\, \dot{\sigma} = q_1 \dot{\varepsilon}$	$\frac{1}{q_1}(t + p_1)$	$\frac{q_1}{p_1} \exp\left(-\frac{1}{p_1} t\right)$

Table 11.1

compliances following from (11.1) with $E = q_0$, from (11.2) with $F = q_1$, and from (11.14) and (11.20). In the last column the relaxation moduli are listed. The one on the first line is evident. To justify the second, we note that differentiation of $\varepsilon = \varepsilon_0 \, \Delta(t)$ yields $\dot{\varepsilon} = \varepsilon_0 \, \delta(t)$. Inserting this in the differential equation of the dashpot, we obtain $\sigma = q_1 \dot{\varepsilon} = \varepsilon_0 q_1 \, \delta(t)$ and hence $J(t) = q_1 \, \delta(t)$. The relaxation moduli of the Kelvin and the Maxwell models are obtained in a similar manner (P1).

Let us assume now that an arbitrary model is unstressed up to the time $t = 0$ and that, from then on, it is subjected to a given stress history described, e.g. by the diagram of Figure 11.9. If the model is linear, the extension ε at

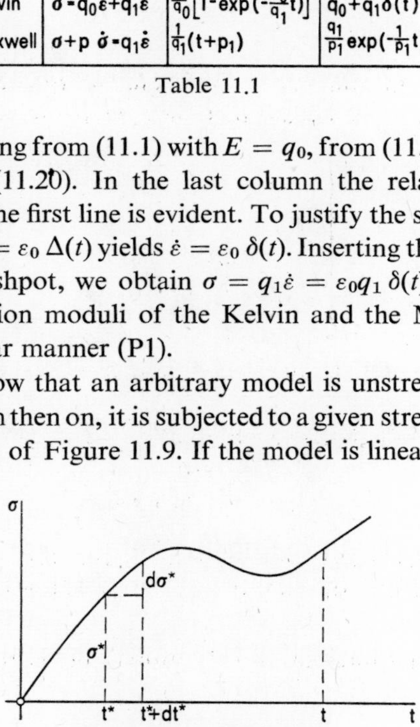

Fig. 11.9 Stress history

time t may be considered as the result of the contributions of all stress increments $d\sigma^*$ during the preceding time elements dt^*. These increments are

$$d\sigma^* = \left(\frac{d\sigma}{dt}\right)_{t^*} dt^* \quad \text{or simply} \quad d\sigma^* = \frac{d\sigma^*}{dt^*}\, dt^* . \qquad (11.27)$$

The curve of Figure 11.9 may be approximated by a sequence of steps, and it is clear that the approximation improves with decreasing size of the steps. The contribution of the step $d\sigma^*$ at time t^* to the extension at time t is

$$d\varepsilon = I(t-t^*)\, d\sigma^* \cdot \tag{11.28}$$

Addition of all contributions during the time interval $0 \leqslant t^* \leqslant t$ yields

$$\varepsilon(t) = \int_0^t I(t-t^*) \frac{d\sigma^*}{dt^*}\, dt^* . \tag{11.29}$$

In the case of an instantaneous stress change the curve of Figure 11.9 becomes discontinuous as shown in Figure 11.10, where $\Delta\sigma'$ represents a

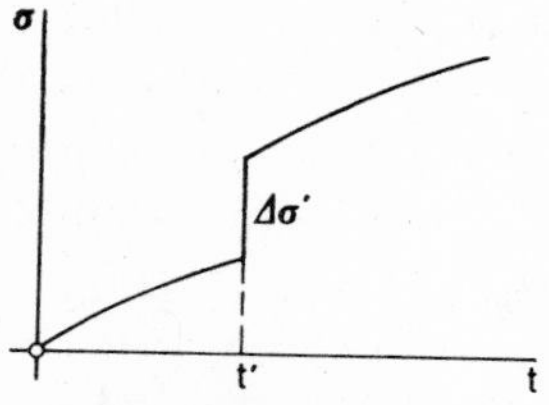

Fig. 11.10 Sudden increase of stress

sudden increase of the stress at time t' $(0 \leqslant t' \leqslant t)$. The corresponding contribution to $\varepsilon(t)$ is

$$\Delta\varepsilon(t) = I(t-t')\, \Delta\sigma' . \tag{11.30}$$

In the evaluation of (11.29) contributions of this type must be taken into account. In other words: the right-hand side of (11.29) is to be interpreted as a *Stieltjes integral.*

By means of partial integration (11.29) may be written in the form

$$\varepsilon(t) = I(t-t^*)\,\sigma(t^*)\Big|_0^t - \int_0^t \frac{dI(t-t^*)}{dt^*}\,\sigma(t^*)\, dt^* . \tag{11.31}$$

If a possible discontinuity of σ at $t = 0$ is included in the integral, we have $\sigma(t^* = 0) = 0$ and hence

$$\varepsilon(t) = I(0)\,\sigma(t) + \int_0^t \sigma(t^*) \frac{dI(t-t^*)}{d(t-t^*)}\, dt^* . \tag{11.32}$$

The first term on the right may be interpreted as the extension at time t in the fictitious case that the stress is maintained zero up to t and that $\sigma(t)$ is instantaneously applied at the time t. The second term is the correction necessary to account for the fact that the stress $\sigma(t)$ is actually built up according to Figure 11.9 during the finite time interval $0 \leqslant t^* \leqslant t$.

There are other ways of modifying (11.29). Since $\sigma(t^*) = 0$ for $t^* < 0$, the lower limit of the integral may be replaced by $-\infty$. Since $I(t-t^*) = 0$ for $t^* > t$, the upper limit may be replaced by $+\infty$.

Let us assume now that the model just considered is unstrained up to the time $t = 0$ and that, from then on, it is subjected to an extension history described by $\varepsilon(t)$ in a manner similar to $\sigma(t)$ in Figure 11.9. On account of the linearity of the model, the stress σ at time t may be considered as the result of the contributions of all extension increments $\mathrm{d}\varepsilon^*$ during the preceding time elements $\mathrm{d}t^*$. These increments are

$$\mathrm{d}\varepsilon^* = \left(\frac{\mathrm{d}\varepsilon}{\mathrm{d}t}\right)_{t^*} \mathrm{d}t^* \quad \text{or simply} \quad \mathrm{d}\varepsilon^* = \frac{\mathrm{d}\varepsilon^*}{\mathrm{d}t^*}\,\mathrm{d}t^*. \tag{11.33}$$

The corresponding contributions to the stress at time t are

$$\mathrm{d}\sigma = J(t-t^*)\,\mathrm{d}\varepsilon^*\,; \tag{11.34}$$

in consequence,

$$\sigma(t) = \int_0^t J(t-t^*)\frac{\mathrm{d}\varepsilon^*}{\mathrm{d}t^*}\,\mathrm{d}t^*. \tag{11.35}$$

This representation is analogous to (11.29), and the remarks made in connection with, and following (11.29) are also valid for (11.35).

As a simple application we consider the Kelvin model subjected to the stress pulse represented in Figure 11.11. According to Table 11.1, the creep compliance is

$$I(t) = \frac{1}{q_0}\left[1-\exp\left(-\frac{q_0}{q_1}t\right)\right]. \tag{11.36}$$

During the time interval $0 \leqslant t^* < t_1$ the derivative $\mathrm{d}\sigma^*/\mathrm{d}t^*$ is zero except for $t^* = 0$ where it is infinite. The integral (11.29) therefore reduces to

$$\varepsilon(t) = I(t)\sigma_0 = \frac{\sigma_0}{q_0}\left[1-\exp\left(-\frac{q_0}{q_1}t\right)\right] \quad (0 \leqslant t < t_1). \tag{11.37}$$

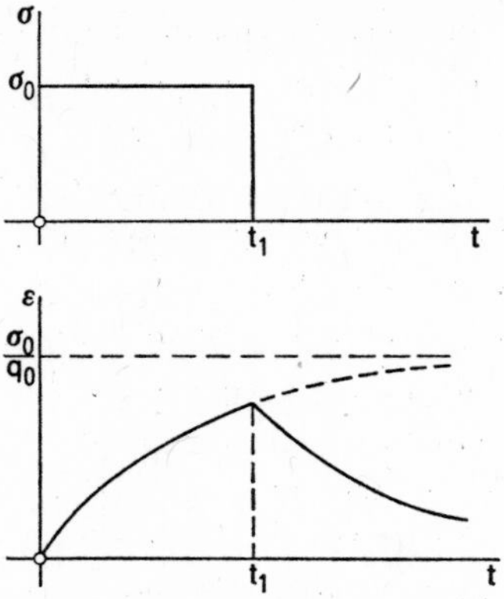

Fig. 11.11 Kelvin model subjected to an extension pulse

This result is identical with (11.14). In fact, for $0 \leqslant t < t_1$ the situation considered here is exactly the same as in the first phase of the standard test. For $t \geqslant t_1$ the integral (11.29) contains a second contribution corresponding to the sudden unloading at $t = t_1$. The extension now becomes

$$\varepsilon(t) = I(t)\sigma_0 - I(t-t_1)\sigma_0 = \frac{\sigma_0}{q_0}\left[\exp\left(\frac{q_0}{q_1}t_1\right)-1\right]\exp\left(-\frac{q_0}{q_1}t\right) \quad (t \geqslant t_1). \tag{11.38}$$

It is easy to verify that the function $\varepsilon(t)$ is continuous at $t = t_1$ as shown in Figure 11.11.

If the Kelvin model is subjected to the extension pulse of Figure 11.12, its response is described by (11.35) in conjunction with the relaxation modulus

$$J(t) = q_0 + q_1\,\delta(t). \tag{11.39}$$

For the first phase we have

$$\sigma(t) = J(t)\varepsilon_0 = \varepsilon_0[q_0 + q_1\,\delta(t)] \quad (0 \leqslant t < t_1). \tag{11.40}$$

This implies that the stress becomes instantaneously infinite at time $t = 0$ (on account of the resistance offered by the dashpot) and drops immediately to the value $q_0\varepsilon_0$. For the second phase (11.35) yields

$$\sigma(t) = J(t)\varepsilon_0 - J(t-t_1)\varepsilon_0 = -\varepsilon_0 q_1\,\delta(t-t_1) \quad (t \geqslant t_1). \tag{11.41}$$

The stress thus becomes $-\infty$ at $t = t_1$ and drops back to zero.

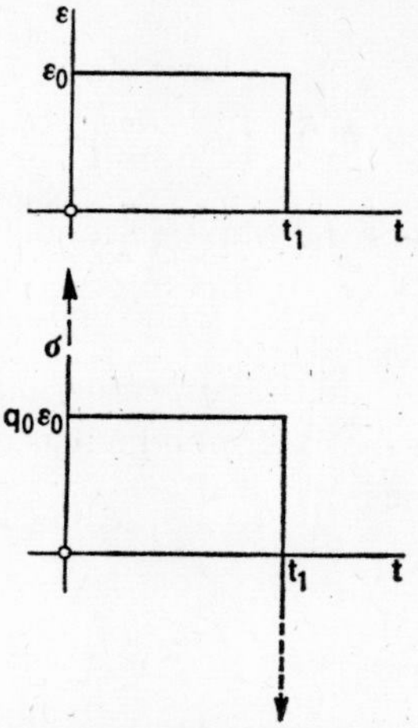

Fig. 11.12 Maxwell model subjected to an extension pulse

Returning to Section 11.1 we note that in the treatment of models on the basis of their differential equations the various models distinguish themselves in the orders of the two sides of (11.24). On the other hand, the integral equations (11.29) and (11.35) established in this section are always the same, whereas the special forms of the functions $I(t)$ and $J(t)$ are characteristic for the various models. Incidentally, it is clear that the creep compliance and the relaxation modulus of one and the same model depend on each other.

Problems

1. Verify the relaxation moduli listed in Table 11.1 for the Kelvin and the Maxwell model.

2. Show that the creep compliance and the relaxation modulus of the model of Figure 11.7 are

$$I(t) = \frac{1}{q_0}\left[1-\left(1-\frac{p_1 q_0}{q_1}\right)\exp\left(-\frac{q_0}{q_1}t\right)\right],$$

$$J(t) = q_0+\left(\frac{q_1}{p_1}-q_0\right)\exp\left(-\frac{1}{p_1}t\right)$$

respectively.

3. Show that for the model of Figure 11.8

$$I(t) = \frac{1}{q_1}\left\{t+\left(p_1-\frac{q_2}{q_1}\right)\left[1-\exp\left(-\frac{q_1}{q_2}t\right)\right]\right\}$$

and

$$J(t) = \frac{1}{p_1}\left[q_2\,\delta(t) + \left(q_1 - \frac{q_2}{p_1}\right)\exp\left(-\frac{1}{p_1}t\right)\right].$$

4. Discuss the response of the Maxwell model subjected to a stress pulse according to Figure 11.11 or to the corresponding extension pulse (Fig. 11.12).

5. Solve the last problem and the corresponding one for the Kelvin model, assuming that the stress or the extension increases linearly from 0 to σ_0 or ε_0 respectively during the time interval $0 \leqslant t \leqslant t_1$, and that it remains constant for $t \geqslant t_1$.

11.3. Constitutive relations

The models introduced in Section 11.1 are of some use in demonstrating the complexity of the response in certain materials. However, it would be wrong to see more in these models than a remote and rather simplified picture of what actually happens in a material element. In the first place, such an element does not consist of a small number of springs and dashpots, nicely separated from each other. In the second place, the models considered here are one-dimensional, and it is in principle impossible to transfer the notions of series or parallel connection to two or three dimensions.

In spite of these shortcomings, the models supply certain indications as to the possibilities of treating elements of viscoelastic materials in real, 3-dimensional, situations. It is plausible, e.g., that the strain and stress tensors of such a body are locally connected by a tensorial differential equation of the general type (11.24). It is equally plausible that the 3-dimenional response may be represented by tensorial generalizations of the integral equations (11.29) and (11.35). Both representations have been developed out of the notion of linear models and thus define materials of the so-called *linear viscoelastic* type.

The mechanical state of an element without internal parameters is determined by the strain components ε_{ij}. It has been shown in Section 4.2 that for the unit of mass the corresponding forces are σ_{ij}/ϱ, where σ_{ij} is the stress tensor and ϱ the density. Provided the stresses are quasiconservative, the body is referred to, according to Section 5.1, as elastic, and if the stressess is not necessarily isotropic, the material is a solid. In the linear case, the constitutive equation of the isotropic elastic solid is given by Hooke's law (5.41):

$$\sigma_{ij} = \lambda\varepsilon_{kk}\delta_{ij} + 2\mu\varepsilon_{ij}. \tag{11.42}$$

Under isothermal conditions the strain energy (5.42) per unit volume is the product of ϱ and the specific free energy,

$$\varrho\psi = \frac{\lambda}{2}\,\varepsilon_{ii}\varepsilon_{jj} + \mu\varepsilon_{ij}\varepsilon_{ij}\,, \tag{11.43}$$

and (11.42) may be obtained by differentiation according to $(4.37)_1$:

$$\sigma_{ij}^{(q)} = \varrho\,\frac{\partial\psi}{\partial\varepsilon_{ij}}\,. \tag{11.44}$$

If, on the other hand, the stresses are dissipative, the body is referred to as purely viscous and is a fluid according to the definition in Section 5.1. Materials of this type are rare. In general, the isotropic part of the stress tensor contains a quasiconservative component in the form of the hydrostatic pressure, given by (5.15) and $(7.14)_1$, i.e. by

$$\sigma_{ij}^{(q)} = -p\delta_{ij} = K\varepsilon_{kk}\delta_{ij}\,. \tag{11.45}$$

If the bulk viscosity is zero, the material, provided it is isotropic and linear, is a Navier–Stokes fluid, and the dissipative stress tensor is given by (5.26),

$$\sigma_{ij}^{(d)} = 2\mu' d'_{ij}\,, \tag{11.46}$$

where the viscosity coefficient is denoted by μ' to distinguish it from Lamé's constant μ in (11.42). In (5.1) we have introduced the specific dissipation function $\varphi = \vartheta\dot{s}^{(i)}$. On account of (4.77), $\varrho\vartheta\dot{s}^{(i)}$ is given in the present case by $\sigma_{ij}^{(d)}d_{ij}$. It follows that

$$\varrho\varphi = 2\mu' d'_{ij}d_{ij} = 2\mu'(\dot{\varepsilon}_{ij} - \tfrac{1}{3}\dot{\varepsilon}_{kk}\delta_{ij})\dot{\varepsilon}_{ij}\,, \tag{11.47}$$

and we note that (11.46) might be obtained by differentiation of the specific dissipation function:

$$\sigma_{ij}^{(d)} = \tfrac{1}{2}\varrho\,\frac{\partial\varphi}{\partial\dot{\varepsilon}_{ij}}\,. \tag{11.48}$$

Strictly speaking, the material just discussed is viscoelastic since the tensors $\sigma_{ij}^{(q)}$ and $\sigma_{ij}^{(d)}$ are different from zero (the first of them being a reaction if the material is incompressible). Let us now consider some truly viscoelastic bodies, still restricting ourselves to the linear, isotropic case and to materials without bulk viscosity. The simplest body of this type is the one without internal parameters. Its quasiconservative stress has the form (11.42) while the dissipative stress is given by (11.46). We thus have

$$\sigma_{ij} = \lambda\varepsilon_{kk}\delta_{ij}+2\mu\varepsilon_{ij}+2\mu'(\dot{\varepsilon}_{ij}-\tfrac{1}{3}\dot{\varepsilon}_{kk}\delta_{ij}), \tag{11.49}$$

and we note again that the quasiconservative part of σ_{ij} follows from (11.43) with (11.44), whereas the dissipative part might be obtained from (11.47) by means of equation (11.48) which, although so far without proof, will be justified in Chapter 18. Comparing (11.49) and (11.6), we observe that the material is of the Kelvin type. Let us emphasize, however, that we have not made use of the Kelvin model in deriving (11.49). Since the quasiconservative stress has a non-vanishing deviatoric part, we call the body a *Kelvin solid.*

Another example is the *Maxwell body*. To define it, let us first be guided by the model of Figure 4.1, assuming that the strain is composed, according to

$$\varepsilon_{ij} = \varepsilon_{ij}^{(e)}+\varepsilon_{ij}^{(v)}, \tag{11.50}$$

of an "elastic" part $\varepsilon_{ij}^{(e)}$, connected with the stress by (11.42),

$$\sigma_{ij} = \lambda\varepsilon_{kk}^{(e)}\delta_{ij}+2\mu\varepsilon_{ij}^{(e)}, \tag{11.51}$$

and a „viscous" part $\varepsilon_{ij}^{(v)}$ determining the same stress by means of (11.45) and (11.46):

$$\sigma_{ij} = K\varepsilon_{kk}^{(v)}\delta_{ij}+2\mu' d_{ij}^{(v)\prime}. \tag{11.52}$$

Combining (11.51) and (11.52), we obtain

$$\sigma_{ij} = \lambda\varepsilon_{kk}^{(e)}\delta_{ij}+2\mu\varepsilon_{ij}^{(e)} = K\varepsilon_{kk}^{(v)}\delta_{ij}+2\mu'(\dot{\varepsilon}_{ij}^{(v)}-\tfrac{1}{3}\dot{\varepsilon}_{kk}^{(v)}\delta_{ij}). \tag{11.53}$$

It is worthwile to write (11.53) in terms of external and internal parameters. According to Figure 4.1 there are two ways of doing so. In the first place, we may use the ε_{ij} as external and the $\alpha_{ij} = \varepsilon_{ij}^{(e)}$ as internal parameters. Equation (11.53) then takes the form

$$\sigma_{ij} = \lambda\alpha_{kk}\delta_{ij}+2\mu\alpha_{ij} = K(\varepsilon_{kk}-\alpha_{kk})\delta_{ij}+2\mu'[\dot{\varepsilon}_{ij}-\dot{\alpha}_{ij}-\tfrac{1}{3}(\dot{\varepsilon}_{kk}-\dot{\alpha}_{kk})\delta_{ij}]. \tag{11.54}$$

It can also be derived from the functions

$$\varrho\psi = \frac{\lambda}{2}\alpha_{ii}\alpha_{jj}+\mu\alpha_{ij}\alpha_{ij}+\frac{K}{2}(\varepsilon_{ii}-\alpha_{ii})(\varepsilon_{jj}-\alpha_{jj}) \tag{11.55}$$

and

$$\varrho\varphi = 2\mu'[(\dot{\varepsilon}_{ij}-\dot{\alpha}_{ij})(\dot{\varepsilon}_{ij}-\dot{\alpha}_{ij})-\tfrac{1}{3}(\dot{\varepsilon}_{ii}-\dot{\alpha}_{ii})(\dot{\varepsilon}_{jj}-\dot{\alpha}_{jj})]. \tag{11.56}$$

In fact, if we apply (11.44), (11.48), and the analogous equations for the internal forces,

$$\beta_{ij}^{(q)} = \varrho \frac{\partial \psi}{\partial \alpha_{ij}}, \qquad \beta_{ij}^{(d)} = \tfrac{1}{2}\varrho \frac{\partial \varphi}{\partial \dot{\alpha}_{ij}}, \tag{11.57}$$

to (11.55) and (11.56), we obtain

$$\begin{aligned} &\sigma_{ij}^{(q)} = K(\varepsilon_{kk}-\alpha_{kk})\delta_{ij}, \qquad \beta_{ij}^{(q)} = \lambda\alpha_{kk}\delta_{ij}+2\mu\alpha_{ij}-K(\varepsilon_{kk}-\alpha_{kk})\delta_{ij}, \\ &\sigma_{ij}^{(d)} = -\beta_{ij}^{(d)} = 2\mu'[\dot{\varepsilon}_{ij}-\dot{\alpha}_{ij}-\tfrac{1}{3}(\dot{\varepsilon}_{kk}-\dot{\alpha}_{kk})\delta_{ij}]. \end{aligned} \tag{11.58}$$

Combining the quasiconservative and the dissipative forces according to (4.36) and (4.42), we in fact obtain (11.54).

If, on the other hand, we use the $\bar{\alpha}_{ij} = \varepsilon_{ij}^{(v)}$ as internal parameters, equation (11.53) assumes the form

$$\sigma_{ij} = \lambda(\varepsilon_{kk}-\bar{\alpha}_{kk})+2\mu(\varepsilon_{ij}-\bar{\alpha}_{ij}) = K\bar{\alpha}_{kk}\delta_{ij}+2\mu'(\dot{\bar{\alpha}}_{ij}-\tfrac{1}{3}\dot{\bar{\alpha}}_{kk}\delta_{ij}). \tag{11.59}$$

It can also be derived by applying (11.44), (11.48), and equations (11.57) (written with bars) to

$$\varrho\psi = \frac{\lambda}{2}(\varepsilon_{ii}-\bar{\alpha}_{ii})(\varepsilon_{jj}-\bar{\alpha}_{jj})+\mu(\varepsilon_{ij}-\bar{\alpha}_{ij})(\varepsilon_{ij}-\bar{\alpha}_{ij})+\frac{K}{2}\bar{\alpha}_{ii}\bar{\alpha}_{jj} \tag{11.60}$$

and

$$\varrho\varphi = 2\mu'(\dot{\bar{\alpha}}_{ij}\dot{\bar{\alpha}}_{ij}-\tfrac{1}{3}\dot{\bar{\alpha}}_{ii}\dot{\bar{\alpha}}_{jj}). \tag{11.61}$$

The corresponding forces are

$$\begin{aligned} \sigma_{ij}^{(q)} &= \lambda(\varepsilon_{kk}-\bar{\alpha}_{kk})\delta_{ij}+2\mu(\varepsilon_{ij}-\bar{\alpha}_{ij}), \\ \bar{\beta}_{ij}^{(q)} &= K\bar{\alpha}_{kk}\delta_{ij}-\lambda(\varepsilon_{kk}-\bar{\alpha}_{kk})\delta_{ij}-2\mu(\varepsilon_{ij}-\bar{\alpha}_{ij}), \\ \sigma_{ij}^{(d)} &= 0, \qquad \bar{\beta}_{ij}^{(d)} = 2\mu'(\dot{\bar{\alpha}}_{ij}-\tfrac{1}{3}\dot{\bar{\alpha}}_{kk}\delta_{ij}), \end{aligned} \tag{11.62}$$

and the combination according to (4.36) and (4.42) in fact yields (11.59).

The analysis just gone through for the simple case of a Maxwell body may appear unnecessarily cumbersome. However, it yields a few results of fundamental importance.

For once, the analysis shows that even for the Maxwell body we can dispense with the model concept, provided we derive the external and internal forces from the free energy and the dissipation function by means of equations like (11.44), (11.48), and (11.57). As far as ψ is concerned, these equations follow from Section 4.2; for φ they will be justified in Chapter 18.

As for the definition of the internal parameters, there are two physically reasonable choices. We accordingly obtain different expressions (11.54) and (11.59) for σ_{ij}. However, it is clear that they are equivalent, and if we eliminate the internal parameters from the two equations (11.54), the result is in fact the same as if we subject (11.59) to the same process: we obtain an equation connecting ε_{kk} with σ_{kk} and a differential equation of the type (11.10) between ε'_{ij} and σ'_{ij} (P1).

Another result of the above analysis confirms a remark already made in Section 5.1: Even though the expressions (11.54) and (11.59) for σ_{ij} are equivalent, a comparison of (11.58) and (11.62) shows that the decomposition of σ_{ij} into $\sigma_{ij}^{(q)}$ and $\sigma_{ij}^{(d)}$ depends on the choice of the internal parameters. This is a result with awkward consequences concerning the distinction between fluids and solids. This distinction has been defined provisionally in Section 5.1 for bodies without internal parameters, and it has been based on the absence or presence of the quasiconservative stress deviator $\sigma_{ij}^{(q)'}$. We obviously cannot generalize it without modification for materials with internal parameters since $(11.58)_1$ would imply that the Maxwell body is a fluid, whereas $(11.62)_1$ supplies

$$\sigma_{ij}^{(q)'} = 2\mu(\varepsilon'_{ij} - \bar{\alpha}'_{ij}) \tag{11.63}$$

and hence would classify the same body as a solid. Now, the definitions in Section 5.1 were meant to imply that, in contrast to the solid, the state of stress in a fluid may be the same before and after a distortion. This is impossible in the case of the constitutive equation (11.49), but it is true according to $(11.58)_1$ and possible in the case of (11.63). It seems therefore reasonable to generalize the former definitions by defining a *fluid* as a body in which, at least for an appropriate choice of the internal parameters, $\sigma_{ij}^{(q)'}$ is identically zero, whereas for a *solid* such a choice is impossible.

The *Maxwell* body, in particular, thus qualifies as a *fluid*.

Problems

1. Eliminate the internal parameters from the two equations (11.54) and subsequently from (11.59). Show that the same equation between ε_{kk} and σ_{kk} and the same differential equation between ε'_{ij} and σ'_{ij} is obtained.

2. Equations (11.51) and (11.52) serve as the basis of the (nonlinear) *creep theory of metals*. The first of them is adopted without modification, also the first term on the right-hand side of (11.52), whereas the second term

is made nonlinear by assuming (cf. [25]) that μ' is a function of the invariant $\sigma'_{(2)}$, to be determined by experiments. Let $\mu'(\sigma'_{(2)})$ be given by

$$\frac{1}{2\mu'} = k(\sigma'_{(2)})^m,$$

where k and m are constants, and establish the connection between σ_{11} and $d_{11}^{(v)}$ for a uniaxial stress state.

CHAPTER 12

GENERAL TENSORS

In Chapter 13 we will deal with large displacements. For this purpose it is convenient to admit curvilinear coordinate systems, and this renders it necessary to generalize the tensor concept introduced in Section 1.1. The present chapter will be concerned with this purely mathematical problem.

12.1. Tensor algebra

The definition of a cartesian tensor has been based in Section 1.1 on the transformation of its components in the transition from an arbitrary cartesian coordinate system to another one. For many purposes, and in particular for the treatment of large deformations, curvilinear coordinate systems are more convenient. It is therefore desirable to use a more general tensor concept, based on transformations between curvilinear coordinate systems.

Figure 12.1 shows a point P with radius vector $\boldsymbol{r}$ in a cartesian coordinate system defined by the unit vectors $\boldsymbol{i}_j$. In the present context it is convenient

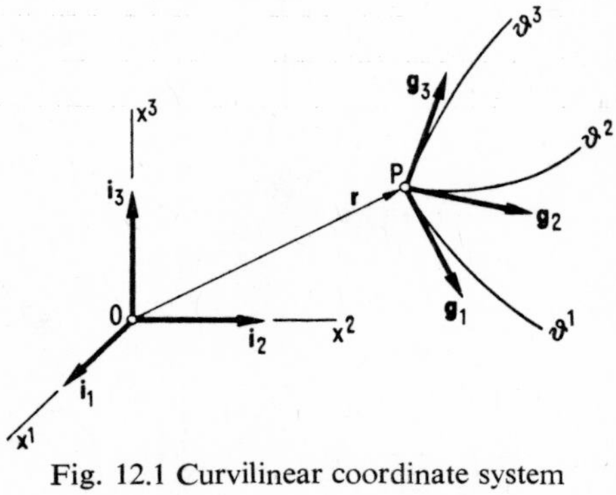

Fig. 12.1 Curvilinear coordinate system

to denote the base vectors by subscripts, the components by superscripts, and to modify the summation convention of Section 1.1 by stipulating that, whenever the same index appears as a subscript and as a superscript in a product, the sum is to be taken over this index. The cartesian coordinates of P are thus x^j, and the radius vector of P is given by

$$\boldsymbol{r} = x^j \boldsymbol{i}_j . \tag{12.1}$$

Since the $\boldsymbol{i}_j$ are constant, (12.1) yields

$$\boldsymbol{i}_j = \frac{\partial \boldsymbol{r}}{\partial x^j} . \tag{12.2}$$

The cartesian base vectors are thus the partial derivatives of the radius vector with respect to the cartesian coordinates.

We now introduce a *curvilinear coordinate system* ϑ^i (e.g. cylindrical or spherical coordinates) represented in Figure 12.1 by the coordinate curves passing through P, i.e. by the curves on which two of the three coordinates ϑ^i are constant. We assume that the line elements corresponding to coordinate increments $d\vartheta^1, \ldots$ on the three curves form a right-handed system, and we refer to the coordinate system itself as right-handed. The curvilinear coordinates may be considered as functions

$$\vartheta^i = \vartheta^i(x^j) \tag{12.3}$$

of the cartesian coordinates, and if we require the Jacobian $\det(\partial\vartheta^i/\partial x^j)$ to be different from zero everywhere, the transformation (12.3) is reversible.

Curvilinear coordinates cannot be characterized by a single system of base vectors. However, we may introduce such a system at every point P, noting that in general at least the directions of the base vectors depend on the position of P. We might use unit vectors, but it is more convenient to define the basis $\boldsymbol{g}_i$ by relations analogous to (12.2), viz. by

$$\boldsymbol{g}_i = \frac{\partial \boldsymbol{r}}{\partial \vartheta^i} \tag{12.4}$$

or on account of (12.1) by

$$\boldsymbol{g}_i = \frac{\partial x^j}{\partial \vartheta^i} \boldsymbol{i}_j . \tag{12.5}$$

It follows that the $\boldsymbol{g}_i$ are tangential to the coordinate curves in P and that also their magnitudes depend on the position of P. They are not unit vectors like the $\boldsymbol{i}_j$, and they may even have different dimensions (as in the case of cylindrical coordinates). By means of the basis (12.4) any locally defined vector may be decomposed. An example is the vectorial line element d$\boldsymbol{r}$, for which (12.4) supplies the decomposition

$$\mathrm{d}\boldsymbol{r} = \frac{\partial \boldsymbol{r}}{\partial \vartheta^i}\,\mathrm{d}\vartheta^i = \boldsymbol{g}_i\,\mathrm{d}\vartheta^i\,. \tag{12.6}$$

Equation (12.5) may be interpreted as the transformation between the unit vectors in the cartesian and the base vectors in the curvilinear coordinate system. On account of (12.2) and (12.4), the inversion of (12.5) is

$$\boldsymbol{i}_j = \frac{\partial \boldsymbol{r}}{\partial x^j} = \frac{\partial \boldsymbol{r}}{\partial \vartheta^i}\,\frac{\partial \vartheta^i}{\partial x^j} = \frac{\partial \vartheta^i}{\partial x^j}\,\boldsymbol{g}_i\,. \tag{12.7}$$

Comparing (12.7) and (12.5) to (1.13), we observe that the single transformation matrix c_{ij} is now replaced by two different ones, $\partial\vartheta^i/\partial x^j$ and $\partial x^j/\partial\vartheta^i$. Besides, the new matrices are dependent on the position of P, whereas c_{ij} was constant.

On account of this duality it is desirable to introduce a second local basis $\boldsymbol{g}^i$, defined by

$$\boldsymbol{g}^i = \frac{\partial \vartheta^i}{\partial x^j}\,\boldsymbol{i}^j\,, \tag{12.8}$$

where $\boldsymbol{i}^j$ is an alternate way of writing $\boldsymbol{i}_j$. Replacing the dummy index j by k and multiplying both sides of (12.8) by $\partial x^j/\partial\vartheta^i$, we have

$$\frac{\partial x^j}{\partial \vartheta^i}\,\boldsymbol{g}^i = \frac{\partial x^j}{\partial \vartheta^i}\,\frac{\partial \vartheta^i}{\partial x^k}\,\boldsymbol{i}^k = \delta^j_k\boldsymbol{i}^k\,, \tag{12.9}$$

where δ^j_k is another form of writing the Kronecker symbol. The inversion of (12.8) is thus given by

$$\boldsymbol{i}^j = \frac{\partial x^j}{\partial \vartheta^i}\,\boldsymbol{g}^i\,. \tag{12.10}$$

On account of (12.8) and (12.5)

$$\boldsymbol{g}^i\!\cdot\!\boldsymbol{g}_k = \frac{\partial \vartheta^i}{\partial x^j}\,\boldsymbol{i}^j\!\cdot\frac{\partial x^l}{\partial \vartheta^k}\,\boldsymbol{i}_l = \frac{\partial \vartheta^i}{\partial x^j}\,\frac{\partial x^l}{\partial \vartheta^k}\,\delta^j_l = \frac{\partial \vartheta^i}{\partial x^j}\,\frac{\partial x^j}{\partial \vartheta^k} = \delta^i_k\,. \tag{12.11}$$

It follows that

$$\boldsymbol{g}^1\!\cdot\!\boldsymbol{g}_1 = 1 \tag{12.12}$$

and that $\boldsymbol{g}^1$ is perpendicular to $\boldsymbol{g}_2$ and $\boldsymbol{g}_3$; similar statements hold for $\boldsymbol{g}^2$ and $\boldsymbol{g}^3$. We distinguish the two bases just defined by denoting the $\boldsymbol{g}_i$ as the *covariant base vectors* of the curvilinear coordinate system and the $\boldsymbol{g}^i$ as its *contravariant basis*. The $\boldsymbol{g}_i$ and the $\boldsymbol{g}^i$ follow from the same cartesian basis by means of two different transformations (12.5) and (12.8) respectively.

In the special case where also the system ϑ^i is cartesian, equations (1.5) show that the matrices $\partial x^j/\partial\vartheta^i$ and $\partial\vartheta^i/\partial x^j$ are identical. It follows that the systems $\boldsymbol{g}_i$ and $\boldsymbol{g}^i$ coincide.

A vector $\boldsymbol{v}$ defined in P be may decomposed with respect to either one of the two bases. The two systems of components are generally different and will be denoted by v^i and v_i respectively. We thus have

$$\boldsymbol{v} = v^i\boldsymbol{g}_i = v_i\boldsymbol{g}^i . \tag{12.13}$$

Combination of (12.13) with (12.5) yields

$$\boldsymbol{v} = \left(\frac{\partial x^j}{\partial\vartheta^i}\, v^i\right)\boldsymbol{i}_j . \tag{12.14}$$

The expressions between parentheses are obviously the cartesian components of $\boldsymbol{v}$. They are connected with the v^i by the same transformation (12.10) as the $\boldsymbol{i}^j$ with the $\boldsymbol{g}^i$. It is therefore reasonable to call the v^i the *contravariant components* of $\boldsymbol{v}$. In a similar manner we obtain

$$\boldsymbol{v} = \left(\frac{\partial\vartheta^i}{\partial x^j}\, v_i\right)\boldsymbol{i}^j \tag{12.15}$$

from (12.13) and (12.8). The expressions between parentheses are again the cartesian components of $\boldsymbol{v}$. Since they follow from the v_i by the same transformation (12.7) as the $\boldsymbol{i}_j$ from the $\boldsymbol{g}_i$, we refer to the v_i as the *covariant components* of $\boldsymbol{v}$. In the special case where the system ϑ^i is cartesian, the contravariant and covariant components of $\boldsymbol{v}$ are the same.

Let us introduce a second curvilinear system by means of

$$\Theta^p = \Theta^p(\vartheta^i), \tag{12.16}$$

and let us assume that the Jacobian det $(\partial\Theta^p/\partial\vartheta^i)$ is different from zero everywhere so that (12.16) is reversible. In analogy to (12.4) the covariant basis $\boldsymbol{G}_p$ corresponding to the system Θ^p is given by

$$\boldsymbol{G}_p = \frac{\partial \boldsymbol{r}}{\partial\Theta^p} = \frac{\partial \boldsymbol{r}}{\partial\vartheta^i}\frac{\partial\vartheta^i}{\partial\Theta^p} = \frac{\partial\vartheta^i}{\partial\Theta^p}\boldsymbol{g}_i , \tag{12.17}$$

and the vectorial line element is

$$\mathrm{d}\boldsymbol{r} = \frac{\partial \boldsymbol{r}}{\partial \Theta^p}\, \mathrm{d}\Theta^p = \boldsymbol{G}_p\, \mathrm{d}\Theta^p . \tag{12.18}$$

Equation (12.17) transforms the base $\boldsymbol{g}_i$ into $\boldsymbol{G}_p$. Replacing the dummy index i by k and multiplying both sides of (12.17) by $\partial\Theta^p/\partial\vartheta^i$, we obtain the inversion

$$\boldsymbol{g}_i = \frac{\partial \Theta^p}{\partial \vartheta^i}\, \boldsymbol{G}_p \tag{12.19}$$

of (12.17). The contravariant base $\boldsymbol{G}^p$ is defined in analogy to (12.8) by

$$\boldsymbol{G}^p = \frac{\partial \Theta^p}{\partial x^j}\, \boldsymbol{i}^j = \frac{\partial \Theta^p}{\partial \vartheta^i}\, \frac{\partial \vartheta^i}{\partial x^j}\, \boldsymbol{i}^j = \frac{\partial \Theta^p}{\partial \vartheta^i}\, \boldsymbol{g}^i , \tag{12.20}$$

and it is easy to verify (P2) that

$$\boldsymbol{g}^i = \frac{\partial \vartheta^i}{\partial \Theta^p}\, \boldsymbol{G}^p \tag{12.21}$$

is the inversion of (12.20).

Let the components of a vector $\boldsymbol{v}$ in the new coordinate system Θ^p be denoted by V^p and V_p respectively. We thus have

$$\boldsymbol{v} = V^p \boldsymbol{G}_p = V_p \boldsymbol{G}^p . \tag{12.22}$$

On account of (12.19), $\boldsymbol{v}$ is also given by

$$\boldsymbol{v} = v^i \boldsymbol{g}_i = \frac{\partial \Theta^p}{\partial \vartheta^i}\, v^i \boldsymbol{G}_p , \tag{12.23}$$

and by comparison of (12.22) and (12.23) we obtain the transformation

$$V^p = \frac{\partial \Theta^p}{\partial \vartheta^i}\, v^i \tag{12.24}$$

between the contravariant components of $\boldsymbol{v}$ in the two curvilinear coordinate systems. It is not difficult to verify (P3) the inversion

$$v^i = \frac{\partial \vartheta^i}{\partial \Theta^p}\, V^p \tag{12.25}$$

of (12.24) and the transformations

$$V_p = \frac{\partial \vartheta^i}{\partial \Theta^p} v_i, \qquad v_i = \frac{\partial \Theta^p}{\partial \vartheta^i} V_p \tag{12.26}$$

between the covariant components of $\boldsymbol{v}$. The contravariant and covariant components of a vector might be defined now as triplets transforming according to (12.24) through (12.26) in the transition from one curvilinear coordinate system to another.

General *tensors* are defined in a similar manner as the cartesian tensors introduced in Section 1.1. A *scalar* λ is independent of the (curvilinear) coordinate system and may be considered as a tensor of order zero. A *vector* or tensor of order one has three contravariant components v^i transforming according to (12.24), (12.25), and three covariant components v_i transforming according to (12.26). Generalizing these transformations, we obtain tensors of higher order. For a *second-order tensor*, e.g., the contravariant components t^{ij} are defined by the transformations

$$T^{pq} = \frac{\partial \Theta^p}{\partial \vartheta^i} \frac{\partial \Theta^q}{\partial \vartheta^j} t^{ij}, \tag{12.27}$$

the covariant components by

$$T_{pq} = \frac{\partial \vartheta^i}{\partial \Theta^p} \frac{\partial \vartheta^j}{\partial \Theta^q} t_{ij}, \tag{12.28}$$

and the so-called mixed components by

$$T^p_q = \frac{\partial \Theta^p}{\partial \vartheta^i} \frac{\partial \vartheta^j}{\partial \Theta^q} t^i_j. \tag{12.29}$$

The inversions of (12.27) through (12.29) are obvious, also the generalizations of these transformations for higher-order tensors.

The rules of tensor algebra discussed in Section 1.2 are easily transferred to general tensors (P4, 5) once we adopt the conventions that the *variance* of a given index (superscript or subscript) must be the same in all terms of an equation and that the summation rule applies to indices appearing with different variance.

Equation (12.6) supplies the vectorial line element in terms of the curvilinear coordinate increments $\mathrm{d}\vartheta^i$. The square of this line element is

$$(\mathrm{d}\boldsymbol{r})^2 = \mathrm{d}x^k\, \mathrm{d}x_k = \boldsymbol{g}_i \cdot \boldsymbol{g}_j\, \mathrm{d}\vartheta^i\, \mathrm{d}\vartheta^j, \tag{12.30}$$

where $\mathrm{d}x_k$ is an alternate way of writing $\mathrm{d}x^k$.

From (12.18) we obtain an analogous expression

$$(\mathrm{d}\boldsymbol{r})^2 = \boldsymbol{G}_p \cdot \boldsymbol{G}_q \, \mathrm{d}\Theta^p \, \mathrm{d}\Theta^q . \tag{12.31}$$

On account of (12.17) and (12.4),

$$\boldsymbol{G}_p \cdot \boldsymbol{G}_q = \frac{\partial \boldsymbol{r}}{\partial \Theta^p} \cdot \frac{\partial \boldsymbol{r}}{\partial \Theta^q} = \frac{\partial \boldsymbol{r}}{\partial \vartheta^i} \frac{\partial \vartheta^i}{\partial \Theta^p} \cdot \frac{\partial \boldsymbol{r}}{\partial \vartheta^j} \frac{\partial \vartheta^j}{\partial \Theta^q} = \frac{\partial \vartheta^i}{\partial \Theta^p} \frac{\partial \vartheta^j}{\partial \Theta^q} \boldsymbol{g}_i \cdot \boldsymbol{g}_j . \tag{12.32}$$

Comparing this to (12.28), we conclude that the scalar products

$$\boldsymbol{g}_i \cdot \boldsymbol{g}_j = g_{ij} \tag{12.33}$$

are the covariant components of a second-order tensor. According to (12.30), (12.33), and (12.31) the square of the line element is given by

$$(\mathrm{d}\boldsymbol{r})^2 = \mathrm{d}x^k \, \mathrm{d}x_k = g_{ij} \, \mathrm{d}\vartheta^i \, \mathrm{d}\vartheta^j = G_{pq} \, \mathrm{d}\Theta^p \, \mathrm{d}\Theta^q . \tag{12.34}$$

In cartesian coordinates (12.34) is the expression of the theorem of Pythagoras; the curvilinear generalization requires the knowledge of the tensor g_{ij} beside the coordinate increments. We denote this tensor as the *metric tensor* of the coordinate system ϑ^i and define its contravariant and mixed components by

$$g^{ij} = \boldsymbol{g}^i \cdot \boldsymbol{g}^j, \qquad g^i_j = \boldsymbol{g}^i \cdot \boldsymbol{g}_j = \delta^i_j \tag{12.35}$$

respectively, where use has been made of (12.11) (P6). Let us finally note that the metric tensor is symmetric in all of its three forms.

Multiplying the second equation (12.13) by $\boldsymbol{g}_j$ and making use of $(12.35)_2$, we obtain

$$v^i g_{ij} = v_j . \tag{12.36}$$

In a similar manner, multiplication by $\boldsymbol{g}^j$ yields

$$v_i g^{ij} = v^j . \tag{12.37}$$

It follows that the metric tensor establishes the connection between the contravariant and covariant components of the same vector: multiplication by g_{ij} lowers the index; multiplication by g^{ij} raises it.

As an example, consider the scalar product of the two vectors $\boldsymbol{v}$ and $\boldsymbol{w}$. It can be written in the various forms

$$\boldsymbol{v} \cdot \boldsymbol{w} = v_i \boldsymbol{g}^i \cdot w^j \boldsymbol{g}_j = v^i \boldsymbol{g}_i \cdot w_j \boldsymbol{g}^j = v^i \boldsymbol{g}_i \cdot w^j \boldsymbol{g}_j = v_i \boldsymbol{g}^i \cdot w_j \boldsymbol{g}^j \tag{12.38}$$

or

$$\boldsymbol{v} \cdot \boldsymbol{w} = v_i w^i = v^i w_i = g_{ij} v^i w^j = g^{ij} v_i w_j . \tag{12.39}$$

For the magnitude of a vector $\boldsymbol{v}$ we thus obtain

$$|\boldsymbol{v}| = \sqrt{v_i v^i} = \sqrt{g_{ij} v^i v^j} = \sqrt{g^{ij} v_i v_j}, \tag{12.40}$$

and the angle φ between $\boldsymbol{v}$ and $\boldsymbol{w}$ is determined by expressions like

$$\cos \varphi = \frac{\boldsymbol{v} \cdot \boldsymbol{w}}{|\boldsymbol{v}| \cdot |\boldsymbol{w}|} = g_{ij} v^i w^j (g_{kl} v^k v^l)^{-1/2} (g_{mn} w^m w^n)^{-1/2}. \tag{12.41}$$

For tensors of the second and higher orders the processes of lowering and raising indices are similar and do not present any problems as long as we restrict ourselves to symmetric tensors.

Let g denote the determinant of the metric tensor, written in covariant matrix form:

$$g = \det g_{ij}. \tag{12.42}$$

On account of the multiplication theorem for determinants we have

$$(\det g_{ij})(\det g^{jk}) = \det(g_{ij} g^{jk}) = \det \delta_i^k = 1. \tag{12.43}$$

It follows that

$$\det g^{jk} = \frac{1}{g}. \tag{12.44}$$

We also have

$$G = \det G_{pq} = \det\left(\frac{\partial \vartheta^i}{\partial \Theta^p} \frac{\partial \vartheta^j}{\partial \Theta^q} g_{ij}\right) = \det\left(\frac{\partial \vartheta^i}{\partial \Theta^p} g_{ij} \frac{\partial \vartheta^j}{\partial \Theta^q}\right). \tag{12.45}$$

On account of the multiplication theorem for determinants and of the fact that the determinant of a matrix and its transpose are equal, (12.45) is equivalent to

$$G = \left(\det \frac{\partial \vartheta^i}{\partial \Theta^p}\right)^2 g. \tag{12.46}$$

Since the transformation (12.46) contains the Jacobian $\det(\partial \vartheta^i / \partial \Theta^p)$, the determinant g is a *pseudo-scalar* as D in (1.27).

In the special case where the coordinate system Θ^p is cartesian ($\Theta^p = x^p$); (12.31) reads

$$(\mathrm{d}\boldsymbol{r})^2 = \delta_{pq}\, \mathrm{d}x^p\, \mathrm{d}x^q. \tag{12.47}$$

It follows that $G = 1$. From (12.46) we obtain

$$\det \frac{\partial \vartheta^i}{\partial x^p} = \frac{1}{\sqrt{g}}, \tag{12.48}$$

where the square root is to be taken with the positive sign, since by convention the coordinate systems x^p and ϑ^i are right-handed. On account of (12.48) we further have

$$\frac{1}{\sqrt{g}} \det \frac{\partial x^p}{\partial \vartheta^j} = \det\left(\frac{\partial \vartheta^i}{\partial x^p} \frac{\partial x^p}{\partial \vartheta^j}\right) = \det \delta^i_j = 1 \tag{12.49}$$

and hence

$$\det \frac{\partial x^p}{\partial \vartheta^i} = \sqrt{g}. \tag{12.50}$$

According to (12.34) the metric tensor g_{ij} assumes the components δ_{ij} in a cartesian coordinate system x^p. This statement might have served to define g_{ij}. In a similar manner we can define a covariant third-order tensor ϵ_{ijk} by the condition that in cartesian coordinates it becomes the permutation tensor e_{ijk} (Section 1.2). Generalizing (12.28), we obtain the transformation

$$e_{pqr} = \frac{\partial \vartheta^i}{\partial x^p} \frac{\partial \vartheta^j}{\partial x^q} \frac{\partial \vartheta^k}{\partial x^r} \epsilon_{ijk} \tag{12.51}$$

with the inversion

$$\epsilon_{ijk} = \frac{\partial x^p}{\partial \vartheta^i} \frac{\partial x^q}{\partial \vartheta^j} \frac{\partial x^r}{\partial \vartheta^k} e_{pqr}\,. \tag{12.52}$$

It is easy to see (P7 of Section 1.2) that the expression on the right is equal to $e_{ijk} \det(\partial x^p/\partial \vartheta^i)$. On account of (12.50), we therefore have

$$\epsilon_{ijk} = \sqrt{g} e_{ijk}\,. \tag{12.53}$$

Defining the contravariant tensor ϵ^{ijk} by the condition that it assumes the form $e^{ijk} = e_{ijk}$ in cartesian coordinates, we find (P8)

$$\epsilon^{ijk} = \frac{1}{\sqrt{g}} e^{ijk}\,. \tag{12.54}$$

The tensor ϵ_{ijk} is particularly useful for the tensorial representation of the vector product. Starting with two covariant base vectors and making use of (12.5) and (12.10), we have

$$\boldsymbol{g}_i \times \boldsymbol{g}_j = \frac{\partial x^p}{\partial \vartheta^i} \boldsymbol{i}_p \times \frac{\partial x^q}{\partial \vartheta^j} \boldsymbol{i}_q = \frac{\partial x^p}{\partial \vartheta^i} \frac{\partial x^q}{\partial \vartheta^j} e_{pqr} \boldsymbol{i}^r$$

$$= e_{pqr} \frac{\partial x^p}{\partial \vartheta^i} \frac{\partial x^q}{\partial \vartheta^j} \frac{\partial x^r}{\partial \vartheta^k} \boldsymbol{g}^k \tag{12.55}$$

or on account of (12.52)

$$\boldsymbol{g}_i \times \boldsymbol{g}_j = \epsilon_{ijk}\boldsymbol{g}^k. \tag{12.56}$$

In a similar manner one obtains (P9)

$$\boldsymbol{g}^i \times \boldsymbol{g}^j = \epsilon^{ijk}\boldsymbol{g}_k. \tag{12.57}$$

The vector product of two arbitrary vectors $\boldsymbol{a}$ and $\boldsymbol{b}$ is

$$\boldsymbol{q} = \boldsymbol{a} \times \boldsymbol{b} = a^i\boldsymbol{g}_i \times b^j\boldsymbol{g}_j = \epsilon_{ijk}a^ib^j\boldsymbol{g}^k, \tag{12.58}$$

where use has been made of (12.56). We thus have

$$q_k = \epsilon_{kij}a^ib^j \quad \text{and (P9)} \quad q^k = \epsilon^{kij}a_ib_j. \tag{12.59}$$

On account of (12.56), (12.11), and (12.53), the triple product of three covariant base vectors is

$$(\boldsymbol{g}_i \times \boldsymbol{g}_j)\cdot\boldsymbol{g}_k = \epsilon_{ijl}\boldsymbol{g}^l\cdot\boldsymbol{g}_k = \epsilon_{ijk} = \sqrt{g}e_{ijk}. \tag{12.60}$$

It follows from (12.58) and (12.59) that the triple product of three arbitrary vectors is

$$(\boldsymbol{a} \times \boldsymbol{b})\cdot\boldsymbol{c} = \boldsymbol{q}\cdot\boldsymbol{c} = q_kc^k = \epsilon_{ijk}a^ib^jc^k. \tag{12.61}$$

Another form of this product (P10) is

$$(\boldsymbol{a} \times \boldsymbol{b})\cdot\boldsymbol{c} = \epsilon^{ijk}a_ib_jc_k. \tag{12.62}$$

Figure 12.2 shows an infinitesimal tetrahedron defined by the vectorial line elements d$\boldsymbol{r}$, $\delta\boldsymbol{r}$, $\Delta\boldsymbol{r}$ of the coordinate curves passing through P. Let dA be the oblique face with the external unit normal $\boldsymbol{\nu}$, and let the other faces

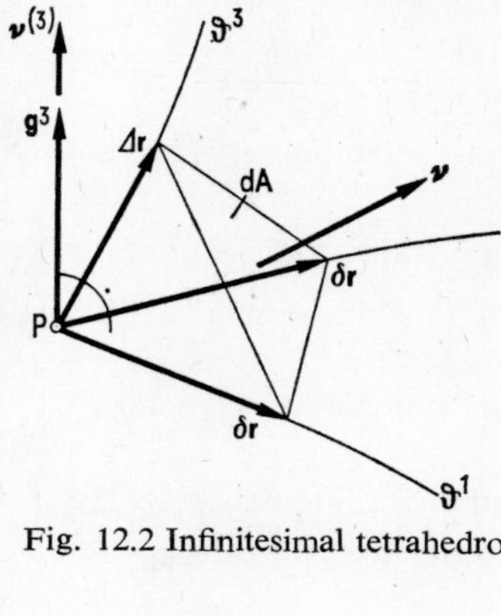

Fig. 12.2 Infinitesimal tetrahedron

and their external unit normals be denoted by $\mathrm{d}A^{(1)}$, ... and $-\boldsymbol{\nu}^{(1)}$, ... respectively. From the figure we deduce

$$\boldsymbol{\nu}\,\mathrm{d}A = \nu_i \boldsymbol{g}^i\,\mathrm{d}A = \tfrac{1}{2}(\delta\boldsymbol{r}-\mathrm{d}\boldsymbol{r})\times(\Delta\boldsymbol{r}-\mathrm{d}\boldsymbol{r})$$
$$= \tfrac{1}{2}(\mathrm{d}\boldsymbol{r}\times\delta\boldsymbol{r}+\delta\boldsymbol{r}\times\Delta\boldsymbol{r}+\Delta\boldsymbol{r}\times\mathrm{d}\boldsymbol{r}) = \boldsymbol{\nu}^{(3)}\,\mathrm{d}A^{(3)}+\ \dots\ . \qquad (12.63)$$

Since $\boldsymbol{\nu}^{(3)}$ is a unit vector with the direction of $\boldsymbol{g}^3$, we further obtain

$$\nu_i \boldsymbol{g}^i\,\mathrm{d}A = \frac{\boldsymbol{g}^3}{\sqrt{\boldsymbol{g}^3\cdot\boldsymbol{g}^3}}\,\mathrm{d}A^{(3)}+\ \dots = \frac{\boldsymbol{g}^1}{\sqrt{g^{11}}}\,\mathrm{d}A^{(1)}+\ \dots\ . \qquad (12.64)$$

The faces $\mathrm{d}A^{(1)}$, ... of the tetrahedron are thus given by

$$\mathrm{d}A^{(1)} = \nu_1\sqrt{g^{11}}\,\mathrm{d}A,\ \dots\ . \qquad (12.65)$$

Let us finally consider the infinitesimal block defined by the line elements $\mathrm{d}\boldsymbol{r} = \boldsymbol{g}_1\,\mathrm{d}\vartheta^1$, $\delta\boldsymbol{r} = \boldsymbol{g}_2\,\delta\vartheta^2$, and $\Delta\boldsymbol{r} = \boldsymbol{g}_3\,\delta\vartheta^3$. On account of (12.60), its volume is

$$\mathrm{d}V = (\mathrm{d}\boldsymbol{r}\times\delta\boldsymbol{r})\cdot\Delta\boldsymbol{r} = (\boldsymbol{g}_1\times\boldsymbol{g}_2)\cdot\boldsymbol{g}_3\,\mathrm{d}\vartheta^1\,\mathrm{d}\vartheta^2\,\mathrm{d}\vartheta^3 = \sqrt{g}e_{123}\,\mathrm{d}\vartheta^1\,\mathrm{d}\vartheta^2\,\mathrm{d}\vartheta^3 \qquad (12.66)$$

or

$$\mathrm{d}V = \sqrt{g}\,\mathrm{d}\vartheta^1\,\mathrm{d}\vartheta^2\,\mathrm{d}\vartheta^3\,. \qquad (12.67)$$

Problems

1. The transformations $\vartheta^1 = 2x^1+x^2$, $\vartheta^2 = x^1+x^2$, applied to the plane cartesian coordinates x^1, x^2, define a skew system ϑ^1, ϑ^2. Find its covariant and contravariant base vectors and verify that $\boldsymbol{g}^1 \perp \boldsymbol{g}_2$ and $\boldsymbol{g}^2 \perp \boldsymbol{g}_1$. Determine the contravariant and covariant components of the vector $\boldsymbol{v} = \boldsymbol{i}_1+2\boldsymbol{i}_2$.
2. Deduce (12.21) from (12.20).
3. Verify (12.25) and (12.26).
4. Show that contraction of the tensor r^{ijk}_{lm} with respect to the indices k and l yields a tensor t^{ij}_{m}.
5. Show that the 27 quantities $t(i, j, k)$ are the components of a tensor t^i_{jk}, provided $t(i, j, k)u_i v^j w^k$ is a scalar for any choice of the vectors u_i, v^j, w^k.
6. Show that $g^{ij} = \boldsymbol{g}^i\cdot\boldsymbol{g}^j$ and $g^i_j = \boldsymbol{g}^i\cdot\boldsymbol{g}_j = \delta^i_j$ are components of a second-order tensor.
7. Find the metric fundamental tensor g_{ij}, g^{ij}, g^i_j for the plane coordinate system ϑ^1, ϑ^2 of P1 and verify that $(\mathrm{d}\boldsymbol{r})^2 = g_{ij}\,\mathrm{d}\vartheta^i\,\mathrm{d}\vartheta^j$.

8. Verify (12.54).
9. Prove (12.57) and $(12.59)_2$.
10. Verify (12.62).

12.2. Tensor analysis

In Section 1.4 we have seen that the partial derivatives of tensors are again tensors, provided we restrict ourselves to cartesian coordinate systems. This is in general not true for the tensors defined in Section 12.1. The process of differentiation implies the comparison of a quantity in adjacent points in space. In the case of a cartesian tensor, only the components are different from point to point, whereas the base vectors are the same. In the case of a general tensor, however, the components *and* the base vectors change.

Let us first consider a scalar φ and let us define its *gradient* as in cartesian coordinates by means of the partial derivatives $\varphi_{,i}$ with respect to the coordinates ϑ^i:

$$\varphi_{,i} = \frac{\partial \varphi}{\partial \vartheta^i}. \tag{12.68}$$

Passing to another curvilinear coordinate system Θ^p, we obtain

$$\frac{\partial \varphi}{\partial \Theta^p} = \frac{\partial \varphi}{\partial \vartheta^i} \frac{\partial \vartheta^i}{\partial \Theta^p}. \tag{12.69}$$

Comparing this to $(12.26)_1$, we see that the gradient of a scalar is a covarian vector.

In a next step let us consider the partial derivative of a vector

$$\boldsymbol{v} = v^k \boldsymbol{g}_k. \tag{12.70}$$

Since either one of the two factors on the right depends on the coordinates ϑ^i, the *gradient* of $\boldsymbol{v}$ has the form

$$\boldsymbol{v}_{,l} = \frac{\partial \boldsymbol{v}}{\partial \vartheta^l} = v^k{}_{,l} \boldsymbol{g}_k + v^k \boldsymbol{g}_{k,l}. \tag{12.71}$$

On account of (12.5) and (12.7),

$$\boldsymbol{g}_{k,l} = \frac{\partial}{\partial \vartheta^l} \left(\frac{\partial x^m}{\partial \vartheta^k} \right) \boldsymbol{i}_m = \frac{\partial^2 x^m}{\partial \vartheta^k \, \partial \vartheta^l} \frac{\partial \vartheta^j}{\partial x^m} \boldsymbol{g}_j = \Gamma^j_{kl} \boldsymbol{g}_j, \tag{12.72}$$

where

$$\Gamma^j_{kl} = \frac{\partial^2 x^m}{\partial\vartheta^k\,\partial\vartheta^l}\,\frac{\partial\vartheta^j}{\partial x^m} \tag{12.73}$$

is the so-called *Christoffel symbol* of the second kind, symmetric in k and l. It can be shown and should be kept in mind that the Christoffel symbol, in spite of the notation used here, is not a tensor. Inserting (12.72) in (12.71), we have

$$\boldsymbol{v}_{,l} = (v^j{}_{,l} + v^k\Gamma^j_{kl})\boldsymbol{g}_j \tag{12.74}$$

or

$$\boldsymbol{v}_{,l} = v^j|_l\,\boldsymbol{g}_j, \tag{12.75}$$

where $v^j|_l$ is defined by

$$v^j|_l = v^j{}_{,l} + v^k\Gamma^j_{kl} \tag{12.76}$$

and is called the *covariant derivative* of v^j.

If the vector $\boldsymbol{v}$ is represented by

$$\boldsymbol{v} = v_k\boldsymbol{g}^k \tag{12.77}$$

in lieu of (12.70), equation (12.71) assumes the form

$$\boldsymbol{v}_{,l} = \frac{\partial\boldsymbol{v}}{\partial\vartheta^l} = v_{k,l}\boldsymbol{g}^k + v_k\boldsymbol{g}^k{}_{,l}. \tag{12.78}$$

It is not difficult to verify (P1) that (12.78) may be written as

$$\boldsymbol{v}_{,l} = v_j|_l\,\boldsymbol{g}^j, \tag{12.79}$$

where

$$v_j|_l = v_{j,l} - v_k\Gamma^k_{jl}. \tag{12.80}$$

Unlike the partial derivatives $v^j{}_{,l}$ and $v_{j,l}$, the covariant derivatives $v^j|_l$ and $v_j|_l$ are tensors. To prove this for $v^j|_l$, let us introduce a second curvilinear coordinate system Θ^p, and let us consider the product $V^q\|_s\boldsymbol{G}_q$, where the vertical double line for the covariant derivative indicates that the Christoffel symbol (12.73) is to be evaluated in the system Θ^p. Using (12.75) we obtain

$$V^q\|_s\boldsymbol{G}_q = \frac{\partial\boldsymbol{v}}{\partial\Theta^s} = \frac{\partial\boldsymbol{v}}{\partial\vartheta^l}\,\frac{\partial\vartheta^l}{\partial\Theta^s} = \frac{\partial\vartheta^l}{\partial\Theta^s}\,v^j|_l\,\boldsymbol{g}_j. \tag{12.81}$$

According to (12.19), this is equivalent to

$$V^q \|_s \boldsymbol{G}_q = \frac{\partial \vartheta^l}{\partial \Theta^s} v^j |_l \frac{\partial \Theta^q}{\partial \vartheta^j} \boldsymbol{G}_q . \tag{12.82}$$

Multiplication of both sides by $\boldsymbol{G}^p$ finally yields the transformation

$$V^p \|_s = \frac{\partial \Theta^p}{\partial \vartheta^j} \frac{\partial \vartheta^l}{\partial \Theta^s} v^j |_l \tag{12.83}$$

for a mixed second-order tensor. For $v_j|_l$ the proof is analogous (P2).

If the coordinate system ϑ^i is cartesian, the second derivatives in (12.73) vanish. The Christoffel symbols are thus zero, and it follows from (12.76) and (12.80) that the covariant and the partial derivatives are identical.

These considerations are readily extended to tensors of higher orders. For a contravariant second-order tensor t^{ij}, e.g., we obtain

$$t^{ij}|_l = t^{ij}{}_{,l} + t^{kj}\Gamma^i_{kl} + t^{ik}\Gamma^j_{kl} . \tag{12.84}$$

Covariant derivatives of this type are again tensors. It follows in particular that the possibility of raising or lowering indices exists. We may define, e.g. a contravariant derivative $t^{ij}|^l$ by means of

$$t^{ij}|^l = g^{lm} t^{ij}|_m . \tag{12.85}$$

For the covariant differentiation of sums, products, etc., the usual rules are valid.

If a vector $\boldsymbol{v}$ is constant, i.e. independent of the position of its point of application, its cartesian components are constant, too. We may express this by the equation $\boldsymbol{v}_{,l} = 0$, but we note that the partial derivatives $v^j{}_{,l}$ and $v_{j,l}$ with respect to curvilinear coordinates need not vanish. However, on account of (12.75) and (12.79), we have $v^j|_l = v_j|_l = 0$. The result can be extended to higher orders: a constant tensor is characterized by components with vanishing covariant derivatives.

An example is the metric tensor. In curvilinear coordinates its covariant, contravariant, and mixed components are g_{ij}, g^{ij}, δ^i_j respectively; the first two systems are functions of ϑ^k. In cartesian coordinates, however, the three systems reduce to $\boldsymbol{i}^i \cdot \boldsymbol{i}_j = \delta^i_j$. The metric tensor is thus constant, and it follows that

$$g_{ij}|_l = g^{ij}|_l = \delta^i_j|_l = 0. \tag{12.86}$$

Another example is the tensor ϵ_{ijk}, since its cartesian components e_{ijk} are constant.

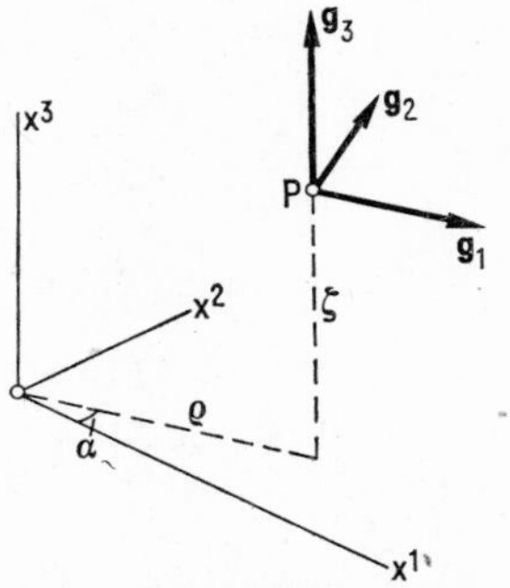

Fig. 12.3 Cylindrical coordinates

As a simple application let us consider the cylindrical coordinate system $\vartheta^1 = \varrho$, $\vartheta^2 = \alpha$, $\vartheta^3 = \zeta$ of Figure 12.3. The transformations $x^i = x^i(\vartheta^j)$ are

$$x^1 = \varrho \cos \alpha, \qquad x^2 = \varrho \sin \alpha, \qquad x^3 = \zeta; \tag{12.87}$$

their inversions are given by

$$\varrho = [(x^1)^2 + (x^2)^2]^{1/2}, \qquad \alpha = \arctan \frac{x^2}{x^1}, \qquad \zeta = x^3. \tag{12.88}$$

The covariant base vectors have radial, circumferential, and axial directions respectively, and their magnitudes are $|\boldsymbol{g}_1| = 1$, $|\boldsymbol{g}_2| = \varrho$, $|\boldsymbol{g}_3| = 1$, since the displacements corresponding to the various coordinate increments are $\mathrm{d}\varrho$, $\varrho\, \mathrm{d}\alpha$, $\mathrm{d}\zeta$. Note that the second base vector is not a unit vector and differs even in dimension from the others. The metric tensor (12.33) is given by

$$g_{11} = 1, \qquad g_{22} = \varrho^2, \qquad g_{33} = 1, \qquad g_{ij} = 0 \quad (i \neq j), \tag{12.89}$$

and the square (12.34) of the vectorial line element is

$$(\mathrm{d}\boldsymbol{r})^2 = (\mathrm{d}\varrho)^2 + \varrho^2(\mathrm{d}\alpha)^2 + (\mathrm{d}\zeta)^2. \tag{12.90}$$

Incidentally, this relation is geometrically obvious, and the metric tensor (12.89) might have been deduced from it.

Evaluating (12.73) in cylindrical coordinates, it is easy to show (P4) that the only non-vanishing Christoffel symbols are

$$\Gamma^1_{22} = -\varrho, \qquad \Gamma^2_{12} = \Gamma^2_{21} = \frac{1}{\varrho}. \tag{12.91}$$

The gradient (12.75) of a vector $\boldsymbol{v}$ is thus given by

$$\begin{aligned}
\boldsymbol{v}_{,1} &= (v^j{}_{,1} + v^k\Gamma^j_{k1})\boldsymbol{g}_j = \qquad v^1{}_{,1}\boldsymbol{g}_1 + \left(v^2{}_{,1} + \frac{v^2}{\varrho}\right)\boldsymbol{g}_2 + v^3{}_{,1}\boldsymbol{g}_3,\\
\boldsymbol{v}_{,2} &= (v^j{}_{,2} + v^k\Gamma^j_{k2})\boldsymbol{g}_j = (v^1{}_{,2} - \varrho v^2)\boldsymbol{g}_1 + \left(v^2{}_{,2} + \frac{v^1}{\varrho}\right)\boldsymbol{g}_2 + v^3{}_{,2}\boldsymbol{g}_3,\\
\boldsymbol{v}_{,3} &= (v^j{}_{,3} + v^k\Gamma^j_{k3})\boldsymbol{g}_j = \qquad v^1{}_{,3}\boldsymbol{g}_1 \qquad + v^2{}_{,3}\boldsymbol{g}_2 + v^3{}_{,3}\boldsymbol{g}_3.
\end{aligned} \tag{12.92}$$

It is sometimes convenient to replace the base vectors $\boldsymbol{g}_i$ by the corresponding unit vectors $\boldsymbol{e}_i$. In the present case they are given by

$$\boldsymbol{e}_1 = \boldsymbol{g}_1, \qquad \boldsymbol{e}_2 = \frac{1}{\varrho}\boldsymbol{g}_2, \qquad \boldsymbol{e}_3 = \boldsymbol{g}_3. \tag{12.93}$$

Equations (12.92) thus assume the forms

$$\begin{aligned}
\boldsymbol{v}_{,1} &= \qquad v^1{}_{,1}\boldsymbol{e}_1 + (\varrho v^2{}_{,1} + v^2)\boldsymbol{e}_2 + v^3{}_{,1}\boldsymbol{e}_3,\\
\boldsymbol{v}_{,2} &= (v^1{}_{,2} - \varrho v^2)\boldsymbol{e}_1 + (\varrho v^2{}_{,2} + v^1)\boldsymbol{e}_2 + v^3{}_{,2}\boldsymbol{e}_3,\\
\boldsymbol{v}_{,3} &= \qquad v^1{}_{,3}\boldsymbol{e}_1 \qquad + \varrho v^2{}_{,3}\boldsymbol{e}_2 + v^3{}_{,3}\boldsymbol{e}_3.
\end{aligned} \tag{12.94}$$

The components of the vector $\boldsymbol{v}$ with respect to the system $\boldsymbol{e}_i$ are referred to as its *physical components*. These are the components we have used so far whenever we considered a problem in cylindrical coordinates, e.g. in (6.27) or in the treatment of Couette flow (Section 9.4). On account of (12.93) they are given by

$$v^{\langle 1\rangle} = v^1 \qquad v^{\langle 2\rangle} = \varrho v^2, \qquad v^{\langle 3\rangle} = v^3; \tag{12.95}$$

hence

$$\boldsymbol{v} = v^1\boldsymbol{e}_1 + \varrho v^2\boldsymbol{e}_2 + v^3\boldsymbol{e}_3. \tag{12.96}$$

Of the partial derivatives of the $\boldsymbol{e}_i$ with respect to the ϑ^j only

$$\boldsymbol{e}_{1,2} = \boldsymbol{e}_2, \qquad \boldsymbol{e}_{2,2} = -\boldsymbol{e}_1 \tag{12.97}$$

are different from zero. Making use of (12.97), we verify equations (12.94) directly by differentiation of (12.96).

Problems

1. Verify (12.79).
2. Prove that $v_j|_l$ is a tensor.
3. Prove (12.86) by transformation onto a cartesian coordinate system.
4. Verify (12.91) and show that the remaining Christoffel symbols are zero.
5. Find the physical components of the vector $\boldsymbol{v} = \boldsymbol{i}_1 + 2\boldsymbol{i}_2$ in the oblique coordinate system introduced in P1 of Section 12.1.

CHAPTER 13

LARGE DISPLACEMENTS

In the treatment of fluids (Chapters 6, 8, and 9) it is generally possible to obtain the answers to all relevant questions by studying the velocity field. The emphasis then lies on the (instantaneous) state of motion or, in other words, on the displacements during an infinitesimal time interval. It is true that in a finite time interval the displacements and the corresponding deformations may become arbitrarily large, but they are of no interest.

In the treatment of solids it is generally necessary to consider the total displacements between a given initial or reference configuration and the instantaneous configuration. As long as the displacements are small (Chapters 7 and 11) the problem is almost the same as the analysis of the velocity field. For large displacements and deformations, however, it is more complicated, but its treatment is simplified by the use of the general tensors defined in Chapter 12.

13.1. Displacements and strains

The state of motion of a continuum has been described so far by the velocity field at a given time; the entire motion is known as soon as the velocity is specified as a function of position and time. In this context it is convenient to use a cartesian coordinate system and to denote the coordinates of spatial points by y_j (in place of the x_j used so far). The velocity field is described by

$$v_i = v_i(y_j, t), \tag{13.1}$$

and the treatment based on (13.1) is sometimes called the *Eulerian approach*.

The v_i in (13.1) are the cartesian velocity components of the material point passing the spatial point y_j at time t. According to (2.14) the deformation rate d_{ij} is the symmetric part $v_{(j,i)}$ of the cartesian velocity gradient obtained by partial differentiation of the velocity components. The d_{ij} describe the deformation per unit time of an infinitesimal cuboid at y_j with edges parallel to the coordinate axes; they apply to different material elements in the course of time.

In this chapter the motion will be described in a different way. We assume that the continuum possesses a reference configuration in which it is free from stresses. We will occasionally refer to it as the initial configuration. The corresponding positions of the various material points will be described by their cartesian coordinates x_j. If the y_j denote the positions of these material points at time t, the motion can be described by the functions

$$y_i = y_i(x_j, t), \tag{13.2}$$

and the treatment based on (13.2) will be referred to as the *Lagrangean approach*. Incidentally, it has been used already (Section 2.2, Chapters 7, 10, and 11) in connection with displacements which are so small that for many purposes the differences between the x_j and the y_j can be disregarded.

A material point cannot assume two different positions at the same time. Conversely, two different material points cannot simultaneously assume the same position. The functions (13.2) are therefore single-valued, and so are the inverse functions

$$x_i = x_i(y_j, t) \tag{13.3}$$

specifying the reference locations of the material points which at time t have the positions y_j. It follows that the Jacobians of (13.2) and (13.3) are different from zero. We will further assume that the functions (13.2) and (13.3) are at least piecewise continuous and differentiable as many times as necessary.

Mathematically, the main difference between the representations (13.1) and (13.2) is the fact that the independent variables are not the same. This is essential in connection with partial differentiation: In the case of (13.1) partial differentiation with respect to t implies that the y_j are kept constant, whereas the x_j are to be fixed in the case of (13.2).

In Figure 13.1 the reference configuration P_0 of a material point and its location P at a fixed time t are referred to the same cartesian coordinate system with unit vectors $\boldsymbol{i}_k$. The coordinates of P_0 are x^k if we use superscripts in view of the transition to curvilinear coordinates, and the radius vector of P_0 is

$$\boldsymbol{r} = x^k \boldsymbol{i}_k\,. \tag{13.4}$$

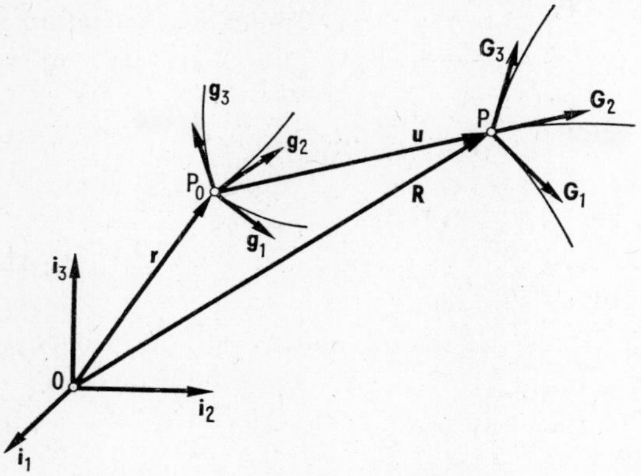

Fig. 13.1 Material coordinate system

At time t the coordinates of the same material point P are y^k, and its radius vector is given by

$$R = y^k i_k . \tag{13.5}$$

The displacement of the point P, referred to the reference configuration P_0, is

$$u = R - r = (y^k - x^k) i_k . \tag{13.6}$$

The reference configuration of the continuum may also be referred to a system of curvilinear coordinates ϑ^j. The x^i are then functions of the form

$$x^i = x^i(\vartheta^j), \tag{13.7}$$

and the y^i may be considered as functions

$$y^i = y^i(\vartheta^j, t). \tag{13.8}$$

The ϑ^j in (13.8) specify the reference configurations. For a given material point the ϑ^j have always the same values: each set of ϑ^j denotes a material point. In the configuration at time t the ϑ^j define a spatially different curvilinear coordinate system; the coordinate curves, however, are material curves insofar as they always consist of the same material points. The ϑ^j may therefore be referred to as *material coordinates.*

Provided we use the ϑ^j and t as independent variables, the radius vector of P_0 is given by $\boldsymbol{r}(\vartheta^j)$, the radius vector of P by $\boldsymbol{R}(\vartheta^j, t)$, and the displacement vector by $\boldsymbol{u}(\vartheta^j, t)$. Of the partial derivatives with respect to ϑ^j and t, the last one is also the material derivative. The base vectors of the material coordinate system ϑ^j are different in the reference configuration and at time t. In the reference configuration they are given by (12.4),

$$g_i(\vartheta^j) = \boldsymbol{r}_{,i}; \tag{13.9}$$

at time t they are

$$\boldsymbol{G}_i(\vartheta^j, t) = \boldsymbol{R}_{,i}. \tag{13.10}$$

The metric tensors (12.33) in the two configurations are

$$g_{ij} = \boldsymbol{g}_i \cdot \boldsymbol{g}_j, \qquad G_{ij} = \boldsymbol{G}_i \cdot \boldsymbol{G}_j, \tag{13.11}$$

and the squares (12.34) of the material line element are given by

$$\mathrm{d}s^2 = g_{ij}\, \mathrm{d}\vartheta^i\, \mathrm{d}\vartheta^j, \qquad \mathrm{d}S^2 = G_{ij}\, \mathrm{d}\vartheta^i\, \mathrm{d}\vartheta^j. \tag{13.12}$$

In order to investigate the deformation of an infinitesimal neighborhood of the point P_0, let us consider (Fig. 13.2) the material points Q_0 and S_0

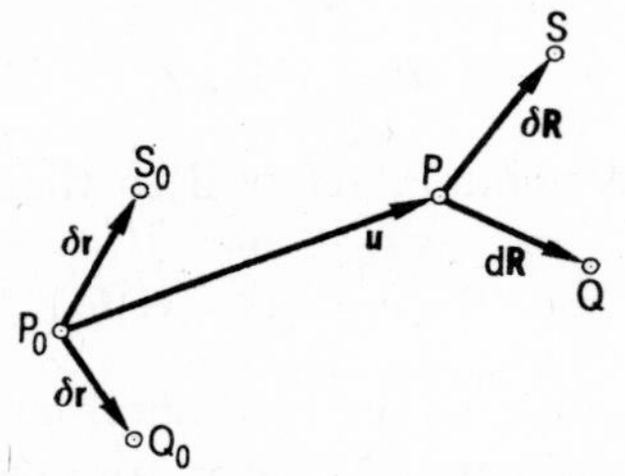

Fig. 13.2 Displacement of a vicinity of P

with radius vectors $\mathrm{d}\boldsymbol{r}$, $\delta\boldsymbol{r}$, from P_0 respectively. Let Q and S denote the positions of these points at time t, and let their radius vectors from P be denoted by $\mathrm{d}\boldsymbol{R}$, $\delta\boldsymbol{R}$ respectively. If $\boldsymbol{u}$ is the displacement of P and if $\boldsymbol{u}+\mathrm{d}\boldsymbol{u}$, $\boldsymbol{u}+\delta\boldsymbol{u}$ are the displacements of Q and S respectively, we conclude from (13.6) that

$$\mathrm{d}\boldsymbol{R} = \mathrm{d}\boldsymbol{r}+\mathrm{d}\boldsymbol{u}, \qquad \delta\boldsymbol{R} = \delta\boldsymbol{r}+\delta\boldsymbol{u}. \tag{13.13}$$

The vicinity of P remains undeformed (P1) if and only if $\mathrm{d}\boldsymbol{R}\cdot\delta\boldsymbol{R} = \mathrm{d}\boldsymbol{r}\cdot\delta\boldsymbol{r}$ for any choice of the vectors $\mathrm{d}\boldsymbol{r}$ and $\delta\boldsymbol{r}$. The difference $\mathrm{d}\boldsymbol{R}\cdot\delta\boldsymbol{R}-\mathrm{d}\boldsymbol{r}\cdot\delta\boldsymbol{r}$ may therefore be used as a measure of the deformation.

According to (13.13) we have

$$\begin{aligned}\mathrm{d}\boldsymbol{R}\cdot\delta\boldsymbol{R}-\mathrm{d}\boldsymbol{r}\cdot\delta\boldsymbol{r} &= (\mathrm{d}\boldsymbol{r}+\mathrm{d}\boldsymbol{u})\cdot(\delta\boldsymbol{r}+\delta\boldsymbol{u})-\mathrm{d}\boldsymbol{r}\cdot\delta\boldsymbol{r}\\ &= \mathrm{d}\boldsymbol{r}\cdot\delta\boldsymbol{u}+\delta\boldsymbol{r}\cdot\mathrm{d}\boldsymbol{u}+\mathrm{d}\boldsymbol{u}\cdot\delta\boldsymbol{u}.\end{aligned} \tag{13.14}$$

On account of (12.6)

$$\mathrm{d}\boldsymbol{r} = \boldsymbol{r}_{,i}\,\mathrm{d}\vartheta^i = \boldsymbol{g}_i\,\mathrm{d}\vartheta^i. \tag{13.15}$$

In a similar manner we obtain

$$\mathrm{d}\boldsymbol{u} = \boldsymbol{u}_{,j}\,\mathrm{d}\vartheta^j. \tag{13.16}$$

Equation (12.75) yields

$$\boldsymbol{u}_{,j} = u^k|_j \boldsymbol{g}_k \quad \text{and hence} \quad \mathrm{d}\boldsymbol{u} = u^k|_j \boldsymbol{g}_k\,\mathrm{d}\vartheta^j. \tag{13.17}$$

From (13.15) and $(13.17)_2$ we further obtain

$$\mathrm{d}\boldsymbol{r}\cdot\delta\boldsymbol{u} = \boldsymbol{g}_i\,\mathrm{d}\vartheta^i u^k|_j \boldsymbol{g}_k\,\delta\vartheta^j = u^k|_j g_{ik}\,\mathrm{d}\vartheta^i\,\delta\vartheta^j = u_i|_j\,\mathrm{d}\vartheta^i\,\delta\vartheta^j \tag{13.18}$$

and similarly

$$\delta\boldsymbol{r}\cdot\mathrm{d}\boldsymbol{u} = u_i|_j\,\delta\vartheta^i\,\mathrm{d}\vartheta^j = u_j|_i\,\mathrm{d}\vartheta^i\,\delta\vartheta^j. \tag{13.19}$$

Equation $(13.17)_2$ finally yields

$$\begin{aligned}\mathrm{d}\boldsymbol{u}\cdot\delta\boldsymbol{u} &= u^k|_i \boldsymbol{g}_k\,\mathrm{d}\vartheta^i u^l|_j \boldsymbol{g}_l\,\delta\vartheta^j = u^k{}_i| u^l|_j g_{kl}\,\mathrm{d}\vartheta^i\,\delta\vartheta^j\\ &= u^k|_i u_k|_j\,\mathrm{d}\vartheta^i\,\delta\vartheta^j,\end{aligned} \tag{13.20}$$

and upon substitution of (13.18) through (13.20) in (13.14) we obtain

$$\mathrm{d}\boldsymbol{R}\cdot\delta\boldsymbol{R} - \mathrm{d}\boldsymbol{r}\cdot\delta\boldsymbol{r} = (u_i|_j + u_j|_i + u^k{}_i| u_k|_j)\,\mathrm{d}\vartheta^i\,\delta\vartheta^j. \tag{13.21}$$

The calculations leading to (13.21) have been carried out with the help of the basis $\boldsymbol{g}_i$. The Christoffel symbols to be used in the evaluation of the covariant derivatives are therefore the ones of (12.73), based on the reference coordinates x^m. The calculations may be repeated (P2) with the basis $\boldsymbol{G}_i$. The result is

$$\mathrm{d}\boldsymbol{R}\cdot\delta\boldsymbol{R} - \mathrm{d}\boldsymbol{r}\cdot\delta\boldsymbol{r} = (U_i\|_j + U_j\|_i - U^k\|_i\, U_k\|_j)\,\mathrm{d}\vartheta^i\,\delta\vartheta^j, \tag{13.22}$$

where the vertical double lines indicate that the Christoffel symbols are to be computed from the instantaneous coordinates y^m in lieu of the x^m. We have noted that the left-hand side of (13.21) represents a measure for the deformation. The expression in parentheses on the right is a covariant tensor of the second order, and it is convenient to define half of it,

$$\gamma_{ij} = \tfrac{1}{2}(u_i|_j + u_j|_i + u^k|_i u_k|_j), \tag{13.23}$$

as the *strain tensor* for large displacements. It is symmetric and completely determined by the displacement field. Inserting (13.23) in (13.21), we have

$$\mathrm{d}\boldsymbol{R}\cdot\delta\boldsymbol{R}-\mathrm{d}\boldsymbol{r}\cdot\delta\boldsymbol{r} = 2\gamma_{ij}\,\mathrm{d}\vartheta^i\,\delta\vartheta^j. \tag{13.24}$$

If we choose $\delta\boldsymbol{r} = \mathrm{d}\boldsymbol{r}$, it is clear that $\delta\boldsymbol{R} = \mathrm{d}\boldsymbol{R}$, so that

$$\mathrm{d}S^2-\mathrm{d}s^2 = 2\gamma_{ij}\,\mathrm{d}\vartheta^i\,\mathrm{d}\vartheta^j. \tag{13.25}$$

Comparing this to (13.12), we obtain another expression for the strain tensor,

$$\gamma_{ij} = \tfrac{1}{2}(G_{ij}-g_{ij}). \tag{13.26}$$

In the special case where the material coordinate system is cartesian in the reference configuration ($\vartheta^i = x^i$), covariant and partial differentiation are the same, and the strain tensor becomes

$$\gamma_{ij} = \tfrac{1}{2}(u_{i,j}+u_{j,i}+u^k{}_{,i}u_{k,j}). \tag{13.27}$$

For small displacement gradients a first approximation of (13.27) is

$$\gamma_{ij} = \tfrac{1}{2}(u_{i,j}+u_{j,i}); \tag{13.28}$$

the strain tensor γ_{ij} is thus a generalization of the tensor ε_{ij} introduced in (2.33) for small displacements.

Returning to the general case and considering a line element emanating from P_0 in the first coordinate curve, we note that according to (13.12) its original and instantaneous lengths are

$$\mathrm{d}s = \sqrt{g_{11}}\,\mathrm{d}\vartheta^1, \qquad \mathrm{d}S = \sqrt{G_{11}}\,\mathrm{d}\vartheta^1 \tag{13.29}$$

respectively. The change in length may be described by the so-called *extension ratio* $\lambda = \mathrm{d}S/\mathrm{d}s$. From (13.29) and (13.26) we obtain the extension ratios

$$\lambda^{(1)} = \sqrt{\frac{G_{11}}{g_{11}}} = \left(1+2\frac{\gamma_{11}}{g_{11}}\right)^{1/2}, \quad \ldots \tag{13.30}$$

for the line elements in the three coordinate curves. The first of them is determined by γ_{11}. For small deformations (13.30) reduces to $\lambda^{(1)} = 1+\gamma_{11}/g_{11}, \ldots$; the quotients $\gamma_{11}/g_{11}, \ldots$ are the extensions $\varepsilon_{11}, \ldots$ in the sense of Section 2.2. The angles $\varphi^{(23)}, \ldots$ between pairs of line elements in the coordinate curves are originally given by

$$\cos\varphi^{(23)} = \frac{\boldsymbol{g}_2\cdot\boldsymbol{g}_3}{|\boldsymbol{g}_2|\cdot|\boldsymbol{g}_3|} = \frac{g_{23}}{\sqrt{g_{22}g_{33}}}, \quad \ldots, \tag{13.31}$$

the instantaneous angles by

$$\cos \Phi^{(23)} = \frac{G_{23}}{\sqrt{G_{22}G_{33}}}, \ldots, \tag{13.32}$$

The changes of the angles can be calculated from (13.31) and (13.32); for $\varphi^{(23)}$, e. g., they are determined by γ_{23}, γ_{22}, and γ_{33}.

The basic invariants of a general tensor are expressions like (1.52); on account of the modified summation convention, however, they must be written in mixed components, i.e. in terms of γ^i_k in the case of the strain tensor. Instead of the γ^i_k, however, it is convenient to use the tensor $\delta^i_k + 2\gamma^i_k$, which may be written as

$$\delta^i_k + 2\gamma^i_k = g^{ij}(g_{jk} + 2\gamma_{jk}) = g^{ij}G_{jk} \tag{13.33}$$

on account of $(12.35)_2$ and (13.26). The basic invariants of (13.33) are referred to as the *strain invariants*; they are given by

$$\begin{aligned} I_{(1)} &= g^{ij}G_{ji}, \\ I_{(2)} &= \tfrac{1}{2}(g^{ij}G_{jk}g^{kl}G_{li} - g^{ij}G_{ji}g^{kl}G_{lk}) = \tfrac{1}{2}(g^{ij}g^{kl}G_{jk}G_{li} - I^2_{(1)}), \\ I_{(3)} &= \det(g^{ij}G_{jk}) = \frac{G}{g}, \end{aligned} \tag{13.34}$$

where use has been made of (12.44).

On account of (12.67) the volume element in the reference configuration is $\mathrm{d}v = \sqrt{g}\,\mathrm{d}\vartheta^1\,\mathrm{d}\vartheta^2\,\mathrm{d}\vartheta^3$. At time t it is $\mathrm{d}V = \sqrt{G}\,\mathrm{d}\vartheta^1\,\mathrm{d}\vartheta^2\,\mathrm{d}\vartheta^3$. Its change may be described by the *dilatation ratio*

$$\frac{\mathrm{d}V}{\mathrm{d}v} = \sqrt{\frac{G}{g}} = \sqrt{I_{(3)}}\,. \tag{13.35}$$

It follows in particular that incompressible materials are characterized by $I_{(3)} = 1$.

According to (12.5) and (12.13), the metric tensor (12.33) in the reference configuration may be written as

$$g_{ij} = \boldsymbol{g}_i \cdot \boldsymbol{g}_j = \frac{\partial x^k}{\partial \vartheta^i}\boldsymbol{i}_k \cdot \frac{\partial x^l}{\partial \vartheta^j}\boldsymbol{i}_l = \frac{\partial x^k}{\partial \vartheta^i}\frac{\partial x_k}{\partial \vartheta^j}\,. \tag{13.36}$$

In a similar manner, $(12.35)_1$, (12.8), and (12.13) yield

$$g^{ij} = \boldsymbol{g}^i \cdot \boldsymbol{g}^j = \frac{\partial \vartheta^i}{\partial x^k}\boldsymbol{i}^k \cdot \frac{\partial \vartheta^j}{\partial x_l}\boldsymbol{i}_l = \frac{\partial \vartheta^i}{\partial x^k}\frac{\partial \vartheta^j}{\partial x_k}\,. \tag{13.37}$$

The corresponding expressions for the metric tensor at time t are

$$G_{ij} = \frac{\partial y^k}{\partial \vartheta^i} \frac{\partial y_k}{\partial \vartheta^j}, \qquad G^{ij} = \frac{\partial \vartheta^i}{\partial y^k} \frac{\partial \vartheta^j}{\partial y_k}. \tag{13.38}$$

The material derivative of the strain tensor is $\dot{\gamma}_{ij} = \gamma_{ij,0}$; it will be referred to as the *strain rate*. On account of (13.26) and $(13.38)_1$ it assumes the form

$$\dot{\gamma}_{ij} = \tfrac{1}{2} G_{ij,0} = \tfrac{1}{2} \left(\frac{\partial y^k}{\partial \vartheta^i} \frac{\partial y_{k,0}}{\partial \vartheta^j} + \frac{\partial y^k{}_{,0}}{\partial \vartheta^i} \frac{\partial y_k}{\partial \vartheta^j} \right). \tag{13.39}$$

The partial derivative $y^k{}_{,0} = y_{k,0}$ is the cartesian component $v^k = v_k$ of the velocity vector. We thus have

$$\dot{\gamma}_{ij} = \tfrac{1}{2} \left(\frac{\partial y^k}{\partial \vartheta^i} \frac{\partial v_k}{\partial \vartheta^j} + \frac{\partial y_k}{\partial \vartheta^j} \frac{\partial v^k}{\partial \vartheta^i} \right) = \tfrac{1}{2} \left(\frac{\partial y^k}{\partial \vartheta^i} \frac{\partial v_k}{\partial y^l} \frac{\partial y^l}{\partial \vartheta^j} + \frac{\partial y_k}{\partial \vartheta^j} \frac{\partial v^k}{\partial y^l} \frac{\partial y^l}{\partial \vartheta^i} \right) \tag{13.40}$$

or

$$\dot{\gamma}_{ij} = \tfrac{1}{2} \left(\frac{\partial v_k}{\partial y^l} + \frac{\partial v_l}{\partial y^k} \right) \frac{\partial y^k}{\partial \vartheta^i} \frac{\partial y^l}{\partial \vartheta^j} = \frac{\partial y^k}{\partial \vartheta^i} \frac{\partial y^l}{\partial \vartheta^j} d_{kl}, \tag{13.41}$$

where d_{kl} is the deformation rate (2.14) at time t. Comparing (13.41) and the inversion of (12.28), written with $\Theta^k = y^k$, we see that the material strain rate $\dot{\gamma}_{ij}$ is the covariant tensor obtained from the deformation rate d_{kl} by transformation from the cartesian coordinates y^k (with unit vectors $\boldsymbol{i}_k$) to the curvilinear material system ϑ^i (with the basis $\boldsymbol{G}_i$).

Problems

1. Show that (Fig. 13.2) the infinitesimal vicinity of a material point P in a continuum remains undeformed if and only if $\mathrm{d}\boldsymbol{R} \cdot \delta \boldsymbol{R} = \mathrm{d}\boldsymbol{r} \cdot \delta \boldsymbol{r}$ for any choice of the vectors $\mathrm{d}\boldsymbol{r}$ and $\delta \boldsymbol{r}$.
2. Verify (13.22).

13.2. Stresses and power of deformation

The deformation rate appearing on the right-hand side of (13.41) corresponds to the Eulerian approach, based on the representation of the velocity as a function $v^i(y^j, t)$ of the instantaneous cartesian coordinates and the time. It describes the deformation per unit time of an infinitesimal cuboid at y^j with edges parallel to the cartesian coordinate axes. If the stress components acting on the faces of this cuboid are denoted by $\sigma^{ij} = \sigma_{ij}$, the specific rate of work (4.35) is given by

$$l = \frac{1}{\varrho}\sigma^{ij}d_{ij}, \tag{13.42}$$

where ϱ is the density of the cuboid at time t.

In the case of small displacements the deformation is described by the strain components ε_{ij} introduced in (2.33). The deformation rate is given by $d_{ij} = \mathrm{d}\varepsilon_{ij}/\mathrm{d}t$, where $\mathrm{d}\varepsilon_{ij}$ may be interpreted as the material strain increment, and the specific elementary work of the stress tensor is

$$\mathrm{d}w = l\,\mathrm{d}t = \frac{1}{\varrho}\sigma^{ij}\,\mathrm{d}\varepsilon_{ij}. \tag{13.43}$$

If we adopt the Lagrangean approach for large displacements, ε_{ij} is to be replaced by the strain tensor γ_{ij} defined by (13.23). Let its material increment be denoted by $\mathrm{d}\gamma_{ij}$. Since the elementary work $\mathrm{d}w$ per unit mass is a scalar, it must have the form

$$\mathrm{d}w = \frac{1}{\varrho}\tau^{ij}\,\mathrm{d}\gamma_{ij}, \tag{13.44}$$

where τ^{ij} is the stress tensor associated with the strain γ_{ij}. The corresponding specific rate of work is

$$l = \frac{1}{\varrho}\tau^{ij}\dot{\gamma}_{ij}. \tag{13.45}$$

We will presently discuss the significance of the stress components τ^{ij}. Before doing so, let us note that besides γ_{ij} numerous other strain measures have been proposed in literature. However, almost none of them admit a representation of $\mathrm{d}w$ of the general form of (13.44). They are unsuitable for the formulation of the energy theorem or the first fundamental law and hence must be discarded [26].

To find the significance of the stress tensor τ^{ij} associated with the strain γ_{ij}, we note that the connection between τ^{ij} and σ^{ij} must be determined by the connection between the strain rates $\dot{\gamma}_{ij}$ and d_{ij}. In fact, on account of (13.45), (13.41), and (13.42) we have

$$\varrho l = \tau^{ij}\dot{\gamma}_{ij} = \tau^{ij}\frac{\partial y^k}{\partial \vartheta^i}\frac{\partial y^l}{\partial \vartheta^j}d_{kl} = \sigma^{kl}d_{kl}. \tag{13.46}$$

It follows that

$$\sigma^{kl} = \frac{\partial y^k}{\partial \vartheta^i} \frac{\partial y^l}{\partial \vartheta^j} \tau^{ij} \tag{13.47}$$

and that, conversely,

$$\tau^{ij} = \frac{\partial \vartheta^i}{\partial y^k} \frac{\partial \vartheta^j}{\partial y^l} \sigma^{kl} . \tag{13.48}$$

Equations (13.48) and (13.47) correspond to the transformation (12.28) and its inverse, written for $\Theta^k = y^k$. It thus turns out that τ^{ij} is the contravariant tensor obtained from σ^{kl} by transformation from the cartesian coordinates y^k (with the unit vectors $\boldsymbol{i}_k$) to the curvilinear system ϑ^i (with the basis $\boldsymbol{G}_i$).

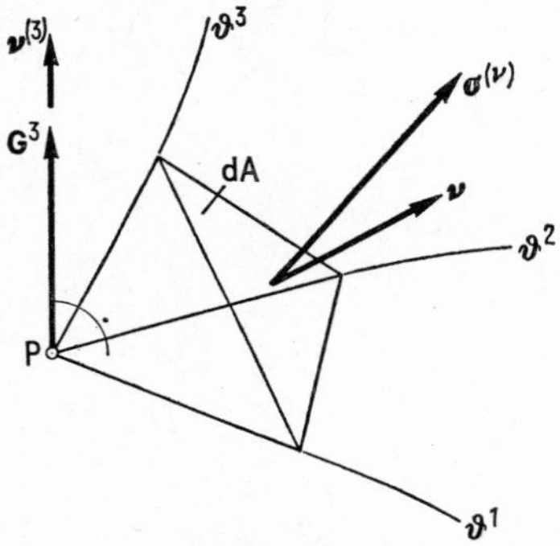

Fig. 13.3 Infinitesimal tetrahedron

Figure 13.3 shows an infinitesimal tetrahedron defined by line elements in the material coordinate curves at time t. As in Figure 12.2, $\mathrm{d}A$ denotes the oblique face with the external unit normal $\boldsymbol{\nu}$, and the external unit normals of the other faces $\mathrm{d}A^{(1)}, \ldots$ will be denoted by $-\boldsymbol{\nu}^{(1)}, \ldots$; the $\boldsymbol{\nu}^{(1)} \ldots$ thus have the directions of the contravariant base vectors $\boldsymbol{G}^1, \ldots$. Let $\boldsymbol{\sigma}^{(\nu)}$ and $-\boldsymbol{\sigma}^{(1)}, \ldots$ denote the stress vectors acting on the faces $\mathrm{d}A$, $\mathrm{d}A^{(1)}, \ldots$ respectively, i.e. the corresponding surface forces per unit area. According to (3.6), the cartesian components of the stress vector $\boldsymbol{\sigma}^{(\nu)}$ acting on the oblique face are given by

$$\sigma^{(\nu)k} = \sigma^{kl} \nu_l , \tag{13.49}$$

where the σ^{kl} are the normal and shear stresses for surface elements perpendicular to the cartesian coordinate axes. On account of (12.26) we have

$$\nu_l = \frac{\partial \vartheta^m}{\partial y^l} N_m , \tag{13.50}$$

where the N_m are the covariant components of $\boldsymbol{\nu}$ in the material coordinate system. Equations (13.49), (13.47), and (13.50) yield

$$\sigma^{(\nu)k} = \sigma^{kl}\nu_l = \frac{\partial y^k}{\partial \vartheta^i}\frac{\partial y^l}{\partial \vartheta^j}\frac{\partial \vartheta^m}{\partial y^l}\tau^{ij}N_m = \frac{\partial y^k}{\partial \vartheta^i}\tau^{ij}N_j . \tag{13.51}$$

Provided the curvilinear components of the vector $\boldsymbol{\sigma}^{(\nu)}$ are denoted by $T^{(\nu)i}$, we further have

$$\sigma^{(\nu)k} = \frac{\partial y^k}{\partial \vartheta^i} T^{(\nu)i} . \tag{13.52}$$

Comparing (13.51) and (13.52), we obtain the relation

$$T^{(\nu)i} = \tau^{ij}N_j , \tag{13.53}$$

which is analogous to (13.49).

Let us apply (13.53) to the stress vector acting on the face $\mathrm{d}A^{(3)}$ of the tetrahedron. Since $\boldsymbol{\nu}^{(3)}$ is a unit vector with the direction of $\boldsymbol{G}^3$, we have

$$\boldsymbol{\nu}^{(3)} = \frac{\boldsymbol{G}^3}{\sqrt{G^{33}}} . \tag{13.54}$$

It follows that the only non-vanishing covariant component of $\boldsymbol{\nu}^{(3)}$ is $N_3^{(3)} = 1/\sqrt{G^{33}}$. Equation (13.53) thus yields

$$T^{(3)i} = \tau^{ij}N_j^{(3)} = \frac{\tau^{i3}}{\sqrt{G^{33}}} \tag{13.55}$$

or, cyclically supplemented,

$$\tau^{i1} = \sqrt{G^{11}}T^{(1)i}, \; \ldots \; . \tag{13.56}$$

The stresses τ^{i1}, ... are thus the contravariant components of the stress vectors $\boldsymbol{\sigma}^{(1)}$, ... acting on surface elements defined by pairs of line elements in the material coordinate curves, multiplied by $\sqrt{G^{11}}$, ... respectively.

If we use unit vectors

$$\boldsymbol{e}_1 = \frac{\boldsymbol{G}_1}{\sqrt{G_{11}}}, \; \ldots \tag{13.57}$$

instead of the covariant basis, (13.55) supplies expressions like

$$T^{(3)1}\boldsymbol{G}_1 = \frac{\tau^{13}}{\sqrt{G^{33}}}\sqrt{G_{11}}\boldsymbol{e}_1 . \tag{13.58}$$

Thus, the vector $\sigma^{(3)} = T^{(3)i}G_i$ assumes the form

$$\sigma^{(3)} = \sqrt{\frac{G_{11}}{G^{33}}}\,\tau^{13}e_1 + \sqrt{\frac{G_{22}}{G^{33}}}\,\tau^{23}e_2 + \sqrt{\frac{G_{33}}{G^{33}}}\,\tau^{33}e_3. \tag{13.59}$$

From (13.59) and the corresponding relations for $\sigma^{(1)}$ and $\sigma^{(2)}$ we deduce the *physical components*

$$\tau^{\langle 11\rangle} = \sqrt{\frac{G_{11}}{G^{11}}}\,\tau^{11}, \quad \ldots, \qquad \tau^{\langle 23\rangle} = \sqrt{\frac{G_{22}}{G^{33}}}\,\tau^{23}, \quad \ldots \tag{13.60}$$

of the tensor τ^{ij}.

With the results obtained in this section and the last one, the general theorems of Chapters 3 and 4 are readily transferred to material coordinates. The equilibrium condition (3.10), e. g., assumes the form

$$\tau^{ij}\,||_j + \varrho f^i = 0, \tag{13.61}$$

where the f^i are the contravariant components of the specific body force and the vertical double line indicates that (13.61) is to be satisfied in the instantaneous configuration of the continuum. In a similar manner the boundary condition (7.21) for the stresses becomes

$$\tau^{ij}N_i = T^{(\nu)i}. \tag{13.62}$$

The principal result to be kept in mind for the following chapters is the necessity to replace expressions (13.42) and (13.43) by (13.45) and (13.44) whenever the Lagrangean approach is used for large displacements.

Problem

What is the mechanical significance of the physical stress components $\tau^{\langle ij\rangle}$?

CHAPTER 14

THERMODYNAMIC ORTHOGONALITY

We have noted in Section 3.3 that in general continuum mechanics cannot be separated from thermodynamics. This has been confirmed by numerous applications, in particular by those concerned with thermoelasticity and with gases. It is true that in other applications it has been possible to separate the mechanical from the thermal problem. However, in the condition that the power of dissipation be nonnegative, an element of thermodynamics persisted even in most of these cases.

In the present chapter we will return to the fundamental aspects of thermodynamics with the intention of gaining some information beyond the fundamental laws. We will in fact establish an orthogonality condition for the dissipative forces and propose a generalization in terms of an orthogonality principle. We will demonstrate the usefulness of these additions to the fundamental laws in setting up constitutive equations. It will turn out that the numerous functions encountered in this connection can be reduced to a single pair: the free energy and the dissipation function.

14.1. The governing functions

It has been pointed out in Section 5.1 that for reversible processes the thermomechanical behavior of a continuum appears completely determined in the classical sense by its caloric equation of state, specifying the free energy $\psi(\varepsilon_{ij}, \vartheta)$ as a function of the independent state variables ε_{ij} and ϑ. In fact, the specific entropy s follows from $(4.37)_2$ and the specific internal energy u from (4.34). The thermal equation of state $(4.37)_1$ supplies the quasiconservative stresses $\sigma_{ij}^{(q)}$, and the dissipative stresses are zero. On account of (4.42) the assumption that the process is reversible precludes the presence of internal parameters, and a possible deviation from reversibility in the form of heat exchange may be accounted for by Fourier's law.

The response of an inviscid fluid, for instance, is governed by the specific free energy $(5.18)_1$ or in the particular case of an ideal gas by the expression given in P1 of Section 5.2. A similar statement holds for the elastic solid. Its specific free energy has the form (5.27), where ε_{ij} would have to be replaced by γ_{ij} in the case of large deformations. In the isothermal case, (5.34) shows that the free energy and the strain energy are essentially the same. On the other hand, the linear theory of thermoelasticity follows entirely from the expression (7.72) for $\varrho\psi$.

It is clear that the particular form of the free energy depends on the material. However, the essential fact is that in all reversible cases the entire reponse of the continuum is determined by a single relation: the caloric equation of state, specifying the free energy as a function of the independent state variables.

The treatment of irreversible processes does not appear that simple. They are characterized by the presence of heat conduction, of dissipative stresses, or of internal parameters, and they seem to be governed by any number of functions beside the free energy. Heat conduction, e. g., requires the addition of a constitutive equation in the form of Fourier's law, and in the anisotropic case the relevant function is the thermal conductivity matrix λ_{kl} appearing in (5.8). A viscous liquid of the quasilinear type is characterized by the viscosity function μ appearing in (9.72), and in the truly nonlinear case (9.65) we need even two such functions, g and h. Plastic bodies require a yield function (10.37), and for viscoelastic bodies we need the appropriate generalizations of the creep compliances and relaxation moduli listed in Table 11.1.

It appears surprising and improbable that the transition from reversible cases to irreversible ones should complicate an originally simple situation to this extent. That the free energy is to be supplemented by something else is clear. One would expect, however, a principle of approximately the same generality as the fundamental laws in place of an almost inexhaustible supply of ad hoc prescriptions covering numerous special cases. In the remainder of this book we will propose, discuss, and apply such a general principle, introduced in two stages and applicable at least to the thermomechanical processes considered here, but possibly also in more general situations. This principle will be based on the assumption that the specific dissipation function, introduced by means of (5.1), plays the central role beside the specific free energy.

According to Section 4.1 a thermomechanical process is irreversible whenever it is accompanied by an entropy production $\dot{S}^{(i)}$. The corresponding dissipation function $\Phi = \vartheta \dot{S}^{(i)}$ has been introduced in (4.29), and by comparison with (4.28) we see that it may be interpreted as a special form of the power of dissipation $L^{(d)}$, obtained by expressing the dissipative forces in terms of the corresponding velocities. In the case of a continuum the situation is slightly complicated by the fact that part of the entropy production is due to heat exchange. The terminology of Section 4.1 may be retained however, also in this case, and the corresponding forces and velocities have been discussed at the end of Section 4.3.

Whenever a process turned out to be irreversible in the chapters concerned with applications, we have established its specific power of dissipation or dissipation rate $l^{(d)}$. Expressing the forces by the corresponding velocities, we obtain the specific dissipation function. In view of the central role of this function, let us first collect its various forms encountered in our applications.

In the case of *heat conduction* the heat flow vector q_k represents the velocity, and the negative temperature gradient $-\vartheta_{,k}$, divided by the temperature, is the corresponding force per unit volume. Provided the connection between the two is not influenced by other processes (e.g. deformation), the specific dissipation function has the form $\varphi(q_k)$ and is obtained from (4.79) by expressing $(\ln \vartheta)_{,k} = \vartheta_{,k}/\vartheta$ in terms of q_k. In the linear case represented by the generalized Fourier law (5.8), $\varrho\varphi = -(\vartheta_{,k}/\vartheta)\, q_k$ assumes the form

$$\varrho\varphi = \frac{\mu_{jk}}{\vartheta} q_j q_k , \tag{14.1}$$

where the matrix μ_{jk} is inverse to λ_{kl}, obeying the relation

$$\mu_{jk}\lambda_{kl} = \delta_{jl} . \tag{14.2}$$

If the material is isotropic, transition from (5.8) to (5.9) shows that $\lambda_{kl} = \lambda\delta_{kl}$. Equation (14.2) yields $\lambda\mu_{jk} = \delta_{jk}$, and the dissipation function per unit volume becomes

$$\varrho\varphi = \frac{1}{\lambda\vartheta} q_j q_j . \tag{14.3}$$

For the deformation of a *Newtonian fluid*, the dissipation function per unit volume follows from (9.3):

$$\varrho\varphi = \lambda d_{ii} d_{jj} + 2\mu d_{ij} d_{ij} , \tag{14.4}$$

where λ and μ are the viscosity coefficients. If the bulk viscosity is zero (or if the fluid is incompressible), (9.16) yields

$$\varrho\varphi = 2\mu d'_{ij}d'_{ij}. \tag{14.5}$$

On account of (9.61) and (9.62), the dissipation function per unit volume of a *non-Newtonian fluid* is

$$\varrho\varphi = fd_{ii}+gd_{ij}d_{ji}+hd_{ij}d_{jk}d_{ki} \tag{14.6}$$

or

$$\varrho\varphi = fd_{(1)}+g(2d_{(2)}+d_{(1)}^2)+h[3d_{(3)}+\tfrac{3}{2}(2d_{(2)}+d_{(1)}^2)d_{(1)}-\tfrac{1}{2}d_{(1)}^3], \tag{14.7}$$

where f, g, and h are functions of the basic invariants of the deformation rate. In the incompressible case, (14.7) reduces to

$$\varrho\varphi = 2gd_{(2)}+3hd_{(3)}, \tag{14.8}$$

and for the quasilinear incompressible liquid (9.72) or (14.8) yield

$$\varrho\varphi = 4\mu d_{(2)}, \tag{14.9}$$

where g, h, and μ are functions of $d_{(2)}$ and $d_{(3)}$.

The dissipation function per unit volume of the *viscoplastic material* defined in Section 10.1 is given by (10.11); it reads

$$\varrho\varphi = 2\left(2a+\frac{k}{\sqrt{d_{(2)}}}\right)d_{(2)}, \tag{14.10}$$

where a and k are constants. For the *perfectly plastic body* obeying v. Mises' yield condition, (10.33) supplies

$$\varrho\varphi = 2k\sqrt{d_{(2)}}, \tag{14.11}$$

and for the *linear viscoelastic body* of the Kelvin type (11.47) yields

$$\varrho\varphi = 2\mu(d_{ij}d_{ij}-\tfrac{1}{3}d_{ii}d_{jj}), \tag{14.12}$$

provided the bulk viscosity is zero and μ denotes the viscosity coefficient. Indicidentally, it is easy to show (P) that (14.5) and (14.12) are equivalent.

Problem

Show that (14.5) and (14.12) are equivalent.

14.2. The thermodynamic forces

In order to establish a principle governing irreversible processes, it is necessary to return to Section 4.1 and to subject the argumentation presented there in connection with the fundamental laws to a critical review. For the sake of clearness the principal arguments will be repeated here in a slightly modified manner.

The object of Section 4.1 was a system whose state is completely determined by the kinematical parameters a_k ($k = 1, 2, \ldots n$) and the absolute temperature $\vartheta > 0$. These are the independent state variables; any function of them is a dependent state variable or state function. Examples are the internal energy $U(a_k, \vartheta)$, the entropy $S(a_k, \vartheta)$, and the free energy, defined by (4.17):

$$\Psi(a_k, \vartheta) = U - \vartheta S. \tag{14.13}$$

The first fundamental law (4.22),

$$\dot{U} = L + Q^* = A_k \dot{a}_k + Q^*, \tag{14.14}$$

states that the increase of the internal energy per unit time is equal to the sum of the rate of work L of the forces A_k corresponding to the parameters a_k, and the heat supply Q^* per unit time. According to the second fundamental law (4.24),

$$\dot{S} = \dot{S}^{(r)} + \dot{S}^{(i)}, \tag{14.15}$$

the rate of entropy increase consists of a reversible contribution $(4.25)_1$,

$$\dot{S}^{(r)} = \frac{Q^*}{\vartheta}, \tag{14.16}$$

called the entropy supply from outside, and a non-negative irreversible contribution $(4.25)_2$,

$$\dot{S}^{(i)} \geqslant 0, \tag{14.17}$$

denoted as the entropy production inside the system. If (14.17) holds with the equality sign, the process is called reversible, otherwise irreversible.

By means of the fundamental laws, the rate of work may be written as

$$L = A_k \dot{a}_k = \dot{U} - Q^* = \dot{U} - \vartheta \dot{S}^{(r)} = \dot{U} - \vartheta \dot{S} + \vartheta \dot{S}^{(i)} \tag{14.18}$$

or, since U and S are state functions, as

$$A_k \dot{a}_k = \left(\frac{\partial U}{\partial a_k} - \vartheta \frac{\partial S}{\partial a_k} \right) \dot{a}_k + \left(\frac{\partial U}{\partial \vartheta} - \vartheta \frac{\partial S}{\partial \vartheta} \right) \dot{\vartheta} + \vartheta \dot{S}^{(i)}. \tag{14.19}$$

Applying (14.19) to pure heating or cooling ($\dot{a}_k = 0$) and noting that (14.17) must be satisfied independent of the sign of $\dot{\vartheta}$, we conclude as in (4.10) that

$$\frac{\partial U}{\partial \vartheta} - \vartheta \frac{\partial S}{\partial \vartheta} = 0. \tag{14.20}$$

Since the expression on the left is a state function, (14.20) holds for arbitrary processes. In consequence, the last term in (14.19) has always the form of a rate of work and may be written as (4.28),

$$\vartheta \dot{S}^{(i)} = A_k^{(d)} \dot{a}_k = L^{(d)} \geqslant 0, \tag{14.21}$$

where $L^{(d)}$ is the dissipation rate and the expressions (4.13),

$$A_k^{(d)} = A_k - \frac{\partial U}{\partial a_k} + \vartheta \frac{\partial S}{\partial a_k}, \tag{14.22}$$

are the dissipative parts of the forces A_k. Equation (14.22) is equivalent to the decomposition (4.15),

$$A_k = A_k^{(d)} + A_k^{(q)}, \tag{14.23}$$

of the actual forces into their dissipative parts and the quasiconservative contributions (4.16),

$$A_k^{(q)} = \frac{\partial U}{\partial a_k} - \vartheta \frac{\partial S}{\partial a_k}. \tag{14.24}$$

On account of (14.13), equation (14.24) reduces to (4.20),

$$A_k^{(q)} = \frac{\partial \Psi}{\partial a_k}, \tag{14.25}$$

and from (14.13) and (14.20) we obtain (4.19),

$$S = -\frac{\partial \Psi}{\partial \vartheta}. \tag{14.26}$$

The quasiconservative forces (14.25) are state functions, whereas the dissipative forces are subject to the inequality (14.21) and hence depend also on the $\dot{a}_k$.

Classical thermodynamics has long been restricted to reversible processes. Here everything appears to be determined by the free energy $\Psi(a_k, \vartheta)$. In fact, the entropy follows from (14.26), and the internal energy from (14.13).

Since the entropy production is zero, there is no reason to introduce dissipative forces; the A_k are therefore considered as quasiconservative and hence are state variables following from (14.25).

At this point classical thermodynamics must be subjected to serious criticism, for our last statement contains a remarkable inaccuracy, which will prove to be significant four our further reasoning. We know from mechanics that in particular situations forces occur that are known as *gyroscopic*. They are dependent on the velocities, and they depend on them in such a way that their rate of work is always zero. Examples are the Coriolis force in a rotating reference frame or part of the Lorentz force in a magnetic field. If now, beside the $A_k^{(q)}$ defined by (14.25), such forces A_k', satisfying the identity

$$A_k' \dot{a}_k = 0 \tag{14.27}$$

were present in a thermodynamic system, they would not affect any of the equations characterizing a reversible process except for the fact that the $A_k^{(q)}$ would have to be replaced by $A_k^{(q)} + A_k'$. In other words: the whole framework of the theory of reversible processes would remain unaltered except for the presence of gyroscopic forces.

The forces A_k', which incidentally may be interpreted as limiting cases of either quasiconservative or dissipative forces, ought to be traceable by experiments. So far, however, no evidence of them has been reported. As a matter of fact, this type of force has been tacitly suppressed in classical thermodynamics. That this has remained without consequence cannot be explained by the current argument that reversible processes are infinitely slow and that gyroscopic forces are therefore negligible. Restriction to infinitely slow processes is not necessary in a thermodynamic field theory; besides, it has never been taken seriously in applications (e.g. in the design of combustion engines). We have to conclude instead that the reversible systems considered in classical thermodynamics are entirely determined by their free energy Ψ and that gyroscopic forces cannot be derived from Ψ in a physically relevant manner. They might occur in rotating reference frames or in magnetic fields since the angular velocity or the magnetic field force constitute new elements besides Ψ. The existence of gyroscopic forces is thus in principle possible, and it seems reasonable therefore to distinguish in the reversible case between *gyroscopic* and *nongyroscopic* systems according to the presence or absence of the A_k'. However, gyroscopic systems have never occurred in classical thermodynamics and thus appear to be extremely scarce.

Retracing the chain of reasoning from (14.13) to (14.26) under the assumption that the system is not reversible, we note that this time $\dot{S}^{(i)} > 0$. As a part of $\dot{S}$, the entropy production $\dot{S}^{(i)}$ is not a state variable but in general a function of $\dot{a}_k$, $\dot{\vartheta}$, a_k, and ϑ. As in (4.29), it is convenient to replace $\dot{S}^{(i)}$ by the product

$$\vartheta \dot{S}^{(i)} = \Phi(\dot{a}_k, \dot{\vartheta}, a_k, \vartheta) \geqslant 0 \tag{14.28}$$

denoted as the dissipation function. The quasiconservative forces have now to be supplemented by the dissipative forces $A_k^{(d)}$, subject on account of (14.21) and (14.28) to the condition

$$L^{(d)} = A_k^{(d)} \dot{a}_k = \Phi. \tag{14.29}$$

It follows that in general also the $A_k^{(d)}$ are functions of $\dot{a}_k$, $\dot{\vartheta}$, a_k, and ϑ.

Let us note that in the condensation given above of the original discussion in Section 4.1 the dissipation function Φ first appears in the form $\vartheta \dot{S}^{(i)}$ in equation (14.18). Being the product of temperature and rate of entropy production, this function has a definite physical meaning. The dissipative forces $A_k^{(d)}$ were introduced subsequently by means of (14.21) in order to interpret Φ as a rate of work. However, these forces are far from being completely defined by (14.21) or (14.29); a gyroscopic component A_k' as discussed in connection with (14.27) remains indetermined. We have already pointed out that such forces A_k' cannot be derived from the free energy. Thus, the simplest possible assumption is that the A_k' and hence the $A_k^{(d)}$ are determined by the only element characterizing the transition from the reversible to the irreversible case: the dissipation function. In fact, any force not determined by Ψ or Φ would have to be derived from a foreign element, e.g. from an angular velocity, a magnetic field force, or a velocity potential as suggested by several authors [27–29]. However, the first two examples have already been recognized as exceptional cases tacitly suppressed in classical thermodynamics, and the third example suffers from the fact that a velocity potential has no physical meaning. Besides, all of these foreign elements define forces which are not typical for irreversible systems; they might as well appear in reversible ones.

Generalizing a concept introduced above, it appears reasonable to denote an irreversible system as *nongyroscopic* or *purely dissipative* whenever the $A_k^{(d)}$ are derivable from the dissipation function. This definition excludes any element except Ψ and Φ for the determination of the forces. The arguments presented above make it extremely probable that the systems which are the objects of this book are purely dissipative; in any event, this assumption is as well justified as the long accepted restriction of classical thermodynamics to nongyroscopic processes.

A model to be treated in Section 15.2 renders it plausible that a particular irreversible process (in fact, the one we are primarily interested in) is purely dissipative. For the present, let us consider the absence of gyroscopic forces as a postulate, and let us note that it reverses the roles of the dissipation function Φ and the dissipative forces $A_k^{(d)}$. The conventional approach considers the $A_k^{(d)}$ as the primary elements and Φ as a function of secondary importance, obtained as a particular form of the dissipation rate. Our own point of view is the exact opposite: we accept Φ as the primary function, and we are now faced with the problem of deriving the $A_k^{(d)}$ from it.

14.3. The orthogonality condition

Let us consider a purely dissipative system, assuming that the variables a_k, ϑ, and $\dot{\vartheta}$ are prescribed, whereas the $\dot{a}_k$ are free. The dissipative forces are thus functions of the $\dot{a}_k$ alone.

For simplicity, we start with the case where the velocities define a vector $\dot{a}_k$ in physical space (e. g. the heat flow vector). The corresponding dissipative forces $A_k^{(d)}$ define another vector to be determined by the dissipation function. So far, we merely have the condition (14.29),

$$A_k^{(d)}(\dot{a}_i)\dot{a}_k = \Phi(\dot{a}_i), \tag{14.30}$$

and we have already observed that this condition does not determine the vector $A_k^{(d)}$. It determines its magnitude once the direction, i.e. the ratio of the components, is known. To obtain this direction, we assume that in the vicinity of any point $\dot{a}_i$ in velocity space the dissipation function admits a power series expansion

$$\Phi(\dot{a}_i+\delta\dot{a}_i) = C+C_k\,\delta\dot{a}_k+C_{kl}\,\delta\dot{a}_k\,\delta\dot{a}_l+\ \ldots . \tag{14.31}$$

Obviously, the scalar $C = \Phi(\dot{a}_i)$ cannot determine the direction of $A_i^{(d)}$. In consequence, the behavior of Φ in the vicinity of $\dot{a}_i$, i.e. the entire expansion (14.31), must be relevant. This expansion is determined by the coefficients $C, C_k, C_{kl}, \ldots$. The only vector among them is the gradient

$$C_k = \frac{\partial\Phi}{\partial\dot{a}_k}(\dot{a}_i). \tag{14.32}$$

It follows that

$$A_k^{(d)} = \nu \frac{\partial \Phi}{\partial \dot{a}_k}, \tag{14.33}$$

where ν is a proportionality factor determined on account of (14.30) by

$$\nu = \Phi\left(\frac{\partial \Phi}{\partial \dot{a}_l}\, \dot{a}_l\right)^{-1}. \tag{14.34}$$

One might be tempted to admit as vectors certain contracted products of the tensors represented by the expansion coefficients, e.g.

$$C_k C_{kl} = \tfrac{1}{2} \frac{\partial \Phi}{\partial \dot{a}_k} \frac{\partial^2 \Phi}{\partial \dot{a}_k\, \partial \dot{a}_l}. \tag{14.35}$$

We have however seen in Chapter 12 that we are not restricted to cartesian coordinate systems when dealing with vectors. We might as well use a skew system. Interpreting the velocity as a contravariant vector, we ought to denote it by $\dot{a}^k$. Since the dissipation function is a scalar, it follows from (14.30) that $A_k^{(d)}$ is a covariant vector. The expansion (14.31) should actually be written

$$\Phi(\dot{a}^i + \delta \dot{a}^i) = C + C_k\, \delta \dot{a}^k + C_{kl}\, \delta \dot{a}^k\, \delta \dot{a}^l + \ldots, \tag{14.36}$$

where the coefficients are covariant tensors. It is obvious that products like (14.35) are not vectors and hence must be discarded.

Returning to the subscript notation, we observe that geometrically (Fig. 14.1) the dissipation function $\Phi(\dot{a}_k)$ may be represented in the velocity

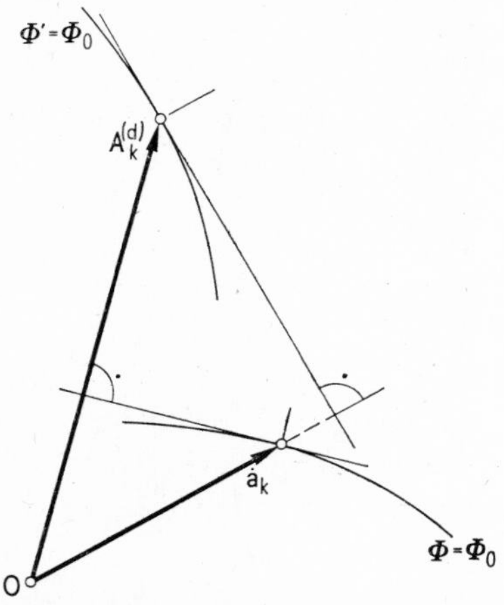

Fig. 14.1 Orthogonality of the dissipative forces and the velocities

space $\dot{a}_k$ by means of *dissipation surfaces* $\Phi = \Phi_0 =$ const and that equation (14.33) may be interpreted as an *orthogonality condition*: the dissipative force $A_k^{(d)}$ corresponding to a velocity $\dot{a}_k$ is orthogonal to the dissipation surface $\Phi = \Phi_0$ in the end point of $\dot{a}_k$.

The solution $A_k^{(d)}(\dot{a}_i)$ of (4.33) and (4.34) is single-valued provided the surface $\Phi = \Phi_0$ is smooth in the end point of $\dot{a}_k$. This is the case if the expansion (14.31) exists. On an arbitrary surface $\Phi = \Phi_0$ corresponding increments of $\dot{a}_k$ and $A_k^{(d)}$ are connected by

$$A_k^{(d)}\,\delta\dot{a}_k + \dot{a}_k\,\delta A_k^{(d)} = 0 \quad (\delta\Phi = 0). \tag{14.37}$$

Besides, it is obvious that the orthogonality condition (14.33) may be written in the form

$$A_k^{(d)}\,\delta\dot{a}_k = 0 \quad (\delta\Phi = 0). \tag{14.38}$$

Thus,

$$\dot{a}_k\,\delta A_k^{(d)} = 0 \quad (\delta\Phi = 0). \tag{14.39}$$

Equations (14.38) and (14.39) suggest a duality between velocities and dissipative forces. In fact, if the function $A_k^{(d)}(\dot{a}_i)$ is reversible, the dissipation function can be written in terms of the $A_k^{(d)}$:

$$\Phi'(A_k^{(d)}) = \Phi(\dot{a}_k). \tag{14.40}$$

Since $\delta\Phi = 0$ is equivalent to $\delta\Phi' = 0$, (14.39) may be replaced by

$$\dot{a}_k\,\delta A_k^{(d)} = 0 \quad (\delta\Phi' = 0). \tag{14.41}$$

We thus obtain a *corollary* (Fig. 14.1) of the orthogonality condition stated above: the velocity $\dot{a}_k$ corresponding to a dissipative force $A_k^{(d)}$ is orthogonal to the dissipation surface $\Phi' = \Phi_0$ in the end point of $A_k^{(d)}$.

Provided $\Phi'(A_k^{(d)})$ is differentiable, this second orthogonality condition may be represented by

$$\dot{a}_k = \nu' \frac{\partial\Phi'}{\partial A_k^{(d)}}, \tag{14.42}$$

where

$$\nu' = \Phi'\left(\frac{\partial\Phi'}{\partial A_l^{(d)}}\, A_l^{(d)}\right)^{-1}. \tag{14.43}$$

The function $\dot{a}_k(A_i^{(d)})$ is single-valued provided the surface $\Phi' = \Phi_0$ is smooth in the end point of $A_k^{(d)}$.

If the velocities define a symmetric tensor of the second order in physical space (e.g. a deformation rate), it is convenient to denote them by $\dot{a}_{kl}$. The corresponding dissipative forces $A_{kl}^{(d)}$ define another symmetric second-order tensor. Equation (14.30) now reads

$$A_{kl}^{(d)}(\dot{a}_{ij})\dot{a}_{kl} = \Phi(\dot{a}_{ij}), \tag{14.44}$$

and the expansion (14.31) has the form

$$\Phi(\dot{a}_{ij}+\delta\dot{a}_{ij}) = C + C_{kl}\,\delta\dot{a}_{kl} + C_{klmn}\,\delta\dot{a}_{kl}\,\delta\dot{a}_{mn} + \dots \,. \tag{14.45}$$

Writing it in terms of covariant and contravariant tensors, we arrive at a similar conclusion as in connection with (14.32): the only second-order tensor defined by the expansion coefficients is

$$C_{kl} = \frac{\partial\Phi}{\partial\dot{a}_{kl}}(\dot{a}_{ij}). \tag{14.46}$$

It follows that

$$A_{kl}^{(d)} = \nu\frac{\partial\Phi}{\partial\dot{a}_{kl}}, \tag{14.47}$$

where on account of (14.44)

$$\nu = \Phi\left(\frac{\partial\Phi}{\partial\dot{a}_{mn}}\dot{a}_{mn}\right)^{-1}. \tag{14.48}$$

If the $\dot{a}_{kl}$ are interpreted as the components of a vector in a space of six dimensions, the function $\Phi(\dot{a}_{kl})$ defines a family of hypersurfaces. They may still be denoted as dissipation surfaces, and it is obvious that (14.47) is the corresponding orthogonality condition for the dissipative forces. The alternate form (14.38) of the condition reads

$$A_{kl}^{(d)}\,\delta\dot{a}_{kl} = 0 \quad (\delta\Phi = 0), \tag{14.49}$$

and the orthogonality condition for the velocities is obtained from (14.47), (14.48), or (14.49) if we exchange the roles of $\dot{a}_{kl}$ and $A_{kl}^{(d)}$, replacing at the same time the function $\Phi(\dot{a}_{kl})$ by the dissipation function $\Phi'(A_{kl}^{(d)})$ in the space of the irreversible forces. In this context it is sometimes convenient to denote the a_{kl} by a single subscript, writing $a_{11} = a_1, \dots, a_{23} = a_4, a_{31} = a_5, a_{12} = a_6$. This has the advantage that the original equations (14.30) through (14.43) can be retained with the understanding that the subscript now assumes all values from 1 through 6. We will make frequent use of this possibility.

In the special case where Φ is a quadratic function of the velocities, it may be written as

$$\Phi = \gamma_{kl}\dot{a}_k\dot{a}_l . \tag{14.50}$$

Decomposing the tensor γ_{kl} into its symmetric and antimetric parts, we note that the second one does not contribute to the function Φ, so that

$$\Phi = \gamma_{(kl)}\dot{a}_k\dot{a}_l . \tag{14.51}$$

The dissipation surfaces are ellipsoids or hyperellipsoids, similar to each other and similarly situated. Equation (14.34) yields $\nu = \frac{1}{2}$, and (14.33) becomes

$$A_k^{(d)} = \frac{1}{2}\frac{\partial \Phi}{\partial \dot{a}_k} = \gamma_{(kl)}\dot{a}_l . \tag{14.52}$$

It thus follows from the orthogonality condition that the tensor establishing the connection between velocities and dissipative forces is symmetric. A possible additional force

$$A_k' = \gamma_{[kl]}\dot{a}_l \tag{14.53}$$

obtained from the antimetric part of γ_{kl} would be gyroscopic. In the three-dimensional case it might be written as

$$A_k' = e_{klm}\omega_m\dot{a}_l = -e_{kml}\omega_m\dot{a}_l , \tag{14.54}$$

where the vector ω_m is dual [cf (1.33)] to the tensor $\gamma_{[kl]}$. The symbolic form of (14.54),

$$\boldsymbol{A}' = -\boldsymbol{\omega} \times \dot{\boldsymbol{a}}, \tag{14.55}$$

shows that the force $\boldsymbol{A}'$ would depend on a vector $\boldsymbol{\omega}$ which itself is independent of Φ. It would appear already in the reversible case and cannot occur in a purely dissipative system.

The orthogonality condition has been proved here under the assumption that the velocities are the components of a single tensor of the first or second order. The extension to tensors of higher order is straightforward. In all these cases the velocities considered are subject to a single transformation law involving all of them. We will refer to systems or processes of this type as *elementary* and we will say that they are characterized by a set of *coherent velocities*.

Let us note that, provided it holds in an n-dimensional space, the orthogonality condition (14.38) is also valid in any linear subspace containing the origin. In fact, starting from

$$A_1^{(d)}\,\delta\dot{a}_1 + \ldots + A_n^{(d)}\,\delta\dot{a}_n = 0 \quad [\delta\Phi(\dot{a}_1, \ldots, \dot{a}_n) = 0] \tag{14.56}$$

and restricting ourselves to the r-dimensional subspace obtained by putting $\dot{a}_{r+1} = \ldots = \dot{a}_n = 0$ $(0 < r < n)$, we have

$$A_1^{(d)}\,\delta\dot{a}_1 + \ldots + A_r^{(d)}\,\delta\dot{a}_r = 0 \quad [\delta\Phi(\dot{a}_1, \ldots, \dot{a}_r, 0, \ldots, 0) = 0] \tag{14.57}$$

where the side condition, by means of the notation

$$\Phi(\dot{a}_1, \ldots, \dot{a}_r, 0, \ldots, 0) = \Phi^*(\dot{a}_1, \ldots, \dot{a}_r), \tag{14.58}$$

may be written as

$$\delta\Phi^*(\dot{a}_1, \ldots, \dot{a}_r) = 0. \tag{14.59}$$

The reduction considered here occurs, e.g., in plane heat flow or in material flows characterized by a plane deformation rate.

If, on the other hand, the velocities characterizing a purely dissipative process are the components of more than one tensor (e.g., the components of the heat flow vector and the deformation rate), the transformation law does not appear as a single set of equations containing all velocity components; it consists of transformations connecting the components of each tensor separately. In this case we will say that the velocities form different coherent sets which as a whole are *incoherent*, and we will refer to the system as *complex*. Its dissipation function Φ depends on all velocities, and it is in general not possible to decompose the system into elementary subsystems with well-defined dissipation functions $\Phi^{(1)}, \Phi^{(2)}, \ldots, \Phi^{(m)}$ so that

$$\Phi = \Phi^{(1)} + \Phi^{(2)} + \ldots + \Phi^{(m)}. \tag{14.60}$$

There is at present no reason to assume that the orthogonality condition holds for the entire set of velocities even if it is valid in case only one of the coherent velocity sets is different from zero. It is to be expected that the dissipative forces depend on all velocities, but in general we have no means of determining them. Cases of this type occur if various elementary processes as, e.g. heat flow and deformation, are *coupled*.

In the particular case of *uncoupled* processes characterized by dissipation functions $\Phi^{(1)}, \Phi^{(2)}, \ldots \Phi^{(m)}$ dependent only on the corresponding velocity sets, the decomposition (14.60) is valid. We will refer to systems or processes of this type as *compound*. The dissipative forces can be obtained by applying the orthogonality condition in turn to each of the partial dissipation functions, and it follows that they depend only on the corresponding velocities. Obviously the orthogonality condition does not generally hold for the entire system. It is true that from

$$A_1^{(d)}\,\delta\dot{a}_1 + \ldots + A_r^{(d)}\,\delta\dot{a}_r = 0 \quad [\delta\Phi^{(1)}(\dot{a}_1, \ldots, \dot{a}_r) = 0] \tag{14.61}$$

and

$$A_{r+1}^{(d)}\,\delta\dot{a}_{r+1} + \ldots + A_n^{(d)}\,\delta\dot{a}_n = 0 \quad [\delta\Phi^{(2)}(\dot{a}_{r+1}, \ldots, \dot{a}_n) = 0] \tag{14.62}$$

one obtains

$$A_1^{(d)}\,\delta\dot{a}_1 + \ldots + A_n^{(d)}\,\delta\dot{a}_n = 0 \quad (\delta\Phi^{(1)} = \delta\Phi^{(2)} = 0). \tag{14.63}$$

However, the two side conditions in (14.63) are more restrictive than the condition

$$\delta\Phi = \delta\Phi^{(1)} + \delta\Phi^{(2)} = 0 \tag{14.64}$$

which would imply orthogonality in the n-dimensional velocity space. Incidentally, since the elementary processes forming part of a compound process are independent of each other, the second fundamental law requires that each of the partial dissipation functions on the right of (14.60) be non-negative.

We thus arrive at the conclusion that the orthogonality condition holds for elementary processes but not necessarily for compound or complex ones.

14.4. Complex processes

Let us consider a *complex* process involving two coherent velocity sets in the form of two vectors $\dot{a}_i$ and $\dot{b}_j$ defined in two vector spaces of possibly different dimension. Denoting the corresponding dissipative forces by $A_i^{(d)}$ and $B_j^{(d)}$ respectively, we have

$$A_k^{(d)}(\dot{a}_i, \dot{b}_j)\dot{a}_k + B_l^{(d)}(\dot{a}_i, \dot{b}_j)\dot{b}_l = \Phi(\dot{a}_i, \dot{b}_j) \tag{14.65}$$

in place of (14.30). Assuming that the dissipation function admits a power series expansion

$$\Phi(\dot{a}_i + \delta\dot{a}_i, \dot{b}_j + \delta\dot{b}_j) = C + C_k\,\delta\dot{a}_k + D_k\,\delta\dot{b}_k + C_{kl}\,\delta\dot{a}_k\,\delta\dot{a}_l + D_{kl}\,\delta\dot{a}_k\,\delta\dot{b}_l + E_{kl}\,\delta\dot{b}_k\,\delta\dot{b}_l + \ldots \tag{14.66}$$

in the vicinity of the point $\dot{a}_i$, $\dot{b}_j$ in the combined velocity space and noting that the expansion is determined by the coefficients C, C_k, D_k, . . ., we conclude as in Section 14.3 that the only vectors among them are the gradients

$$C_i = \frac{\partial \Phi}{\partial \dot{a}_i}, \qquad D_j = \frac{\partial \Phi}{\partial \dot{b}_j}. \tag{14.67}$$

The numbers of their components are generally different. Even if they are equal, the velocities $\dot{a}_i$ and $\dot{b}_j$ may be referred to two different coordinate systems. It follows that

$$A_i^{(d)} = \nu^{(1)} \frac{\partial \Phi}{\partial \dot{a}_i}, \qquad B_j^{(d)} = \nu^{(2)} \frac{\partial \Phi}{\partial \dot{b}_j}, \tag{14.68}$$

where $\nu^{(1)}$ and $\nu^{(2)}$ are two scalars, generally dependent on all velocities and connected according to (14.65) by

$$\nu^{(1)} \frac{\partial \Phi}{\partial \dot{a}_i} \dot{a}_i + \nu^{(2)} \frac{\partial \Phi}{\partial \dot{b}_j} \dot{b}_j = \Phi. \tag{14.69}$$

Since (14.69) does not determine $\nu^{(1)}$ and $\nu^{(2)}$ separately, this confirms the statement made in Section 14.3 that in general the dissipation function provides no sufficient means of determining the dissipative forces in a complex process.

In the special case of a *compound* process the dissipation function has the form

$$\Phi(\dot{a}_i, \dot{b}_j) = \Phi^{(1)}(\dot{a}_i) + \Phi^{(2)}(\dot{b}_j). \tag{14.70}$$

Equation (14.69) therefore decomposes into

$$\nu^{(1)} \frac{\partial \Phi^{(1)}}{\partial \dot{a}_i} \dot{a}_i = \Phi^{(1)}, \qquad \nu^{(2)} = \frac{\partial \Phi^{(2)}}{\partial \dot{b}_j} \dot{b}_j = \Phi^{(2)}. \tag{14.71}$$

Thus, the forces $A_i^{(d)}$ and $B_j^{(d)}$ follow from the orthogonality condition, applied in turn to each of the two elementary processes. However, since in general $\nu^{(1)} \neq \nu^{(2)}$, our observation in Section 14.3 that the orthogonality condition need not hold for the entire process is confirmed.

The results just obtained are readily extended to more than two coherent velocity sets. Moreover, there are a few special cases where the dissipative forces are determined even in a complex process and where the orthogonality condition remains valid for the entire set of velocities. If, e.g., Φ is a *homogeneous function* of degree s in the arguments $\dot{a}_i$ and $\dot{b}_j$, Euler's equation reads

$$\frac{\partial \Phi}{\partial \dot{a}_i}\dot{a}_i+\frac{\partial \Phi}{\partial \dot{b}_j}\dot{b}_j = s\Phi, \tag{14.72}$$

and by comparing this to (14.69), we obtain the constant values $\nu^{(1)} = \nu^{(2)} = 1/s$. From (14.68) we get

$$A_i^{(d)} = \frac{1}{s}\frac{\partial \Phi}{\partial \dot{a}_i}, \qquad B_j^{(d)} = \frac{1}{s}\frac{\partial \Phi}{\partial \dot{b}_j}. \tag{14.73}$$

It should be noted, however, that Φ depends on all velocities $\dot{a}_i$ and $\dot{b}_j$ except for the case of a compound process, where $\Phi(\dot{a}_i, \dot{b}_j)$ may be replaced by $\Phi^{(1)}(\dot{a}_i)$ in $(14.73)_1$ and by $\Phi^{(2)}(\dot{b}_j)$ in $(14.73)_2$. In any case, equations (14.73) represent the orthogonality condition for the entire system. Again, the result is easily extended to more than two coherent velocity sets.

The most important case of this type is the one where the dissipation function is a quadratic form in the velocities $\dot{a}_i$ and $\dot{b}_j$,

$$\Phi = \gamma_{ij}\dot{a}_i\dot{a}_j+2\gamma'_{ij}\dot{a}_i\dot{b}_j+\gamma''_{ij}\dot{b}_i\dot{b}_j, \tag{14.74}$$

where $\gamma_{ji} = \gamma_{ij}$ and $\gamma''_{ji} = \gamma''_{ij}$. Here (14.68) yields

$$A_i^{(d)} = 2\nu^{(1)}(\gamma_{ij}\dot{a}_j+\gamma'_{ij}\dot{b}_j), \qquad B_j^{(d)} = 2\nu^{(2)}(\gamma'_{ij}\dot{a}_i+\gamma''_{ij}\dot{b}_i). \tag{14.75}$$

Upon substitution of (14.75) and (14.74) in (14.65) we obtain $\nu^{(1)} = \nu^{(2)} = \frac{1}{2}$ and hence

$$A_i^{(d)} = \gamma_{ij}\dot{a}_j+\gamma'_{ij}\dot{b}_j, \qquad B_j^{(d)} = \gamma'_{ij}\dot{a}_i+\gamma''_{ij}\dot{b}_i. \tag{14.76}$$

The connection between velocities and dissipative forces is thus linear, and it is easy to see that the corresponding matrix is symmetric. Incidentally, if the process is compound, the γ'_{ij} are zero.

As an example, let us consider the case where the vectors a_i and b_j have two and three components respectively. It occurs, e.g. in the plane problem of a heat flow combined with a deformation, where $a_1 = q_1$, $a_2 = q_2$ are the components of the heat flow vector and $b_1 = d_{11}$, $b_2 = d_{12}$, $b_3 = d_{22}$ those of the deformation rate. The corresponding dissipative forces per unit volume are $A_1^{(d)} = -\vartheta_{,1}/\vartheta$, $A_2^{(d)} = -\vartheta_{,2}/\vartheta$, and $B_1^{(d)} = \sigma_{11}^{(d)}$, $B_2^{(d)} = \sigma_{12}^{(d)}$, $B_3^{(d)} = \sigma_{22}^{(d)}$.

On account of (14.76) they are given by

$$\begin{aligned}
A_1^{(d)} &= \gamma_{11}\dot{a}_1+\gamma_{12}\dot{a}_2+\gamma'_{11}\dot{b}_1+\gamma'_{12}\dot{b}_2+\gamma'_{13}\dot{b}_3\,,\\
A_2^{(d)} &= \gamma_{21}\dot{a}_1+\gamma_{22}\dot{a}_2+\gamma'_{21}\dot{b}_1+\gamma'_{22}\dot{b}_2+\gamma'_{23}\dot{b}_3\,,\\
B_1^{(d)} &= \gamma'_{11}\dot{a}_1+\gamma'_{21}\dot{a}_2+\gamma''_{11}\dot{b}_1+\gamma''_{21}\dot{b}_2+\gamma''_{31}\dot{b}_3\,,\\
B_2^{(d)} &= \gamma'_{12}\dot{a}_1+\gamma'_{22}\dot{a}_2+\gamma''_{12}\dot{b}_1+\gamma''_{22}\dot{b}_2+\gamma''_{32}\dot{b}_3\,,\\
B_3^{(d)} &= \gamma'_{13}\dot{a}_1+\gamma'_{23}\dot{a}_2+\gamma''_{13}\dot{b}_1+\gamma''_{23}\dot{b}_2+\gamma''_{33}\dot{b}_3\,.
\end{aligned} \tag{14.77}$$

In view of the relations $\gamma_{ji} = \gamma_{ij}$ and $\gamma''_{ji} = \gamma''_{ij}$ the symmetry of the matrix is obvious. In a compound process the matrix elements γ'_{ij} are zero.

In thermodynamics the symmetry of the matrix connecting the velocities and the dissipative forces in (14.76) or, in particular, in (14.77) is known as Onsager's theory [30]; it is proved there by statistical means. The orthogonality condition (14.33), (14.34) is a generalization of the so-called reciprocity relations of Onsager for nonlinear elementary processes. So far, however, we have no indication that it also applies to complex systems unless the dissipation function is homogeneous in its arguments. For compound systems, where there is no coupling between the elementary partial processes, it is not generally valid. This can be shown (P3) by combining two elementary processes the dissipation functions of which are homogeneous of different degrees.

Problems

1. Apply (14.76) to two vectors a_i and b_j in a three-dimensional space and verify the symmetry of the matrix connecting the $A_i^{(d)}$ and $B_j^{(d)}$ with $\dot{a}_i$ and $\dot{b}_j$.

2. Supplement the dissipation function (14.74) by the third-degree term

$$\gamma_{ijk}\dot{a}_i\dot{a}_j\dot{a}_k+\gamma'_{ijk}\dot{a}_i\dot{a}_j\dot{b}_k+\gamma''_{ijk}\dot{a}_i\dot{b}_j\dot{b}_k+\gamma'''_{ijk}\dot{b}_i\dot{b}_j\dot{b}_k\,,$$

where γ_{ijk} and γ'''_{ijk} are symmetric with respect to any pair of subscripts, while $\gamma'_{jik} = \gamma'_{ijk}$ and $\gamma''_{ikj} = \gamma''_{ijk}$. Confirm that $\nu^{(1)}$ and $\nu^{(2)}$ cannot be determined separately.

3. Use the result of the last problem to show that a compound process with elementary dissipation functions that are homogeneous of the second and third degree respectively, does not obey the orthogonality condition as a whole.

14.5. Dissipation surfaces

In this section we will restrict ourselves to systems obeying the orthogonality condition, e.g. to elementary systems or to complex ones with a homogeneous dissipation function. Thus, equations (14.33), (14.34) are valid and map the velocity space onto the space of the dissipative forces. The inversion of this mapping is given by (14.42) and (14.43). Provided both mappings are single-valued, the dissipation surfaces in the two spaces have a few remarkable properties, which are usually introduced as postulates but are actually consequences of the orthogonality condition [31].

On account of (14.30) and the analogous equation

$$\dot{a}_k(A_i^{(d)})A_k^{(d)} = \Phi'(A_i^{(d)}) \tag{14.78}$$

the assumption that the functions $A_k^{(d)}(\dot{a}_i)$ and $\dot{a}_k(A_i^{(d)})$ are single-valued implies that also the dissipation functions $\Phi(\dot{a}_i)$ and $\Phi'(A_i^{(d)})$ are single-valued in their respective arguments; according to (14.33) and (14.42) they are also differentiable and hence continuous.

Let us restrict ourselves at present to three dimensions; let s in Figure 14.2 denote an arbitrary ray emanating from the origin O in velocity space,

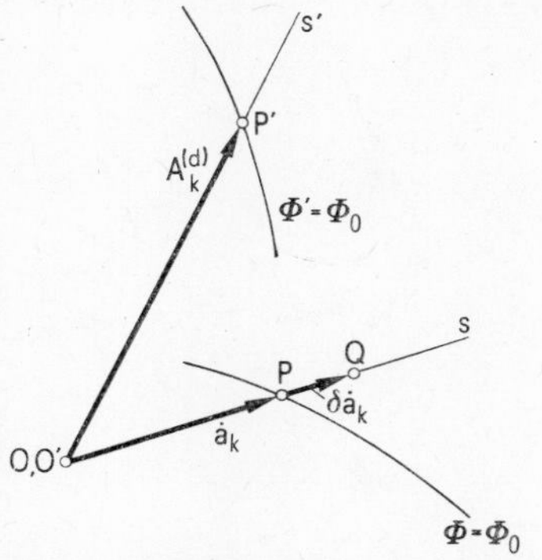

Fig. 14.2 Increase of the dissipation functions on rays s and s'

and let s' denote an arbitrary ray from O′ in the space of the dissipative forces. The functions $\Phi(\dot{a}_k)$ and $\Phi'(A_k^{(d)})$ are nonnegative and vanish at O and O′ respectively. Within sufficiently small vicinities of these points they must be positive since the mapping $A_k^{(d)}(\dot{a}_i)$ and its inverse are assumed to be single-valued. Both functions are thus positive definite, and it follows that surfaces $\Phi(\dot{a}_k) = \Phi_0 > 0$ and $\Phi'(A_k^{(d)}) = \Phi_0 > 0$ can be found containing the origin and with the additional property that inside these surfaces the dissipation function has a positive gradient on any ray s or s' respectively, and hence increases in a strongly monotonic manner. In consequence these surfaces and in fact all dissipation surfaces contained by them have each a single point in common with any ray s, s' respectively. We refer to this property by calling the surfaces *star-shaped* with respect to the origin, and we note that they are also *ordered* in the sense that each of them encloses those with smaller values of the dissipation function.

In Section 10.3 we have defined the supporting lines of a closed curve. Generalizing this concept, we define a *supporting plane* of a closed surface as a plane containing at least one point of the surface but none of the points in its interior. Since the dissipation function is differentiable, the dissipation surfaces are smooth; thus, any supporting plane is a tangential plane. If also the reverse is true, i.e. if any tangential plane is a supporting plane, the surface is referred to as *convex*. Provided the surface has only a single point in common with every supporting plane, it is called *strongly*, otherwise *weakly convex*.

On account of the orthogonality condition, the dissipation surfaces $\Phi = \Phi_0$ and $\Phi' = \Phi_0$ discussed above are convex. If this were not true, e.g. for $\Phi' = \Phi_0$, the surface (Fig. 14.3) would have different tangential planes

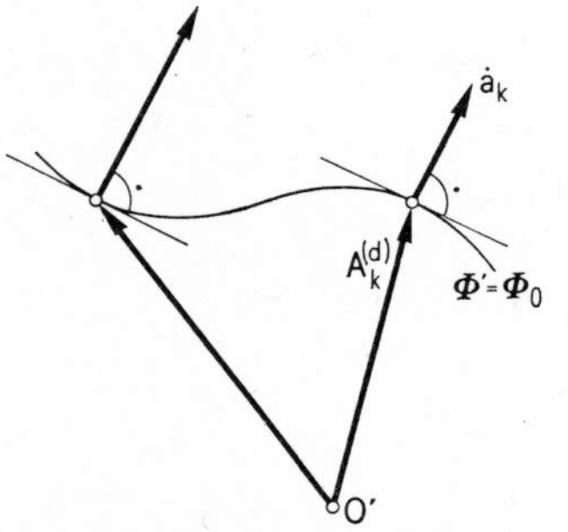

Fig. 14.3 Convexity of the dissipation surface

with the same outward normal. According to (14.42) and (14.43), the corresponding dissipative forces would belong to velocities of equal direction but different magnitude. This is inconsistent with the fact that the surface $\Phi = \Phi_0$ in velocity space is star-shaped. The dissipation surfaces are even strongly convex. If, e.g. $\Phi' = \Phi_0$ were merely weakly convex, the various points of contact with a supporting plane would define different forces $A_k^{(d)}$

corresponding to the same velocity $\dot{a}_k$. This is inconsistent with the fact that the function $A_k^{(d)}(\dot{a}_k)$ is single-valued.

So far our discussion was restricted to surfaces $\Phi = \Phi_0$ and $\Phi' = \Phi_0$ lying within sufficiently small vicinities of the origins O or O′ respectively. In order to see that the results remain valid for arbitrary dissipation surfaces, let us increase the value of Φ_0. As long as $\Phi(\dot{a}_k)$ and $\Phi'(A_k^{(d)})$ increase monotonically on every ray s or s' respectively, our results still hold. In particular, the dissipation surfaces are strongly convex. Let us tentatively assume that for a certain finite value Φ_0^* of Φ_0 the dissipation function $\Phi(\dot{a}_k)$ ceases to increase on a certain ray s. The surface $\Phi = \Phi_0^*$ is then still convex, but it contains at least one point P so that $\delta\Phi = 0$ for an infinitesimal step $\delta\dot{a}_k$ away from O on s. The vector $\delta\dot{a}_k$ thus lies on the dissipation surface $\Phi = \Phi_0^*$, and it follows that its tangential plane contains the origin O. However, a tangential plane passing through O cannot be a supporting plane of the surface $\Phi = \Phi_0^*$; in consequence, the surface $\Phi = \Phi_0^*$ cannot be convex as stated above. It follows that $\delta\Phi \neq 0$ everywhere on s outside O, and the continuity of Φ finally requires $\delta\Phi > 0$.

A similar argument applies for the function $\Phi'(A_k^{(d)})$. We have thus proved that the dissipation surfaces in both spaces are star-shaped with respect to the respective origins, that they are convex and ordered in the sense that each one of them encloses those with smaller values of the dissipation function. Combining this result with the second fundamental law we see that, with reference to the dissipation surface passing through the end point of $\dot{a}_k$, the vector $A_k^{(d)}$ has the direction of the outward normal and vice versa.

In Figure 14.2 $\delta\dot{a}_k$ denotes an infinitesimal step on s leading from the point P on the surface $\Phi = \Phi_0$ to an exterior point Q. In P (14.30) is valid in the form

$$A_k^{(d)}\dot{a}_k = \Phi(\dot{a}_k) > 0. \tag{14.79}$$

We have shown that the step from P to Q corresponds to an increase of Φ; thus,

$$\delta\Phi = A_k^{(d)}\,\delta\dot{a}_k + \dot{a}_k\,\delta A_k^{(d)} > 0. \tag{14.80}$$

Since $\delta\dot{a}_k$ has the direction of $\dot{a}_k$, (14.79) implies that the first term in (14.80) is positive. The same is true for the second term since on account of the monotonic increase of $\Phi'(A_k^{(d)})$ the image Q′ of Q lies outside the surface $\Phi' = \Phi_0$. We thus conclude that

$$\delta\Phi > A_k^{(d)}\,\delta\dot{a}_k\,, \tag{14.81}$$

i.e. that the increase of $\Phi(\dot{a}_k)$ on any ray s is stronger than linear. An analogous result holds for the increase of $\Phi'(A_k^{(d)})$ on s'.

The arguments presented here are based on the concepts of three-dimensional geometry ($n = 3$). However, the results obtained remain valid for arbitrary positive values of n. To see this, we merely have to interpret the surfaces in the velocity space and the space of the irreversible forces as *hypersurfaces* and their tangential planes as *hyperplanes* having n-fold points in common with the hypersurfaces.

Closing this section, let us repeat that the results are based on the orthogonality condition and that they are restricted to single-valued mappings $A_k^{(d)}(\dot{a}_i)$ and $\dot{a}_k(A_i^{(d)})$. There is a case of considerable practical interest where the last condition is not satisfied. It occurs in plasticity and will be treated in Section 17.1.

CHAPTER 15

MAXIMAL DISSIPATION

In this chapter we will first show that the orthogonality condition established in Section 14.3 is equivalent to certain extremum principles. The most interesting among them is a principle of maximal dissipation rate. It may also be stated as a principle of maximal rate of entropy production, and it suggests a generalization of the orthogonality condition.

In a second step we will study the dissipation mechanism in a microsystem serving as a model for plastic bodies. This model makes it plausible that the extremum principles to be established in Section 15.1 are direct consequences of micro-motion.

Finally, the implications of the orthogonality principle for the constitutive equations of some of the bodies treated in the preceding chapters will be discussed.

15.1. Extremum principles

Physical statements assume their invariant and hence most general forms when they are expressed in terms of extremum principles. The statement in which we are interested here is the orthogonality condition, valid in particular for elementary systems of the purely dissipative type. There are various ways of expressing this statement in the form of an extremum principle. To discuss them, let us start from the observation made at the end of Section 14.2 that from our point of view the dissipation function is the key to the dissipative forces. We therefore assume that the function $\Phi(\dot{a}_k)$ is known and that the regularity conditions mentioned in Section 14.3 are satisfied.

Figure 15.1 shows a dissipation surface $\Phi = \Phi_0$ in a three-dimensional velocity space with various velocity vectors ending on it. On account of

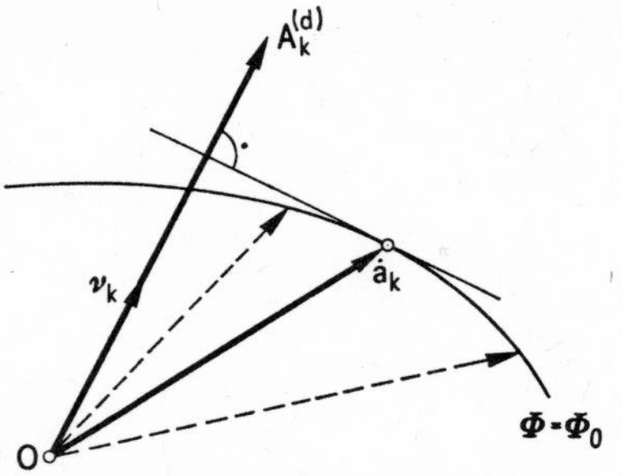

Fig. 15.1 Proof of the principle of least dissipative force

(14.29), these vectors and their corresponding irreversible forces are subject to the condition

$$A_k^{(d)}\dot{a}_k = \Phi_0 > 0. \tag{15.1}$$

Let us now prescribe the direction of the irreversible force $A_k^{(d)}$ by means of the unit vector ν_k, so that

$$A_k^{(d)} = M\nu_k\,, \tag{15.2}$$

where M denotes the magnitude of the irreversible force. It follows from (15.1) and (15.2) that

$$M\nu_k\dot{a}_k = \Phi_0 > 0. \tag{15.3}$$

On account of the strong convexity of the surface $\Phi = \Phi_0$, each velocity vector with end point on it defines a supporting plane, and according to the orthogonality condition (14.33) the supporting plane of the actual vector $\dot{a}_k$ corresponding to a dissipative force $A_k^{(d)}$ with the direction ν_k is perpendicular to $A_k^{(d)}$. Since the remaining points of the surface $\Phi = \Phi_0$ lie on the same side of this supporting plane as the origin O, the actual velocity $\dot{a}_k$ maximizes the scalar product $\nu_k\dot{a}_k$ and hence minimizes M on account of (15.3). It follows that the orthogonality condition is equivalent to a *principle of least dissipative force*: Provided the value Φ_0 of the dissipation function Φ and the direction ν_k of the dissipative force $A_k^{(d)}$ are prescribed, the actual velocity $\dot{a}_k$ minimizes the magnitude of $A_k^{(d)}$ subject to the condition (15.1).

It is obvious how the proof of this principle is to be extended to more than three velocities: The dissipation surface $\Phi = \Phi_0$ has to be interpreted as a hypersurface, and its supporting planes are hyperplanes. In analytical terms the statement that the actual velocity $\dot{a}_k$ maximizes the scalar product

$\nu_k \dot{a}_k$ for given values of ν_k and Φ takes the form

$$\frac{\partial}{\partial \dot{a}_k} [\nu_i \dot{a}_i - \mu \Phi(\dot{a}_i)] = 0, \tag{15.4}$$

where μ is a Lagrangean multiplier. The solution of (15.4),

$$\nu_k = \mu \frac{\partial \Phi}{\partial \dot{a}_k}, \tag{15.5}$$

is clearly equivalent to the orthogonality condition (14.33).

It has been shown in Section 14.3 that the orthogonality condition remains valid if the roles of velocities and irreversible forces are exchanged. It follows that the principle just discussed has a corollary in the form of a *principle of least velocity*: Provided the value Φ_0 of the dissipation function Φ' and the direction ν_k of the velocity $\dot{a}_k$ are prescribed, the actual dissipative force $A_k^{(d)}$ minimizes the magnitude of $\dot{a}_k$ subject to the condition (15.1).

To establish another extremum principle, let us start once more from the three-dimensional case and let us prescribe the dissipative force $A_k^{(d)}$. For any given value Φ_0, the first of the two equations

$$L^{(d)} = A_k^{(d)} \dot{a}_k = \Phi_0, \qquad \Phi(\dot{a}_k) = \Phi_0 \tag{15.6}$$

defines a plane E perpendicular to $A_k^{(d)}$ (Fig. 15.2), the second a dissipation surface F. The curve in which E and F intersect contains the end points of

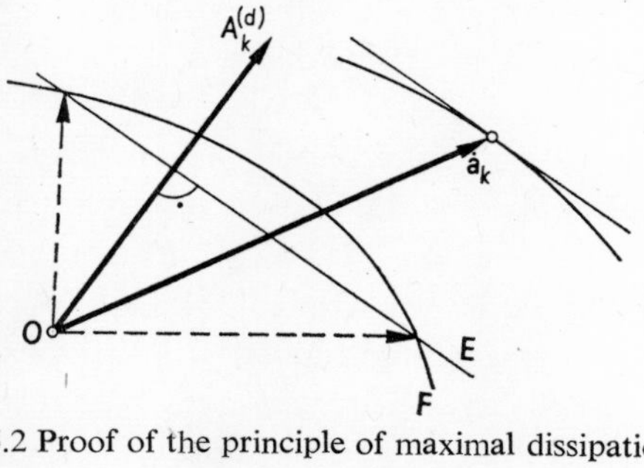

Fig. 15.2 Proof of the principle of maximal dissipation rate

velocity vectors obeying both equations (15.6). By increasing Φ_0, we pass from the dissipation surface F to others enclosing it. At the same time the plane E moves away from O. We have noted in connection with (14.81) that the increase of $\Phi(\dot{a}_k)$ is stronger than linear on any ray from O. On the other hand, with $A_k^{(d)}$ prescribed, the increase of $L^{(d)}$ on any such ray is linear. It follows that E moves faster away from O than F expands, and it is therefore to be expected that for a certain value of Φ_0 the plane E becomes tangential to the surface F. The radius vector $\dot{a}_k$ of the point of contact satisfies the orthogonality condition (14.33) and at the same time maximizes the value Φ_0 for which both equations (15.6) and hence condition (14.30) are satisfied. In consequence, the orthogonality condition is equivalent to a *principle of maximal dissipation rate*: Provided the dissipative force $A_k^{(d)}$ is prescribed, the actual velocity $\dot{a}_k$ maximizes the dissipation rate $L^{(d)} = A_k \dot{a}_k$ subject to the side condition

$$\Phi(\dot{a}_k) = A_k^{(d)} \dot{a}_k = L^{(d)} > 0. \tag{15.7}$$

The extension of this principle to more than three velocities is obvious. The analytical formulation of the extremum problem is given by

$$\frac{\partial}{\partial \dot{a}_k} \{A_i^{(d)} \dot{a}_i - \mu[\Phi(\dot{a}_i) - A_i^{(d)} \dot{a}_i]\} = 0, \tag{15.8}$$

where μ is a Lagrangean multiplier. The solution of (15.8),

$$A_k^{(d)} = \frac{\mu}{1+\mu} \frac{\partial \Phi}{\partial \dot{a}_k}, \tag{15.9}$$

is again equivalent to the orthogonality condition (14.33).

Exchanging the roles of velocities and irreversible forces, we obtain a *corollary* of the principle of maximal dissipation rate: Provided the velocity $\dot{a}_k$ is prescribed, the actual dissipative force $A_k^{(d)}$ maximizes the dissipation rate $A_k^{(d)} \dot{a}_k$ subject to the side condition (15.7).

In the linear case, i.e. for quadratic dissipation functions, the dissipation surfaces are hyperellipsoids. For this special case other extremum principles have been proposed by Onsager [30], Biot [32] and by Prigogine and De Groot [33]. They have been compared in [38]. Of all these possibilities the principle of maximal dissipation rate in its two forms seems to be the most interesting. In the first place, it is not restricted to linear problems. Moreover, it is closely connected with the principle already discussed for plastic bodies in connection with (10.49). Finally, (14.28) shows that division of the dissipation function by the temperature yields the rate of entropy production. It follows that the principle of maximal dissipation rate may also be formulated as a *principle of maximal rate of entropy production*: Provided the dissipative force $A_k^{(d)}$ (or the velocity $\dot{a}_k$) is prescribed, the actual velocity $\dot{a}_k$ (or dissipative force $A_k^{(d)}$) maximizes the rate of entropy production $\dot{S}^{(i)}$ subject to the side condition

$$\dot{S}^{(i)} = \frac{1}{\vartheta} A_k^{(d)} \dot{a}_k > 0. \tag{15.10}$$

From a physical point of view, this last principle is particularly appealing since it may be considered as an extension of the second fundamental law. In fact, if a closed system tends towards its state of maximal entropy, it seems reasonable that the rate of entropy increase under prescribed forces be a maximum, i.e. that the system should approach its final state on the shortest possible path.

The principle of maximal rate of entropy production and the fact just mentioned point far beyond the range of application for which the orthogonality condition has been established in Section 14.3. Not only does it lend itself to processes of a more general type; it actually strongly suggests such a generalization. In fact, we have seen already in Section 14.4 that the orthogonality condition, even though it has been proved only for elementary systems, retains its validity for complex systems with homogeneous dissipation functions and in particular for arbitrary linear processes. This fact suggests the tentative formulation of the following *orthogonality principle*: The orthogonality condition and the corresponding extremum principles remain valid for truly complex processes, i.e. for cases where the various elementary processes are coupled with each other.

Although we have no proof for this generalization, Section 14.4 provides at least a strong argument in its favor: In those cases where the orthogonality condition has been found to apply to complex systems the factors $\nu^{(1)}$ and $\nu^{(2)}$ introduced in (14.68) were equal; in the linear case, e.g., they had the common value $\frac{1}{2}$. In the remaining cases (14.69) is not sufficient to determine $\nu^{(1)}$ and $\nu^{(2)}$ separately; the dissipative forces thus remain indetermined. This situation is physically unsatisfactory. To render the problem well-defined, we need an additional equation between $\nu^{(1)}$ and $\nu^{(2)}$, and the obvious condition that provides the proper connection with the solvable cases is the postulate that $\nu^{(1)}$ and $\nu^{(2)}$ be equal. This postulate is clearly equivalent to the extension of the orthogonality condition to complex systems and hence to the orthogonality principle formulated above. Some consequences of this principle will be discussed in Section 15.3.

It is important to note that the orthogonality principle does not hold for compound processes, i.e. for cases where the elementary processes are independent. They are singular cases, characterized by dissipation functions each of which depends on a single coherent velocity set, and it is clear that, if these functions are maximized separately for given dissipative forces by their actual coherent velocities, this does not necessarily imply that also their sum is maximized by the entire set of velocities.

15.2. A deformation mechanism

The proof of the orthogonality condition in Section 14.3 was based on the assumption that the considered system is purely dissipative. It has been made clear in Section 14.2 that in all probability the processes treated in this book are nongyroscopic, i.e. purely dissipative; so far, however, an entirely conclusive proof has not been provided. In an attempt to bridge this gap at least for the case of entropy production by deformation and to obtain some insight into the mechanism of dissipation for this particular case, we now return to the micro-motion already discussed to a certain extent in Section 3.3.

Let us consider an element of a solid which from a macroscopic point of view is at rest and free of internal parameters. Let the kinematical parameters, i.e. the strain components be denoted by a_k and let us assume that they vary sufficiently slowly. The corresponding macroforces A_k are the stresses acting on the element, and the elementary work done on it is given by

$$\mathrm{d}W = A_k \,\mathrm{d}a_k \tag{15.11}$$

as in (4.1). From a microscopic viewpoint, the element consists of a large number of molecules with coordinates q_k, the a_k representing the external constraints of the whole system and the A_k the corresponding forces. Let us assume that the microsystem has purely mechanical properties and in particular that it obeys the energy principle of mechanics. Thus, the sum of the work (15.11) of the external forces and the work of the internal forces equals the increase of the total kinetic energy of all molecules.

If we tentatively assume that the internal forces have a single-valued potential and that the kinetic energy of the molecules is zero, the work of the external forces is given by

$$\mathrm{d}W = \mathrm{d}U, \tag{15.12}$$

where U is the potential energy of the internal forces. The equilibrium positions q_k of the molecules are entirely determined by the a_k, and the same is true for U. It follows that the macrosystem is conservative and that $-U(a_k)$ is the potential of the macroforces A_k.

We know, however, that except for a few special cases (elastic bodies under isothermal or adiabatic conditions) the macrosystem is not conservative. At least one of the tentative assumptions just introduced is therefore to be dropped. The customary and obvious solution is to sacrifice the last one and to admit that the molecules move and hence have a certain kinetic energy. This has the following consequences: In the first place, the quantity U in the energy principle (15.12) now denotes the sum of the potential energy of the internal forces and the kinetic energy of the molecules. In the second place, U does not depend any more on the a_k alone but also on the configuration and the state of motion of the molecules. In other words, it becomes possible to change the energy U while the a_k have fixed values. Consequently, the energy principle (15.12) must be re-written in the form of the first law (4.2),

$$\mathrm{d}U = \mathrm{d}W + \mathrm{d}Q. \tag{15.13}$$

Here $\mathrm{d}Q$ denotes the energy supplied while the a_k are fixed. From the macroscopic point of view, this is an energy input by extramechanical means; microscopically, it is due to rapid molecular interaction along the boundary of the system, taking place even if the a_k are constant. Finally, it is obvious that all terms in (15.13) are functions of the configuration of the microsystem, of its state of motion and the change of this state. Since the molecular motion is not perceptible on the macroscopic level, it becomes necessary to interpret the quantities in (15.13) as averages over time intervals that are long from the microscopic viewpoint, but macroscopically short. In this sense the external forces become the macroforces A_k, while U becomes the internal energy of the macrosystem, and $\mathrm{d}Q$ represents what is thermodynamically called the heat supply. Since U, even though interpreted as an average, does not depend on the a_k alone, it is necessary to introduce an additional macroscopic parameter in the form of the temperature ϑ, providing the macroscopic observer with an overall information concerning the state of the microsystem.

Let us restrict ourselves from now on to adiabatic deformations of the element, characterized by $\mathrm{d}Q = 0$, and let us define the equilibrium values $a_k = 0$ of the external parameters by the absence of external forces. The corresponding equilibrium positions about which the molecules oscillate will be denoted by $q_k^{(0)}$. If the a_k are made nonzero by application of external forces A_k, the equilibrium positions of the molecules will be modified too. Let us denote them by q_k, and let us assume that on account of the molecular interaction the potential and kinetic energies are equally distributed over the entire microsystem.

Provided the equilibrium configuration of the unloaded system, defined by the kinematical parameters $a_k = 0$ and the corresponding microcoordinates $q_k^{(0)}$, is always the same, the motion of the microsystem in an arbitrary loading process is reversible. More precisely: it is conceivable that in the corresponding unloading process, i.e. if the functions $a_k(t)$ are replaced by $a_k(-t)$, the motion of the molecules is also inverted. Since we have no means of controlling the initial conditions of the molecules, this will never exactly be the case, but it may at least be assumed that the average values of the potential and kinetic energies of the microsystem decrease in the same way during the unloading process as they increased in loading. From the macroscopic point of view the process is thus reversible. The macroforces A_k are quasiconservative and hence derivable according to (4.20) from a potential in the form of the free energy. The system is free of dissipation and hence is elastic; in fact, according to the definitions given in Section 5.1 it represents an element of an elastic solid.

The question now arises which one of the assumptions made above has to be dropped to explain the dissipative response of a viscous material, e.g. a plastic body. The answer, suggested by dislocation theory, is obvious: in an element of a viscous solid the loading process is apt to modify the equilibrium position described by $a_k = 0$ (and the corresponding $q_k^{(0)}$) of the unloaded system, with the effect that the system runs through a large number of subsequent equilibrium configurations $a_k^{(1)}$, $a_k^{(2)}$, ... in the course of time. In a configuration space a_k these equilibrium positions correspond to points in which the potential of the internal forces is minimal. The elastic case may be roughly illustrated by a single potential trough in which the representative point performs small oscillations around the bottom in the absence of external forces and somewhere up a slope if external forces are present. In a similar manner, the viscous case may be represented by a whole pattern of potential troughs separated by ridges. If the external forces are increased proportional to each other, the point representing the equilibrium configuration climbs towards a saddle point in an adjacent ridge and subsequently drops into the next trough. To keep the whole displacement slow, it would be necessary to apply negative external forces. However, such forces are not available, and since the motion leading from the saddle to the bottom of the next trough is extremely fast, the external forces drop to zero during this phase of the process, and the potential difference between saddle and trough is converted into kinetic energy.

The example of moving dislocations shows that in physical space the process just considered consists in an exchange of positions of only a small number of molecules. On account of the molecular interaction, however, the kinetic energy gained in the second stage is rapidly dissipated over the entire element, representing a temperature increase from the viewpoint of thermodynamics. The irreversible character of the entire two-stage process is obvious: although an inversion of the process is micro-mechanically conceivable, it is extremely improbable that part of the kinetic energy distributed over the element becomes concentrated in a few molecules and enables them to return to their original positions in such a way that the representative point in configuration space climbs back onto the saddle without external assistance and reaches its initial position under the influence of a slowly decreasing external force.

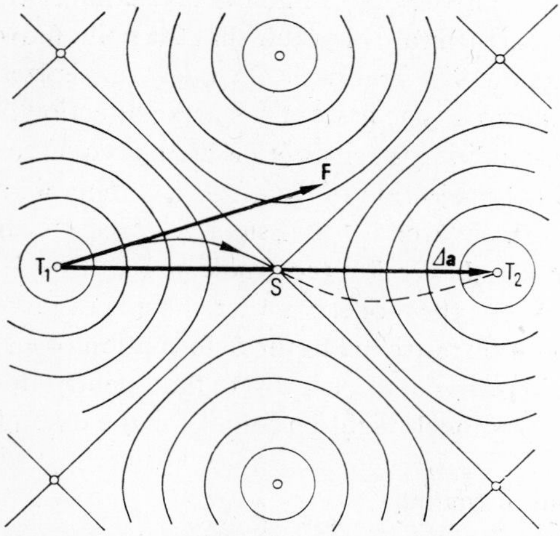

Fig. 15.3 Potential pattern and displacement of the representative point

We have described the process in terms of equilibrium positions. Actually, the molecules oscillate about them. However, we know from experience that under normal conditions the temperature increase due to deformation is rather limited, and we may assume that, except for the second phase, the deviations of the representative point in configuration space from its equilibrium positions are negligibly small. To illustrate the process in some more detail, let us restrict ourselves to the two-dimensional case and let us assume for simplicity that the minimal potential energies in the various troughs as well as those in the saddle points are the same respectively. Figure 15.3 shows a section of the potential pattern. The equilibrium position at time t is represented by the point T_1, where the potential is minimal. Under the influence of an increasing external force $\boldsymbol{F}$ with prescribed direction, the representative point in the configuration plane climbs in such a manner that the potential gradient in this point has always a direction opposite to $\boldsymbol{F}$. When it arrives at the saddle point S, the magnitude of $\boldsymbol{F}$ has become zero and stays zero during the rapid descent from S to the next potential minimum at T_2. The total work of the force $\boldsymbol{F}$ during the whole process is equal to the potential increase $\Delta\Phi$ from T_1 to S, and the energy gained in this first phase of the process is dissipated in the second phase, i.e. between S and T_2.

It is clear that the actual potential pattern for a three-dimensional material element is far more complicated than the idealization of Figure 15.3. There are more troughs adjacent to T_1 than shown in this Figure. However, the path of the representative point is initially tangential to the line of action of the force $\boldsymbol{F}$, and it is therefore plausible that the point moves to a trough next to this line of action. The real path obviously maximizes the projection of $\Delta\boldsymbol{a}$ onto $\boldsymbol{F}$. The macroscopic force $\boldsymbol{A}^{(d)}$ has the direction of $\boldsymbol{F}$; its magnitude follows from the condition that its work $\boldsymbol{A}^{(d)}\,\Delta\boldsymbol{a}$ is equal to $\Delta\Phi$. Since the real path maximizes the projection of $\Delta\boldsymbol{a}$ on $\boldsymbol{A}^{(d)}$, it minimizes the magnitude of $\boldsymbol{A}^{(d)}$. Thus the principle of least dissipative force is satisfied for the step considered here and hence for the whole process.

Needless to say that the model considered here is a rather crude one, apart from the fact that it is restricted to the deformation of a solid. However, the model might be improved and extended to fluids, and it does not seem impossible that a similar model might be found for the case of heat conduction.

15.3. Application to continua

At the end of Section 14.1 we have listed the dissipation functions for some of the processes treated in Chapters 5 and 9 through 11. Now the question arises whether the constitutive relations established in these chapters actually obey the orthogonality condition. We will see that for certain processes this is apparently the case due to the isotropy of the material. In other cases it will turn out that the orthogonality condition simplifies the constitutive relations. The situation is similar to the one already encountered in elasticity:

The linear isotropic elastic solid obeys Hooke's law (5.41) and hence has the strain energy (5.42); in the case of generalization (7.60) for anisotropic solids, the existence of a strain energy requires the symmetry condition (7.65) to be satisfied.

The dissipation functions collected in Section 14.1 are referred to the unit volume. With $\Phi = \varrho\varphi$ the orthogonality condition (14.33), (14.34) assumes the form

$$A_i^{(d)} = \nu\varrho \frac{\partial \varphi}{\partial \dot{a}_i}, \qquad \nu = \varphi \left(\frac{\partial \varphi}{\partial \dot{a}_j} \dot{a}_j \right)^{-1}, \tag{15.14}$$

where the $\dot{a}_i$ denote the velocities and the $A_i^{(d)}$ the dissipative forces per unit volume.

A simple and very instructive example is *heat conduction*. Here the velocities are the components of the heat flow vector q_i, and the dissipative forces per unit volume are given by (4.78),

$$A_i^{(d)} = -\frac{\vartheta_{,i}}{\vartheta}. \tag{15.15}$$

For an *isotropic* material the dissipation function (14.3) per unit volume is

$$\varrho\varphi = \frac{1}{\lambda\vartheta} q_i q_i, \tag{15.16}$$

where λ denotes the thermal conductivity. Applying (15.14) to (15.16), we obtain $\nu = \frac{1}{2}$ and hence

$$A_i^{(d)} = -\frac{\vartheta_{,i}}{\vartheta} = \frac{1}{\lambda\vartheta} q_i, \tag{15.17}$$

i.e. Fourier's law (5.9),

$$q_i = -\lambda\vartheta_{,i}. \tag{15.18}$$

The result appears almost trivial, for it seems obvious that in an anisotropic medium the vectors q_i and $\vartheta_{,i}$ must have the same line of action for reasons of symmetry. However, if the dissipative force $A_i^{(d)}$ had a gyroscopic component, defined according to Section 14.3 by means of a vector $\boldsymbol{\omega}$ independent of φ, the lines of action of q_i and $\vartheta_{,i}$ would be different. The assumption that heat flow is a purely dissipative process is therefore essential even if one does not want to use the orthogonality condition. The example confirms that the restriction to purely dissipative systems in Chapter 14 is not new; although not mentioned, it has always been an important element in thermodynamics.

If the material is *anisotropic*, the dissipation function for heat conduction is the quadratic form (14.1), where μ_{jk} may be assumed to be symmetric. We thus have

$$\varrho\varphi = \frac{\mu_{ij}}{\vartheta} q_i q_j \qquad (\mu_{ji} = \mu_{ij}). \tag{15.19}$$

Inserting this in (15.14), we obtain $\nu = \frac{1}{2}$ and

$$A_i^{(d)} = -\frac{\vartheta_{,i}}{\vartheta} = \frac{\mu_{ij}}{\vartheta} q_j \tag{15.20}$$

or in short

$$-\vartheta_{,i} = \mu_{ik} q_k . \tag{15.21}$$

Multiplying both sides by the tensor λ_{ij}, defined by means of (14.2),

$$\mu_{ki}\lambda_{ij} = \delta_{kj}, \tag{15.22}$$

and making use of the symmetry of μ_{jk}, we obtain

$$q_j = -\lambda_{ij}\vartheta_{,i} . \tag{15.23}$$

Thus, (15.23) becomes the generalization (5.8) of Fourier's law, supplemented by the symmetry condition $\lambda_{ji} = \lambda_{ij}$. In fact, the symmetry of μ_{jk} implies that also the thermal conductivity tensor λ_{ij} is symmetric, for if we exploit (15.22) in turn for $k = 1, 2, 3$, using the system of principal axes of μ_{ik}, we obtain $\lambda_{23} = \ldots = 0$ and $\lambda_{11} = \lambda_{\mathrm{I}} = 1/\mu_{\mathrm{I}}, \ldots$. Again, the result is not trivial but a consequence of the orthogonality condition or, equivalently, of the assumption that the process is purely dissipative. In thermodynamics the statement that $\lambda_{ji} = \lambda_{ij}$ is considered as a special case of Onsager's theory [30] and is referred to as a reciprocity relation.

As another example, let us consider the deformation of a *Newtonian fluid*. Here the velocities are the components of the deformation rate d_{ij}, and the dissipative forces per unit volume are the stress components $\sigma_{ij}^{(d)}$. The dissipation function per unit volume is given by (14.4),

$$\varrho\varphi = \lambda d_{ii} d_{jj} + 2\mu d_{ij} d_{ij}, \tag{15.24}$$

where λ and μ denote the viscosity coefficients. Applying (15.14) to (15.24), we obtain $\nu = \frac{1}{2}$ and

$$\sigma_{ij}^{(d)} = \lambda d_{kk}\delta_{ij} + 2\mu d_{ij} . \tag{15.25}$$

This is in fact the dissipative stress already defined by $(9.2)_2$. The total stress follows from it by addition of the reversible stress $(9.2)_1$, which is hydrostatic. If the bulk viscosity $\lambda+\frac{2}{3}\mu$ is zero, (15.25) reduces to

$$\sigma_{ij}^{(d)} = 2\mu(d_{ij}-\tfrac{1}{3}d_{kk}\delta_{ij}) = 2\mu d'_{ij}, \tag{15.26}$$

which corresponds to the dissipation function (14.5). The same result (with $d'_{ij} = d_{ij}$) follows from (15.25) if the fluid is incompressible.

The remaining materials for which the dissipation functions have been listed in Section 14.1 will be treated separately in the next chapters. Before doing so, let us consider the case where a process is determined by two different dissipation mechanisms.

The simplest and most common example of a complex process is the combination of *flow of heat* and *deformation of the material*. The corresponding velocities are the components of the heat flow vector q_i and the deformation rate d_{kl}. They form two coherent sets in the sense of Section 14.3. The corresponding forces per unit volume are given by the negative gradient of the logarithmic temperature, $-\vartheta_{,i}/\vartheta$, and the dissipative stress tensor $\sigma_{kl}^{(d)}$. The dissipation function, referred to the unit volume, has the form $\Phi = \varrho\varphi(q_i, d_{kl})$, and in place of (14.30) we obtain

$$-\frac{\vartheta_{,i}}{\vartheta}q_i+\sigma_{kl}^{(d)}d_{kl} = \varrho\varphi(q_p, d_{rs}). \tag{15.27}$$

If the material is *isotropic*, the dissipation function has the form

$$\Phi = \varrho\varphi(q_{(1)}, d_{(1)}, d_{(2)}, d_{(3)}), \tag{15.28}$$

where

$$q_{(1)} = \tfrac{1}{2}q_iq_i \tag{15.29}$$

is the only basic invariant of the vector q_i.

In the *linear* case φ is a quadratic form in the arguments q_i and d_{kl}. It follows from (1.52) that $d_{(3)}$ cannot be an argument and that Φ must have the form

$$\Phi = Aq_{(1)}+Bd_{(1)}^2+Cd_{(2)}, \tag{15.30}$$

where A, B, and C are constants. Since (15.30) is of the form (14.60) with $\Phi^{(1)}(q_{(1)})$ and $\Phi^{(2)}(d_{(1)}, d_{(2)})$, the special case considered here is compound. In consequence, there is no coupling between heat flow and deformation, and it is easy to see (P1) that the orthogonality condition, applied in turn to

$\Phi^{(1)}$ and $\Phi^{(2)}$, yields Fourier's law (15.18) and the constitutive equation (15.25) of the Newtonian fluid. In view of the remark made in connection with (14.64) we observe that, at least in the linear case, the separation of the Second fundamental law indicated in (4.80) is justified.

Comparing the various terms in (15.30) to (15.16) and (152.4), we find that the constants are given by

$$A = \frac{2}{\lambda' \vartheta}, \qquad B = \lambda + 2\mu, \qquad C = 4\mu, \tag{15.31}$$

where λ' denotes the thermal conductivity and λ, μ are Lamé's constants. We thus have

$$\Phi^{(1)} = -\frac{\vartheta_{,i}}{\vartheta} q_i = \frac{2}{\lambda'\vartheta} q_{(1)} \tag{15.32}$$

and

$$\Phi^{(2)} = \sigma_{kl}^{(d)} d_{kl} = (\lambda + 2\mu) d_{(1)}^2 + 4\mu d_{(2)}. \tag{15.33}$$

Choosing the coordinate system so that the axis x_1 has the direction of the heat flow, we obtain

$$\Phi^{(1)} = -\frac{\vartheta_{,1}}{\vartheta} q_1 = \frac{1}{\lambda'\vartheta} q_1^2 \tag{15.34}$$

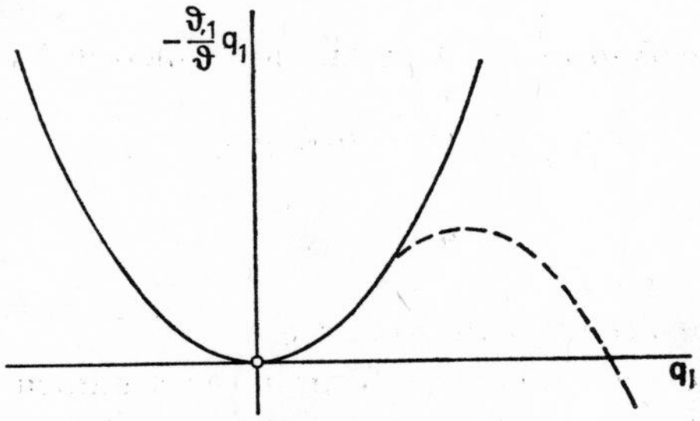

Fig. 15.4 Dissipation rate as a function of heat flow

in place of (15.32). Figure 15.4 shows the corresponding parabola. A similar parabola would be obtained by considering, e.g. the case of simple shear d_{12}, where (15.33) reduces to

$$\Phi^{(2)} = 2\sigma_{12}^{(d)} d_{12} = 4\mu d_{12}^2. \tag{15.35}$$

In the general *nonlinear* case all arguments are present in (15.28). From (15.29) and (1.52) we calculate

$$\frac{\partial q_{(1)}}{\partial q_i} = q_i, \qquad \frac{\partial d_{(1)}}{\partial d_{ij}} = \delta_{ij}, \qquad \frac{\partial d_{(2)}}{\partial d_{ij}} = d_{ij} - d_{(1)}\delta_{ij},$$
$$\frac{\partial d_{(3)}}{\partial d_{ij}} = d_{ik}d_{kj} - d_{(1)}d_{ij} - d_{(2)}\delta_{ij}, \tag{15.36}$$

and we further note that

$$\delta_{ij}d_{ij} = d_{(1)}, \qquad (d_{ij} - d_{(1)}\delta_{ij})d_{ij} = 2d_{(2)},$$
$$(d_{ik}d_{kj} - d_{(1)}d_{ij} - d_{(2)}\delta_{ij})d_{ij} = 3d_{(3)}. \tag{15.37}$$

According to (15.36) we have

$$\frac{\partial \Phi}{\partial q_i} = \frac{\partial \Phi}{\partial q_{(1)}} q_i,$$
$$\frac{\partial \Phi}{\partial d_{ij}} = \frac{\partial \Phi}{\partial d_{(1)}} \delta_{ij} + \frac{\partial \Phi}{\partial d_{(2)}} (d_{ij} - d_{(1)}\delta_{ij}) + \frac{\partial \Phi}{\partial d_{(3)}} (d_{ik}d_{kj} - d_{(1)}d_{ij} - d_{(2)}\delta_{ij}). \tag{15.38}$$

Equations (14.68) assume the forms

$$-\frac{\vartheta_{,i}}{\vartheta} = \nu^{(1)} \frac{\partial \Phi}{\partial q_i}, \qquad \sigma_{ij}^{(d)} = \nu^{(2)} \frac{\partial \Phi}{\partial d_{ij}} \tag{15.39}$$

or, on account of (15.38),

$$-\frac{\vartheta_{,i}}{\vartheta} = \nu^{(1)} \frac{\partial \Phi}{\partial q_{(1)}} q_i,$$
$$\sigma_{ij}^{(d)} = \nu^{(2)} \left[\frac{\partial \Phi}{\partial d_{(1)}} \delta_{ij} + \frac{\partial \Phi}{\partial d_{(2)}} (d_{ij} - d_{(1)}\delta_{ij}) + \frac{\partial \Phi}{\partial d_{(3)}} (d_{ik}d_{kj} - d_{(1)}d_{ij} - d_{(2)}\delta_{ij}) \right]. \tag{15.40}$$

Making use of (15.37), we finally obtain

$$2\nu^{(1)} \frac{\partial \Phi}{\partial q_{(1)}} q_{(1)} + \nu^{(2)} \left(\frac{\partial \Phi}{\partial d_{(1)}} d_{(1)} + 2\frac{\partial \Phi}{\partial d_{(2)}} d_{(2)} + 3\frac{\partial \Phi}{\partial d_{(3)}} d_{(3)} \right) = \Phi. \tag{15.41}$$

As in Section 14.4, it is in general impossible to determine the two factors $\nu^{(1)}$ and $\nu^{(2)}$ by means of the single condition (15.41). The dissipative forces thus remain undetermined. However, there exist two exceptions:

Provided we were allowed to assume that the two elementary processes of heat conduction and deformation are uncoupled, Φ would be of the form $\Phi^{(1)}(q_{(1)})+\Phi^{(2)}(d_{(1)}, d_{(2)}, d_{(3)})$, and (15.41) would decay into the two equations

$$
\begin{aligned}
2\nu^{(1)}\frac{\partial\Phi^{(1)}}{\partial q_{(1)}}q_{(1)} &= \Phi^{(1)},\\
\nu^{(2)}\left(\frac{\partial\Phi^{(2)}}{\partial d_{(1)}}d_{(1)}+2\frac{\partial\Phi^{(2)}}{\partial d_{(2)}}d_{(2)}+3\frac{\partial\Phi^{(2)}}{\partial d_{(3)}}d_{(3)}\right) &= \Phi^{(2)}
\end{aligned}
\tag{15.42}
$$

determining $\nu^{(1)}$ and $\nu^{(2)}$ separately.

However, if we renounce the simplifying assumption that the processes are not coupled, the only means of obtaining well-defined dissipative forces for arbitrary dissipation functions is to accept the orthogonality principle stated in Section 15.1. It requires that $\nu^{(1)} = \nu^{(2)}$, and it follows that (15.40) reduces to

$$
\begin{aligned}
-\frac{\vartheta_{,i}}{\vartheta} &= \nu\frac{\partial\Phi}{\partial q_{(1)}}q_i,\\
\sigma_{ij}^{(d)} &= \nu\left[\frac{\partial\Phi}{\partial d_{(1)}}\delta_{ij}+\frac{\partial\Phi}{\partial d_{(2)}}(d_{ij}-d_{(1)}\delta_{ij})+\frac{\partial\Phi}{\partial d_{(3)}}(d_{ik}d_{kj}-d_{(1)}d_{ij}-d_{(2)}\delta_{ij})\right],
\end{aligned}
\tag{15.43}
$$

where ν is determined by

$$
\nu\left(2\frac{\partial\Phi}{\partial q_{(1)}}q_{(1)}+\frac{\partial\Phi}{\partial d_{(1)}}d_{(1)}+2\frac{\partial\Phi}{\partial d_{(2)}}d_{(2)}+3\frac{\partial\Phi}{\partial d_{(3)}}d_{(3)}\right) = \Phi. \tag{15.44}
$$

Equation $(15.43)_1$ is a generalization of Fourier's law (15.18). The lines of action of the heat flow vector and the temperature gradient still coincide, but the thermal conductivity is now given by

$$
\lambda = \left(\vartheta\nu\frac{\partial\Phi}{\partial q_{(1)}}\right)^{-1}, \tag{15.45}
$$

where Φ is a generalization of (15.16). It follows that λ is now a function of the four invariants appearing in (15.28). As was to be expected, the dissipative stress defined by $(15.43)_2$ has the general form (9.60) with the exception that the functions f, g, and h now depend on $q_{(1)}$ as well as on $d_{(1)}, \ldots$. In short, the constitutive equations considered so far for isotropic materials remain essentially valid. The only modification is that the scalars present in the relations between velocities and dissipative forces are now functions of the four invariants $q_{(1)}$, $d_{(1)}$, $d_{(2)}$, and $d_{(3)}$.

We are particularly interested in constitutive equations that are close to the linear ones and hence may be obtained by addition of higher-order terms. Since we consider Φ as the fundamental function, this means that we have to add terms of the third and higher degrees to the quadratic forms used so far. In the present case, the next approximation is obtained by adding the third-order expression

$$Dq_{(1)}d_{(1)}+Ed_{(1)}^3+Fd_{(1)}d_{(2)}+Gd_{(3)} \tag{15.46}$$

to the function Φ given by (15.30). The first term in (15.46) establishes the coupling between heat flow and deformation.

In principle the possibility exists that the thermal conductivity (15.45) and hence the first term on the left-hand side of (15.27) become negative for certain magnitudes of q_i. The same is true for the second term in (15.27) but not for both of them simultaneously since Φ is positive definite. In one or the other case the separation of the second fundamental law indicated by (4.80) would not be justified. Figure 15.4 however, which is also typical for deformation, shows that the deviations from the linear response represented by a parabolic increase of the dissipation rate must be quite considerable for this to happen. We thus conclude that, at least for isotropic materials without internal parameters, the spearation (4.80) is justified as long as the deviations from the linear response are not extremely high.

Problems

1. Show that the orthogonality condition, applied to the two parts of (15.30), yields (15.18) and (15.25).
2. Add the expression (15.46) to the dissipation function (15.30). Find the dissipative forces $-\vartheta_{,i}/\vartheta$ and $\sigma_{ij}^{(d)}$ and establish the relation determining ν.
3. Solve the last problem for the case where Φ also contains fourth-order terms.

CHAPTER 16

NON-NEWTONIAN LIQUIDS

In this and the following chapters we will study the implications of the orthogonality condition for the constitutive relations of the more complicated materials listed at the end of Section 14.1. Provided they are linear, there is no coupling between the mechanical response and heat conduction, and for nonlinear isotropic materials we have seen in Section 15.3 that a possible coupling only affects the scalar functions appearing in the constitutive equations, provided we accept the orthogonality principle stated at the end of Section 15.1.

We start here with the discussion of the *isotropic* and *incompressible* non-Newtonian liquid already studied in Section 9.4 without making use of orthogonality.

16.1. Constitutive equations

In an incompressible liquid the deformation rate d_{ij} is a deviator. The basic invariant $d_{(1)}$ is therefore zero, and the dissipation function (15.28) for the isotropic case reduces to

$$\Phi = \varrho\varphi(q_{(1)}, d_{(2)}, d_{(3)}). \tag{16.1}$$

However, to obtain the connection between the deformation rate and the dissipative stress it is not sufficient to set $d_{(1)} = 0$ in $(15.43)_2$. On account of the side condition $d_{(1)} = 0$, the relation $\sigma_{ij}^{(d)} = \nu\, \partial\Phi/\partial d_{ij}$ must be replaced by

$$\sigma_{ij}^{(d)} = \nu \frac{\partial}{\partial d_{ij}} (\Phi - \gamma d_{(1)}), \tag{16.2}$$

where γ is a Lagrangean multiplier. Consequently the second equation

(15.43) now assumes the form

$$\sigma_{ij}^{(d)} = \nu\left[-\gamma\delta_{ij}+\frac{\partial\Phi}{\partial d_{(2)}}d_{ij}+\frac{\partial\Phi}{\partial d_{(3)}}(d_{ik}d_{kj}-d_{(2)}\delta_{ij})\right]. \tag{16.3}$$

Noting that not only d_{ij} but also $\sigma_{ij}^{(d)}$ must be a deviator, we obtain the equation

$$-3\gamma+\frac{\partial\Phi}{\partial d_{(3)}}(d_{ik}d_{ki}-3d_{(2)}) = 0. \tag{16.4}$$

The factor γ is thus given by

$$\gamma = -\tfrac{1}{3}\frac{\partial\Phi}{\partial d_{(3)}}d_{(2)}. \tag{16.5}$$

inserting this in (16.3), we have

$$\sigma_{ij}^{(d)} = \nu\left[\frac{\partial\Phi}{\partial d_{(2)}}d_{ij}+\frac{\partial\Phi}{\partial d_{(3)}}(d_{ik}d_{kj}-\tfrac{2}{3}d_{(2)}\delta_{ij})\right] \tag{16.6}$$

In place of $(15.43)_2$, and ν is determined by

$$\nu\left(2\frac{\partial\Phi}{\partial q_{(1)}}q_{(1)}+2\frac{\partial\Phi}{\partial d_{(2)}}d_{(2)}+3\frac{\partial\Phi}{\partial d_{(3)}}d_{(3)}\right) = \Phi. \tag{16.7}$$

Comparing (16.6) and (9.65), we observe that the material considered here is essentially a *Reiner–Rivlin liquid.* However, its properties are determined by a single function Φ of the basic invariants in place of the two functions g and h. The comparison shows that

$$g = \nu\frac{\partial\Phi}{\partial d_{(2)}}, \qquad h = \nu\frac{\partial\Phi}{\partial d_{(3)}}, \tag{16.8}$$

and it follows that the orthogonality condition narrows the field of acceptable Reiner–Rivlin fluids. On the other hand, we have admitted dissipation functions (16.1) dependent on $q_{(1)}$. Insofar as g and h are now functions of $q_{(1)}$, $d_{(2)}$, and $d_{(3)}$, our constitutive equation (16.6) appears more general than (9.65).

An important special case is the *quasilinear* liquid defined by (9.72). Comparing (9.72) and (16.6), we note that the simplification of the constitutive equation resulting in (9.72) requires that $\partial\Phi/\partial d_{(3)}$ be zero. The viscosity coefficient μ is given by

$$2\mu = \nu\frac{\partial\Phi}{\partial d_{(2)}} \tag{16.9}$$

and hence is independent of $d_{(3)}$. We have made use of this result in the definition of the viscoplastic material considered in Section 10.1, where we assumed that μ is a function of $d_{(2)}$ alone. It should be noted however that in the present context μ may also depend on $q_{(1)}$ as long as coupling between deformation and heat conduction cannot be excluded.

16.2. Approximations

We have noted at the end of Section 15.3 that we are particularly interested in constitutive equations that are close to the linear ones and hence may be obtained by addition of higher-order terms. Since Φ is the fundamental function, the proper manner of constructing simple nonlinear constitutive equations is to start with a quadratic form for Φ, which may be successively generalized by the addition of higher-order terms.

According to (15.30), the *linear* case of an incompressible isotropic liquid is characterized by the dissipation function

$$\Phi = Aq_{(1)} + Bd_{(2)}, \tag{16.10}$$

where A and B are constants. Inserting this in (16.7), we obtain $\nu = \frac{1}{2}$, and (16.6) reduces to

$$\sigma_{ij}^{(d)} = \tfrac{1}{2}Bd_{ij}. \tag{16.11}$$

This is the approximation to be expected on account of $(9.2)_2$; it represents a Newtonian liquid of viscosity $\mu = B/4$.

To obtain a *second approximation*, we tentatively add the only third-order term $Cd_{(3)}$ to the right-hand side of (16.10), obtaining

$$\Phi = Aq_{(1)} + Bd_{(2)} + Cd_{(3)}. \tag{16.12}$$

Equation (16.7) now reads

$$\nu(2Aq_{(1)} + 2Bd_{(2)} + 3Cd_{(3)}) = Aq_{(1)} + Bd_{(2)} + Cd_{(3)}, \tag{16.13}$$

and (16.6) becomes

$$\sigma_{ij}^{(d)} = \nu[Bd_{ij} + C(d_{ik}d_{kj} - \tfrac{2}{3}d_{(2)}\delta_{ij})]. \tag{16.14}$$

It is important to note that ν, calculated from (16.13), is no longer a constant, and that therefore the expression (16.14) is not an approximation of the second order in d_{ij}.

At first glance this result is unexpected. Being used to dealing with potential functions, e.g. in the form of polynomials, we expect the forces to be polynomials of the next lower degree. It is clear though that the factor ν in the orthogonality condition complicates matters and that a polynomial for Φ generally supplies an infinite series for $\sigma_{ij}^{(d)}$. However, in practical applications we need approximations of various orders of the constitutive equation. To obtain them, it is necessary to expand also the function $\nu(q_{(1)}, d_{(2)}, d_{(3)})$ and to truncate the corresponding expression for $\sigma_{ij}^{(d)}$. In other words, we have to consider Φ and $\sigma_{ij}^{(d)}$ as infinite series. The n-th order approximation for the constitutive equation is obtained by truncating the expression for $\sigma_{ij}^{(d)}$ after the terms of order n in q_i and d_{ij}; on account of (15.27),

$$-\frac{\vartheta_{,i}}{\vartheta} q_i+\sigma_{ij}^{(d)} d_{ij} = \Phi(q_j, d_{kl}), \tag{16.15}$$

the corresponding expression for Φ must be truncated after the term of order $n+1$.

In the second approximation (16.14), ν is still to be considered as a constant since the next term in its expansion would already be of the second order. Equation (16.13) thus yields $\nu = \frac{1}{2}$ and $C = 0$. This means that on account of the orthogonality condition the second approximation is still a Newtonian liquid. The last condition imposes a restriction on Φ; this, incidentally, was to be expected since with $C \neq 0$ the dissipation function could not be positive definite.

To treat the next two approximations in a single step, let us add 4th and 5th order terms to the dissipation function defined by (16.12). We thus obtain

$$\Phi = (Aq_{(1)}+Bd_{(2)})+Cd_{(3)}+(Dq_{(1)}^2+Eq_{(1)}d_{(2)}+Fd_{(2)}^2)+(Gd_{(1)}d_{(3)}+Hd_{(2)}d_{(3)}), \tag{16.16}$$

where $A, \ldots, H$ are constants. The expression (16.16) is of the 5th degree in q_i and d_{ij}, and the parentheses contain terms of equal degree. The partial derivatives needed in (16.7) and (16.6) are

$$\begin{aligned} \frac{\partial \Phi}{\partial q_{(1)}} &= A+(2Dq_{(1)}+Ed_{(2)})+Gd_{(3)}, \\ \frac{\partial \Phi}{\partial d_{(2)}} &= B+(Eq_{(1)}+2Fd_{(2)})+Hd_{(3)}, \\ \frac{\partial \Phi}{\partial d_{(3)}} &= C+(Gq_{(1)}+Hd_{(2)}). \end{aligned} \tag{16.17}$$

The expansion of ν has the form

$$\nu = \tfrac{1}{2}+(A'q_{(1)}+B'd_{(2)})+C'd_{(3)}+ \ldots, \tag{16.18}$$

where A', B', C', ... are constants. Calculating the left-hand side of (16.7) by means of (16.18) and (16.17) and restricting ourselves to terms including the order 5, we obtain

$$(Aq_{(1)}+Bd_{(2)})+\tfrac{3}{2}Cd_{(3)}+2[(AA'+D)q_{(1)}^2+(AB'+BA'+E)q_{(1)}d_{(2)}$$
$$+(BB'+F)d_{(2)}^2]+[(2AC'+3CA'+\tfrac{5}{2}G)q_{(1)}d_{(3)}$$
$$+(2BC'+3CB'+\tfrac{5}{2}H)d_{(2)}d_{(3)}]. \tag{16.19}$$

On account of (16.7), the expressions (16.19) and (16.16) must be equal. Comparing the coefficients, we get

$$C = 0, \qquad E = \frac{AF}{B}+\frac{BD}{A}, \qquad H = \frac{BG}{A}, \tag{16.20}$$

$$A' = -\tfrac{1}{2}\frac{D}{A}, \qquad B' = -\tfrac{1}{2}\frac{F}{B}, \qquad C' = -\tfrac{3}{4}\frac{G}{A}. \tag{16.21}$$

We note that the condition $C = 0$ is carried over into the higher-order approximations. Inserting the coefficients (16.20) in (16.16), we have

$$\Phi = (Aq_{(1)}+Bd_{(2)})+\left[Dq_{(1)}^2+\left(\frac{AF}{B}+\frac{BD}{A}\right)q_{(1)}d_{(2)}+Fd_{(2)}^2\right]$$
$$+\left[Gq_{(1)}d_{(3)}+\frac{BG}{A}d_{(2)}d_{(3)}\right]. \tag{16.22}$$

Equations (16.17) take the forms

$$\frac{\partial\Phi}{\partial q_{(1)}} = A+\left[2Dq_{(1)}+\left(\frac{AF}{B}+\frac{BD}{A}\right)d_{(2)}\right]+Gd_{(3)},$$
$$\frac{\partial\Phi}{\partial d_{(2)}} = B+\left[\left(\frac{AF}{B}+\frac{BD}{A}\right)q_{(1)}+2Fd_{(2)}\right]+\frac{BG}{A}d_{(3)}, \tag{16.23}$$
$$\frac{\partial\Phi}{\partial d_{(3)}} = \left(Gq_{(1)}+\frac{BG}{A}d_{(2)}\right),$$

and in place of (16.18) we now have

$$v = \tfrac{1}{2}\left[1-\left(\frac{D}{A}q_{(1)}+\frac{F}{B}d_{(2)}\right)-\tfrac{3}{2}\frac{G}{A}d_{(3)}\right]. \tag{16.24}$$

Equation (16.6) eventually becomes

$$\sigma_{ij}^{(d)} = \tfrac{1}{2}\left\{\left[B+\left(\frac{AF}{B}q_{(1)}+Fd_{(2)}\right)-\tfrac{1}{2}\frac{BG}{A}d_{(3)}\right]d_{ij}\right.$$

$$\left.+\left(Gq_{(1)}+\frac{BG}{A}d_{(2)}\right)(d_{ik}d_{kj}-\tfrac{2}{3}d_{(2)}\delta_{ij})\right\}. \tag{16.25}$$

Reducing (16.25) to the first-order term, we once more obtain (16.11) or

$$\sigma_{ij} = 2\mu d_{ij}, \tag{16.26}$$

where the notation

$$4\mu = B \tag{16.27}$$

has been used. This is the contitutive equation of the Newtonian liquid. Since (16.25) does not contain any second-order terms, our observation that the second approximation is still a Newtonian liquid is confirmed. With the notations

$$4\lambda = \frac{AF}{B}, \qquad 4\mu' = F \tag{16.28}$$

the *third approximation* reads

$$\sigma_{ij}^{(d)} = 2(\mu+\lambda q_{(1)}+\mu' d_{(2)})d_{ij}. \tag{16.29}$$

It represents a quasilinear liquid as defined in Section 9.4, and we observe that, as was to be expected in view of the last remarks in Section 16.1, the coefficient of d_{ij} is independent of $d_{(3)}$ but dependent on $q_{(1)}$. With the additional notations

$$4\lambda' = G, \qquad 8\mu'' = \frac{BG}{A} \tag{16.30}$$

the *fourth approximation* becomes

$$\sigma_{ij}^{(d)} = 2(\mu+\lambda q_{(1)}+\mu' d_{(2)}-\mu'' d_{(3)})d_{ij}+2(\lambda' q_{(1)}+2\mu'' d_{(2)})(d_{ik}d_{kj}-\tfrac{2}{3}d_{(2)}\delta_{ij}). \tag{16.31}$$

It represents a Reiner–Rivlin liquid.

Comparing the various approximations, we note that the transition from the Newtonian to the Reiner–Rivlin liquid occurs gradually. The fourth approximation is the first one representing a truly nonlinear liquid. It is easy to see that, starting with the second approximation, the orthogonality principle reduces the number of constants that would otherwise appear in the various expansions. The process can be carried on to supply higher-order approximations. The Reiner–Rivlin character of the liquid persists beyond the fourth-order approximation, but the scalar functions in (16.31) become more complicated.

We have restricted ourselves to the discussion of the dissipative stress $\sigma_{ij}^{(d)}$. A similar process supplies the thermal forces $-\vartheta_{,i}/\vartheta$. The result (P2) is a generalization of Fourier's law, characterized by expressions dependent on $q_{(1)}$, $d_{(2)}$, and $d_{(3)}$ for the thermal conductivity. On the other hand, the expression for the dissipative stress is considerably simplified if we assume that there is no heat flow ($q_{(1)} = 0$) or that its influence on $\sigma_{ij}^{(d)}$ is negligibly small ($\lambda = \lambda' = \ldots = 0$). We have treated this case in Section 9.4 without making use of the orthogonality condition, and it is interesting to compare the results. In Section 9.4 we started from the constitutive equation (9.65) for the incompressible Reiner–Rivlin liquid, and we obtained approximations by truncating the expansions (9.67) and (9.68) for the functions g and h respectively. In view of the orthogonality condition, these functions are now connected by the two equations (16.8), and the approximations are the ones just discussed. Comparing the first approximations (16.26) and (9.69), we note that they are the same (with $\mu = g^{(0)}$). Our second approximation still corresponds to the Newtonian liquid, whereas (9.70) already contains the square of the dissipation rate. Our third approximation is quasilinear, and the fourth approximation (16.31), which reduces to

$$\sigma_{ij}^{(d)} = 2(\mu + \mu' d_{(2)} - \mu'' d_{(3)}) d_{ij} + 4\mu'' d_{(2)} (d_{ik} d_{kj} - \tfrac{2}{3} d_{(2)} \delta_{ij}), \qquad (16.32)$$

is the first that is truly nonlinear.

The discussion of the Weissenberg effect in *Couette flow* was based in Section 9.4 on the constitutive equation (9.70). Since in simple shear $d_{(3)} = 0$, (9.70) formally corresponds to (16.32) if we set $\mu = g^{(0)}$, $\mu' = 0$, and $\mu'' d_{(2)} = \mu'' d_{r\alpha}^2 = h^{(0)}$. We note, however, that $h^{(0)}$ is no longer a constant. In consequence, $h^{(0)} d_{r\alpha}^2$ must be replaced by $\mu'' d_{r\alpha}^4$ in (9.88), (9.90), and (9.91). In lieu of (9.92) we now obtain

$$p = \tfrac{4}{3}\mu'' \frac{A^4}{r^8} + \gamma z + C, \qquad (16.33)$$

and the vertical normal stress $(9.93)_2$ must be replaced by

$$\sigma_{zz} = -4\mu'' \frac{A^4}{r^8} - \gamma z - C. \qquad (16.34)$$

These modifications are of a quantitative nature and do not affect the general conclusions drawn in Section 9.4 concerning the *Weissenberg effect*.

Problems

1. Show that the dissipation function (16.12) is not positive definite for $A > 0$, $B > 0$, $C \neq 0$.

2. Apply the process used in this section to show that the fourth-order approximation of Fourier's law is

$$-\frac{\vartheta_{,i}}{\vartheta} = \tfrac{1}{2}\left[A + \left(Dq_{(1)} + \frac{BD}{A}\, d_{(2)}\right) - \tfrac{1}{2}Gd_{(3)}\right] q_i .$$

Verify that the result, together with (16.31), satisfies (16.15).

16.3. The Green–Rivlin effect

In Section 9.2 we have briefly mentioned the steady *Poiseuille flow* of an incompressible Newtonian liquid in a tube of circular cross section. The corresponding problem for a Reiner–Rivlin liquid has been treated by Green and Rivlin [13]. They have shown that, provided the cross section is not a circle but, for instance an ellipse, the parallel flow is accompanied by a secondary motion of circular character as illustrated in Figure 9.6. Their calculations will be reproduced here for an incompressible non-Newtonian liquid obeying the orthogonality condition.

The motion is governed by the continuity condition (9.20),

$$v_{i,\,i} = 0, \tag{16.35}$$

the kinematic relation (2.14),

$$d_{ij} = \tfrac{1}{2}(v_{j,\,i} + v_{i,\,j}), \tag{16.36}$$

and the theorem of linear momentum (3.9),

$$\varrho v_{i,\,j} v_j = \sigma_{ij,\,j} . \tag{16.37}$$

The quasiconservative stress is given by $(9.2)_1$, and the fourth approximation for the dissipative stress is represented by (16.32) provided we exclude thermal effects. Adding the two stresses, we obtain

$$\sigma_{ij} = -p\delta_{ij} + gd_{ij} + h(d_{ik}d_{kj} - \tfrac{2}{3}d_{(2)}\delta_{ij}), \tag{16.38}$$

where

$$g = 2(\mu + \mu' d_{(2)} - \mu'' d_{(3)}), \qquad h = 4\mu'' d_{(2)} . \tag{16.39}$$

If x_3 is the axis of the tube, the elliptic cross section is given by

$$\frac{x_1^2}{a^2}+\frac{x_2^2}{b^2} = 1, \tag{16.40}$$

and the boundary condition to be satisfied on it is $v_i = 0$.

Let us represent the velocity field in the form

$$v_1 = -\psi_{,2}, \qquad v_2 = \psi_{,1}, \qquad v_3 = 2f \tag{16.41}$$

in terms of two functions $\psi(x_1, x_2)$ and $f(x_1, x_2)$. The flow (16.41) is independent of x_3 and obeys the continuity condition (16.35). From the kinematic relations (16.36) we obtain the rate of deformation tensor

$$d_{ij} = \begin{pmatrix} -\psi_{,12} & \frac{1}{2}(\psi_{,11}-\psi_{,22}) & f_{,1} \\ \frac{1}{2}(\psi_{,11}-\psi_{,22}) & \psi_{,12} & f_{,2} \\ f_{,1} & f_{,2} & 0 \end{pmatrix} \tag{16.42}$$

with the basic invariants

$$\begin{aligned} d_{(2)} &= f_{,1}^2+f_{,2}^2+\psi_{,12}^2+\tfrac{1}{4}(\psi_{,11}-\psi_{,22})^2, \\ d_{(3)} &= (f_{,2}^2-f_{,1}^2)\psi_{,12}+f_{,1}f_{,2}(\psi_{,11}-\psi_{,22}). \end{aligned} \tag{16.43}$$

The stress components (16.38) become

$$\begin{aligned} \sigma_{11} &= -g\psi_{,12} + h[\psi_{,12}^2+\tfrac{1}{4}(\psi_{,11}-\psi_{,22})^2+f_{,1}^2]-q, \\ \sigma_{22} &= g\psi_{,12} + h[\psi_{,12}^2+\tfrac{1}{4}(\psi_{,11}-\psi_{,22})^2+f_{,2}^2]-q, \\ \sigma_{33} &= h(f_{,1}^2+f_{,2}^2) - q, \\ \sigma_{23} &= gf_{,2} + h[f_{,2}\psi_{,12}+\tfrac{1}{2}f_{,1}(\psi_{,11}-\psi_{,22})], \\ \sigma_{31} &= gf_{,1} + h[-f_{,1}\psi_{,12}+\tfrac{1}{2}f_{,2}(\psi_{,11}-\psi_{,22})], \\ \sigma_{12} &= \tfrac{1}{2}g(\psi_{,11}-\psi_{,22})+hf_{,1}f_{,2}, \end{aligned} \tag{16.44}$$

where

$$q = p+\tfrac{2}{3}hd_{(2)}. \tag{16.45}$$

In complex variables

$$z = x_1+ix_2, \qquad \bar{z} = x_1-ix_2 \tag{16.46}$$

the stress components combine in the form

$$\begin{aligned} \sigma_{11}+\sigma_{22} &= 4h(f_{,z}f_{,\bar{z}}+2\psi_{,zz}\psi_{,\bar{z}\bar{z}})-2q, \\ \sigma_{11}-\sigma_{22}+2i\sigma_{12} &= 4ig\psi_{,\bar{z}\bar{z}}+4hf_{,\bar{z}}^2, \\ \sigma_{13}+i\sigma_{23} &= 2gf_{,\bar{z}}+4ihf_{,z}\psi_{,\bar{z}\bar{z}}, \\ \sigma_{23} &= 4hf_{,z}f_{,\bar{z}}-q, \end{aligned} \tag{16.47}$$

where the partial derivatives with respect to z and $\bar{z}$ have to be interpreted in a purely formal way. The invariants (16.43) become

$$d_{(2)} = 4(f_{,z}f_{,\bar{z}}+\psi_{,zz}\psi_{,\bar{z}\bar{z}}), \qquad d_{(3)} = 4i(f_{,z}^2\psi_{,\bar{z}\bar{z}}-f_{,\bar{z}}^2\psi_{,zz}), \tag{16.48}$$

and the equations of motion (16.37) reduce to

$$\begin{aligned} -q_{,\bar{z}}+2[h(f_{,z}f_{,\bar{z}}+2\psi_{,zz}\psi_{,\bar{z}\bar{z}})]_{,\bar{z}}+2(ig\psi_{,\bar{z}\bar{z}}+hf_{,\bar{z}}^2)_{,z} \\ = 2\varrho(\psi_{,z}\psi_{,\bar{z}\bar{z}}-\psi_{,\bar{z}}\psi_{,z\bar{z}}), \\ -q_{,3}+2(gf_{,\bar{z}}+2ihf_{,z}\psi_{,\bar{z}\bar{z}})_{,z}+2(gf_{,z}-2ihf_{,\bar{z}}\psi_{,zz})_{,\bar{z}} \\ = 4i\varrho(f_{,z}\psi_{,\bar{z}}-f_{,\bar{z}}\psi_{,z}). \end{aligned} \tag{16.49}$$

It follows that

$$-q = kx_3+r(z, \bar{z}), \tag{16.50}$$

where

$$\begin{aligned} k+2(gf_{,\bar{z}}+2ihf_{,z}\psi_{,\bar{z}\bar{z}})_{,z}+2(gf_{,z}-2ihf_{,\bar{z}}\psi_{,zz})_{,\bar{z}} \\ =4i\varrho(f_{,z}\psi_{,\bar{z}}-f_{,\bar{z}}\psi_{,z}), \end{aligned} \tag{16.51}$$

and that

$$\begin{aligned} r_{,\bar{z}}+2[h(f_{,z}f_{,\bar{z}}+2\psi_{,zz}\psi_{,\bar{z}\bar{z}})]_{,\bar{z}}+2(ig\psi_{,\bar{z}\bar{z}}+hf_{,\bar{z}}^2)_{\,z} \\ = 2\varrho(\psi_{,z}\psi_{,\bar{z}\bar{z}}-\psi_{,\bar{z}}\psi_{,z\bar{z}}). \end{aligned} \tag{16.52}$$

The calculations reproduced here from [13] result in the relations (16.51) and (16.52). Separating the real and imaginary parts, we easily see that there are three differential equations for the functions ψ, f, and r. The constant k is the negative pressure gradient, and the boundary conditions require that $\psi_{,z} = 0$ and $f = 0$ on the ellipse (16.40).

In the *first two* approximations the liquid is Newtonian, and (16.39) reduces to $g = 2\mu$ and $h = 0$. Denoting the corresponding functions by the superscript 2, we readily see that the problem can be solved under the assumption that

$$\psi = \psi^{(2)} = 0, \tag{16.53}$$

i.e. that the flow is rectilinear. Equations (16.51) and (16.52) reduce to

$$k+8\mu f_{,z\bar{z}} = 0, \qquad r_{,\bar{z}} = 0. \tag{16.54}$$

Thus, $r = r^{(2)}$ is constant, and the function f is subject to the differential equation

$$\Delta f = -\frac{k}{2\mu} \tag{16.55}$$

and to the boundary condition $f = 0$ on the ellipse. The solution is

$$f = f^{(2)} = c\left(1 - \frac{x_1^2}{a^2} - \frac{x_2^2}{b^2}\right), \tag{16.56}$$

where

$$c = \frac{k}{4\mu} \frac{a^2 b^2}{a^2 + b^2}. \tag{16.57}$$

In the *third* approximation the liquid is quasilinear with

$$g = 2(\mu + \mu' d_{(2)}), \qquad h = 0 \tag{16.58}$$

in place of (16.39). We can still assume that

$$\psi = \psi^{(3)} = 0. \tag{16.59}$$

It follows from (16.48) that

$$d_{(2)} = 4f_{,z}f_{,\bar{z}}, \qquad d_{(3)} = 0, \tag{16.60}$$

and from (16.58) we obtain

$$g = 2(\mu + 4\mu' f_{,z}f_{,\bar{z}}). \tag{16.61}$$

Equations (16.51) and (16.52) reduce to

$$k + 8\mu f_{,z\bar{z}} + 16\mu'(f_{,z}^2 f_{,\bar{z}\bar{z}} + 4f_{,z}f_{,\bar{z}}f_{,z\bar{z}} + f_{,\bar{z}}^2 f_{,zz}) = 0 \tag{16.62}$$

and $r_{,\bar{z}} = 0$. Thus, $r = r^{(3)}$ is still constant, and the differential equation for f, written in terms of x_1 and x_2, becomes

$$k + 2\mu\,\Delta f + 2\mu'[(3f_{,1}^2 + f_{,2}^2)f_{,11} + 4f_{,1}f_{,2}f_{,12} + (f_{,1}^2 + 3f_{,2}^2)f_{,22}] = 0. \quad (16.63$$

Let us content ourselves with an approximate solution, valid for small values of μ' and based on the expansions

$$f = f^{(2)} + \mu' f^* + \ldots, \qquad k = k^{(2)} + \mu' k^* + \ldots, \tag{16.64}$$

where $f^{(2)}$ and $k^{(2)}$ denote the second-order approximations, satisfying (16.56), i.e.

$$f^{(2)} = c^{(2)}\left(1 - \frac{x_1^2}{a^2} - \frac{x_2^2}{b^2}\right) \quad \text{with} \quad c^{(2)} = \frac{k^{(2)}}{4\mu} \frac{a^2 b^2}{a^2 + b^2}. \tag{16.65}$$

Equation (16.63) yields

$$k^* + 2\mu\,\Delta f^* + 2(3f_{,1}^{(2)2} + f_{,2}^{(2)2})f_{,11}^{(2)} + 8f_{,1}^{(2)}f_{,2}^{(2)}f_{,12}^{(2)} + 2(f_{,1}^{(2)2} + 3f_{,2}^{(2)2})f_{,22}^{(2)} = 0 \tag{16.66}$$

or, on account of (16.65),

$$k^* + 2\mu\,\Delta f^* - \frac{16c^{(2)3}}{a^2b^2}\left(\frac{a^2+3b^2}{a^4}\,x_1^2 + \frac{3a^2+b^2}{b^4}\,x_2^2\right) = 0. \tag{16.67}$$

The solution of (16.67) satisfying the boundary condition $f^* = 0$ is

$$f^* = (A'x_1^2 + B'x_2^2 + C')\left(1 - \frac{x_1^2}{a^2} - \frac{x_2^2}{b^2}\right), \tag{16.68}$$

where

$$\begin{aligned} A' &= \frac{2c^{(2)3}}{3\mu}\,\frac{3a^6 - 5a^4b^2 - 19a^2b^4 - 3b^6}{a^4b^2(a^4+6a^2b^2+b^4)}, \\ B' &= \frac{2c^{(2)3}}{3\mu}\,\frac{-3a^6 - 19a^4b^2 - 5a^2b^4 + 3b^6}{a^2b^4(a^4+6a^2b^2+b^4)}, \\ C' &= -\frac{2c^{(2)3}}{3\mu}\,\frac{3a^8 + 16a^6b^2 + 10a^4b^4 + 16a^2b^6 + 3b^8}{a^2b^2(a^2+b^2)(a^4+6a^2b^2+b^4)} + \frac{k^*}{4\mu}\,\frac{a^2b^2}{a^2+b^2}. \end{aligned} \tag{16.69}$$

Inserting (16.68) and (16.69) in (16.64), we note that $c^{(2)}$ may be replaced by c and that the third approximation finally becomes

$$f^{(3)} = [c + \mu'(Ax_1^2 + Bx_2^2 + C)]\left(1 - \frac{x_1^2}{a^2} - \frac{x_2^2}{b^2}\right), \tag{16.70}$$

where c is still given by (16.57) and

$$\begin{aligned} A &= \frac{2c^3}{3\mu}\,\frac{3a^6 - 5a^4b^2 - 19a^2b^4 - 3b^6}{a^4b^2(a^4+6a^2b^2+b^4)}, \\ B &= \frac{2c^3}{3\mu}\,\frac{-3a^6 - 19a^4b^2 - 5a^2b^4 + 3b^6}{a^2b^4(a^4+6a^2b^2+b^4)}, \\ C &= -\frac{2c^3}{3\mu}\,\frac{3a^8 + 16a^6b^2 + 10a^4b^4 + 16a^2b^6 + 3b^8}{a^2b^2(a^2+b^2)(a^4+6a^2b^2+b^4)}. \end{aligned} \tag{16.71}$$

It thus turns out that the only difference with respect to the first two approximations is a modification of the velocity component v_3.

To stress the influence of the function h, we simplify the *fourth* approximation representing a truly nonlinear liquid by confining ourselves to the case $\mu' = 0$. We thus obtain

$$g = 2(\mu - \mu''d_{(3)}), \qquad h = 4\mu''d_{(2)} \tag{16.72}$$

in place of (16.39). Approximating for small values of μ'', we have

$$f = f^{(2)} + \mu'' f^* + \ldots, \qquad \psi = \mu'' \psi^* + \ldots, \qquad r = r^{(2)} + \mu'' r^* + \ldots \tag{16.73}$$

with

$$k = k^{(2)} + \mu'' k^* + \ldots . \tag{16.74}$$

Equations (16.48) now yield

$$\begin{aligned} d_{(2)} &= 4(f^{(2)}_{,z} f^{(2)}_{,\bar{z}} + \mu'' f^{(2)}_{,z} f^{*}_{,\bar{z}} + \mu'' f^{(2)}_{,\bar{z}} f^{*}_{,z}), \\ d_{(3)} &= 4i\mu''(f^{(2)}_{,z} \psi^{*}_{,\bar{z}\bar{z}} - f^{(2)}_{,\bar{z}} \psi^{*}_{,zz}), \end{aligned} \tag{16.75}$$

and from (16.51), (16.52) we obtain

$$k^* + 8\mu f^{*}_{,z\bar{z}} = 4i\varrho(f^{(2)}_{,z} \psi^{*}_{,\bar{z}} - f^{(2)}_{,\bar{z}} \psi^{*}_{,z}), \tag{16.76}$$

$$r^{*}_{,\bar{z}} + 4i\mu\psi^{*}_{,z\bar{z}\bar{z}} + 32(f^{(2)2}_{,z} f^{(2)2}_{,\bar{z}})_{,\bar{z}} + 32(f^{(2)}_{,z} f^{(2)3}_{,\bar{z}})_{,z} = 0. \tag{16.77}$$

The last two relations, supplemented by the boundary conditions $\psi^{*}_{,z} = 0$ and $f^* = 0$, determine the functions ψ^*, f^*, and r^*. The last one may be eliminated by differentiating (16.77) with respect to z and subtracting the result from its conjugate complex. The resulting differential equation is

$$i\mu\psi^{*}_{,zz\bar{z}\bar{z}} + 8(f^{(2)}_{,z} f^{(2)3}_{,\bar{z}})_{,zz} - 8(f^{(2)3}_{,z} f^{(2)}_{,\bar{z}})_{,\bar{z}\bar{z}} = 0. \tag{16.78}$$

In complex coordinates the equation of the ellipse (16.40) reads

$$(a^2 + b^2)z\bar{z} - \tfrac{1}{2}(a^2 - b^2)(z^2 + \bar{z}^2) - 1 = 0. \tag{16.79}$$

According to (16.65), the function $f^{(2)}$ is thus given by

$$f^{(2)} = \frac{c^{(2)}}{2a^2b^2}\,[2a^2b^2 - (a^2 + b^2)z\bar{z} + \tfrac{1}{2}(a^2 - b^2)(z^2 + \bar{z}^2)]. \tag{16.80}$$

Calculating the partial derivatives with respect to z and $\bar{z}$ and inserting them in (16.78), we obtain

$$i\mu\psi^{*}_{,zz\bar{z}\bar{z}} = 12c^{(2)4}\,\frac{a^4 - b^4}{a^6b^6}\,(z^2 - \bar{z}^2). \tag{16.81}$$

Let us try a solution of the form

$$i\mu\psi^* = A[(a^2 + b^2)z\bar{z} - \tfrac{1}{2}(a^2 - b^2)(z^2 + \bar{z}^2) - 2a^2b^2]\,(z^2 - \bar{z}^2), \tag{16.82}$$

where A is a constant. It satisfies the boundary condition for ψ^* and also the differential equation (16.81) provided

$$A = \frac{2c^{(2)4}}{a^6b^6} \frac{a^4-b^4}{4(a^2+b^2)^2-(a^2-b^2)^2}. \tag{16.83}$$

Inserting this in (16.82), we obtain the function ψ^*, and if we use $(16.73)_2$ and transform the result back to cartesian coordinates, we finally have

$$\psi = 16 \frac{\mu'' c^{(2)4}}{\mu} \frac{a^4-b^4}{a^6b^6} \frac{(b^2x_1^2+a^2x_2^2-a^2b^2)x_1x_2}{4(a^2+b^2)^2+(a^2-b^2)^2}. \tag{16.84}$$

The corresponding approximation of the function f could now be obtained by integration of (16.76). Since it merely results in a modification of the velocity component v_3, we omit its calculation, and we content ourselves with the observation that the secondary flow in the cross section as represented by (16.84) has the circulatory character sketched in Figure 9.6.

Problem

Carry out the intermediate calculations omitted in the text.

CHAPTER 17

PLASTICITY

The perfectly plastic body has been defined in Chapter 10 as a limiting case of the isotropic incompressible viscous liquid. Its quasiconservative stress is hydrostatic and has the character of a reaction. The deformation rate and the dissipative stress are deviators with coinciding principal axes. Their connection is such that for stresses below the yield limit the deformation rate is zero, whereas for stress states at the yield limit it is determined by the theory of the plastic potential.

It has been noted in Section 10.2 that the theory of the plastic potential was originally a mere postulate. Using the transition from the viscous liquid to the plastic body we will show in this chapter that the theory of the plastic potential is a consequence of the orthogonality condition established in Section 14.3. Moreover, we will discuss a generalization of the plastic potential.

17.1. The orthogonality condition

In Section 10.2 we have seen that in the space of the principal stresses $\sigma_{\mathrm{I}}, \ldots$ the yield surface of a perfectly plastic body is a cylinder (Fig. 10.3) the axis g of which forms equal angles with the coordinate axes $\sigma_{\mathrm{I}}, \ldots$. The equation of this cylinder is given by (10.37), where X is the yield function, dependent on the basic invariants of the stress deviator or, in other words, of the dissipative stress. The hydrostatic or quasi-conservative stress is represented by the projection of the vector $\boldsymbol{\sigma} = (\sigma_{\mathrm{I}}, \ldots)$ on the axis g; the deviatoric or dissipative stress is its projection $\boldsymbol{\sigma}'$ onto the plane $\sigma_{\mathrm{I}} + \ldots = 0$.

Since the yield function depends on the basic invariants of the stress deviator alone, the following discussion may be limited, as far as the stress is concerned, to the plane of Figure 10.7 or 10.12, containing the vector $\boldsymbol{\sigma}'$.

For convenience we will denote it briefly as the *stress plane*, and the intersection of the yield cylinder with this plane will be referred to as the *yield locus*. Since the deformation rate is deviatoric, its components $d_{\mathrm{I}}, \ldots$ may be represented in a similar manner by a vector $\boldsymbol{d} = (d_{\mathrm{I}}, \ldots)$ lying in the plane $d_{\mathrm{I}} + \ldots = 0$, which will be referred to as the *plane of the deformation rate*.

In order to obtain the yield locus by means of a limiting process applied to an incompressible viscous liquid, let us first consider the quasilinear liquid with the constitutive equations $(9.2)_1$ and (9.72), where p is a reaction and μ the viscosity. Acording to Section 16.1 the orthogonality condition requires that μ, provided we neglect its dependence on $q_{(1)}$, is a function of $d_{(2)}$ alone. The stress deviator is thus given by

$$\sigma'_{ij} = \sigma^{(d)}_{ij} = 2\mu(d_{(2)})d_{ij}, \tag{17.1}$$

and the dissipation function per unit volume becomes

$$\varrho\varphi = \sigma'_{ij}d_{ij} = 4\mu(d_{(2)})d_{(2)} \tag{17.2}$$

as already indicated in (14.9).

Since d_{ij} is a deviator, it follows from $(1.59)_2$ that

$$d_{(2)} = -d_{\mathrm{I}}d_{\mathrm{II}} - \ldots = -\tfrac{1}{2}[(d_{\mathrm{I}} + \ldots)^2 - d_{\mathrm{I}}^2 - \ldots] = \tfrac{1}{2}(d_{\mathrm{I}}^2 + \ldots). \tag{17.3}$$

This equation shows that in the plane of the deformation rate $d_{(2)}$ and hence the function $\Phi = \varrho\varphi$ are constant on circles around the origin. On account of (17.1) the vector $\boldsymbol{\sigma}'$ has the direction of $\boldsymbol{d}$. The orthogonality condition is thus satisfied in the plane of the deformation rate, and it is easy to verify that the circles just mentioned have the properties proved in Section 14.5 for dissipation surfaces of arbitrary dimension, provided the function $\mu(d_{(2)})$ is chosen in such a manner that the connection between the vectors $\boldsymbol{d}$ and $\boldsymbol{\sigma}'$ and its inverse are single-valued.

On account of (17.1)

$$\sigma'_{(2)} = 4\mu^2(d_{(2)})d_{(2)}\,. \tag{17.4}$$

Solving this equation for $d_{(2)}$ and substituting the result in (17.2), we obtain $\Phi' = \varrho\varphi$ as a function of $\sigma'_{(2)}$. Equation $(1.59)_2$, applied to $\boldsymbol{\sigma}'$, yields

$$\sigma'_{(2)} = \tfrac{1}{2}(\sigma_{\mathrm{I}}'^2 + \ldots). \tag{17.5}$$

Thus, Φ' is constant on circles in the stress plane with centers at the origin. Again, it is easy to verify that the orthogonality condition is satisfied in the stress plane and that the circles have the properties proved in Section 14.5.

To define the liquid completely it is necessary to specify μ as a function of $d_{(2)}$ or $\sigma'_{(2)}$. Let us assume that this is done in such a way that the mapping between the planes of the stress and the deformation rate is single-valued, and let us represent the dissipation function in the stress plane by means of equidistant curves, i.e. by circles on which the values of Φ' form an algebraic progression starting from zero. Modifying the viscosity function μ, it is obviously possible without violating the results of Section 14.5 to concentrate the entire family of these circles more and more in the vicinity of a single circle with the equation

$$X(\sigma'_{(2)}) = \sigma'_{(2)} - k^2 = 0, \tag{17.6}$$

where k is an arbitrary real constant. Proceeding to the limit, we obtain a dissipation function that is zero inside this circle, assumes all positive values on the circle and is not defined outside of it. In this manner the liquid becomes a perfectly plastic body. Equation (17.6) represents its yield condition (10.35) in the sense of Maxwell, and the orthogonality condition, which is preserved in the process, becomes the principal statement of the theory of the plastic potential, denoted by (*b*) in Section 10.2 and represented analytically by (10.42). Since the circle is a convex curve, the other statement of the theory of the plastic potential, denoted by (*a*) in Section 10.2, is also satisfied. Moreover, the results remain valid in the entire stress space. Here (17.6) represents the circular and hence convex yield cylinder of axis g, and the vector $\boldsymbol{d}$ is orthogonal to it at the end point of $\boldsymbol{\sigma}$.

The first equality (17.2) shows that the deformation rate $\boldsymbol{d}$ is zero for stresses below the yield limit and that the magnitude of the vector $\boldsymbol{d}$ corresponding to a given stress at the yield limit remains indetermined. In the plane of the deformation rate the function Φ is zero at the origin and increases linearly on any ray emanating from it. The origin in the plane of the deformation rate corresponds to the entire interior of the yield locus in the stress plane, and each ray emanating from the origin corresponds to a single point on this circle. It follows that in the limit the connection between the vectors $\boldsymbol{d}$ and $\boldsymbol{\sigma}'$ is no longer single-valued.

The limiting process just considered has been based on a quasilinear viscous liquid. In the general case of a truly nonlinear liquid the constitutive equation (17.1) has to be replaced by (16.6), where ν is determined by (16.7). If we again neglect the influence of $q_{(1)}$, the dissipation function now depends on $d_{(2)}$ and $d_{(3)}$. The dissipation curves in the planes of the deformation rate and of the stress are not circles any longer, but they still have the properties proved in Section 14.5, provided the connection between $\boldsymbol{d}$ and $\boldsymbol{\sigma}'$ and its inverse are single-valued. By means of a limiting process analogous to the one considered above it is possible to obtain a perfectly plastic body with an arbitrary convex yield locus [e.g. the hexagon (Fig. 10.8) of Tresca]. The corresponding yield surface in the entire stress space is a (noncircular) cylinder or prism with axis g, and it is again obvious that in this way statement (b) of the theory of the plastic potential is obtained as a consequence of the orthogonality condition.

The argumentation presented here leaves two questions open. In the first place it is not applicable to nonconvex yield surfaces; it therefore remains to be shown that the orthogonality condition precludes their existence. In the second place we have restricted ourselves to the representation of $\boldsymbol{d}$ and $\boldsymbol{\sigma}$ in the (3-dimensional) spaces of the principal values $d_{\mathrm{I}}, \ldots$ and $\sigma_{\mathrm{I}}, \ldots$. It thus remains to be shown that the theory of the plastic potential remains valid in an arbitrary (6- or 9-dimensional) stress space σ_{ij}. Both problems will be dealt with in Section 17.2.

17.2. The yield surface

In order to show that the yield surface of a perfectly plastic body is always convex, let us start with a coordinate system coinciding with the principal axes, and let us represent the deformation rate by a vector $\boldsymbol{d}$ in the plane $d_{\mathrm{I}}+\ldots=0$, in the way it has been done in Section 17.1. We have shown in connection with (10.48) that the dissipation rate is uniquely determined by the deformation rate; the dissipation function is thus a single-valued function in this plane, and it is reasonable to assume that it is continuous.

The stress deviator may be represented by a vector $\boldsymbol{\sigma}'$ in the plane $\sigma_{\mathrm{I}}+\ldots=0$. The yield locus is a continuous curve F (Fig. 17.1) in this plane, closed and star-shaped with respect to the origin O′. We know from Tresca's yield condition that this curve may contain corners. Let us assume however that it is piecewise smooth. The vector $\boldsymbol{d}$ corresponding to a stress state at the yield limit obeys the orthogonality condition, i.e. it is normal to the tangent in regular points P′ of the yield locus. Since $\boldsymbol{\sigma}'\cdot\boldsymbol{d}>0$, the vector $\boldsymbol{d}$ points to the outside of F, and its magnitude is indetermined.

In the plane of the deformation rate (Fig. 17.2) the vectors $\boldsymbol{d}$ corresponding to a stress state $\boldsymbol{\sigma}'$ at the yield limit lie on a ray s emanating from the origin O. On any such ray the dissipation function

$$\Phi=\varrho\varphi=\boldsymbol{\sigma}'\cdot\boldsymbol{d}\geqslant 0 \tag{17.7}$$

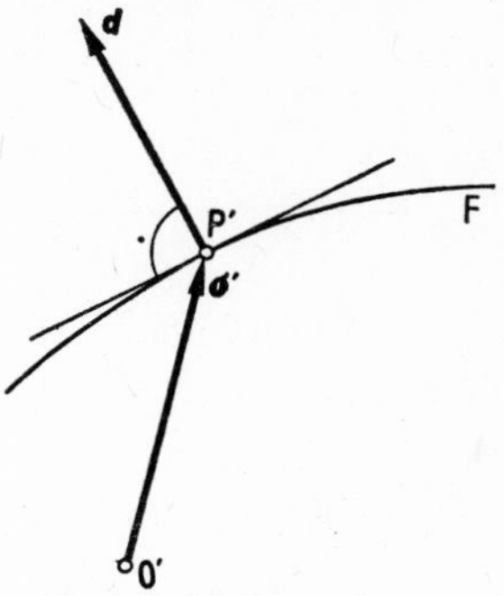

Fig. 17.1 Yield surface

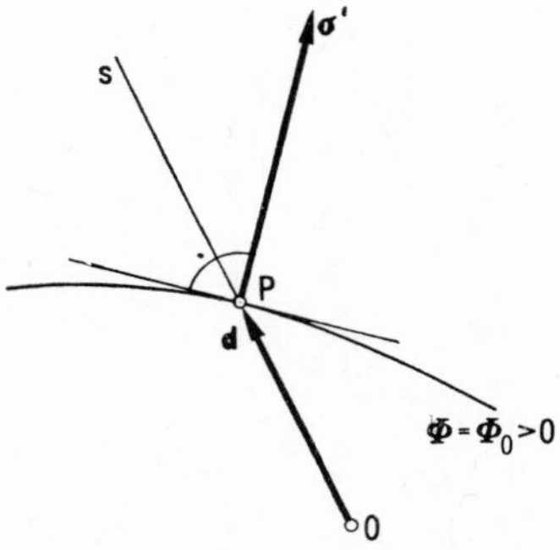

Eig. 17.2 Dissipation surface

increases proportional to the magnitude of $\boldsymbol{d}$. There is thus a single point P on any ray s in which the dissipation function assumes a given value $\Phi(\boldsymbol{d}) = \Phi_0 > 0$. It follows that the dissipation function may be represented by a family of similar and finite dissipation curves each of which is star-shaped with respect to the origin O and is an image of the yield locus F.

On account of (17.7) corresponding increments of $\boldsymbol{d}$ and $\boldsymbol{\sigma}'$ are connected by the relation

$$\boldsymbol{\sigma}' \cdot \delta \boldsymbol{d} + \boldsymbol{d} \cdot \delta \boldsymbol{\sigma}' = \delta \Phi(\boldsymbol{d}). \tag{17.8}$$

Provided the vector $\delta \boldsymbol{d}$ connects two adjacent points on the curve $\Phi = \Phi_0$, the vector $\delta \boldsymbol{\sigma}'$ connects two points on its image F. In this case the right-hand side of (17.8) is zero, and the second product on the left vanishes on account of the orthogonality condition in the stress plane. We thus have $\boldsymbol{\sigma}' \cdot \delta \boldsymbol{d} = 0$, and this implies that the orthogonality condition is also valid in the plane of the deformation rate: in any regular point P of a curve $\Phi(\boldsymbol{d}) = \Phi_0$ the vector $\boldsymbol{\sigma}'$ is normal to the tangent; besides, it is directed away from O on account of (17.7).

Let us assume for the moment that the yield locus F is smooth and thus has a single tangent at each of its points. If F were not convex, at least one of these tangents t would not be a supporting line (Fig. 17.3). The tangent

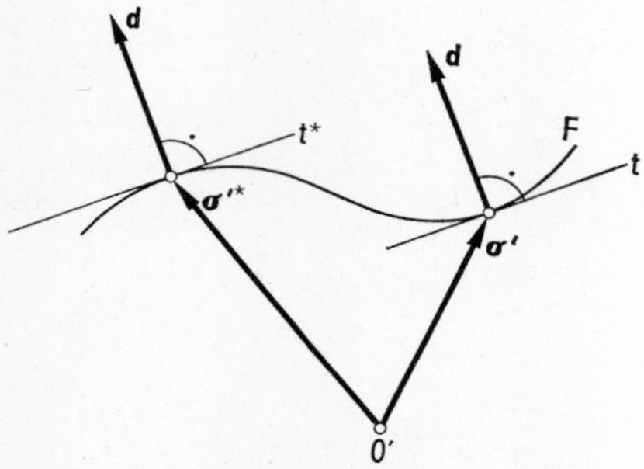

Fig. 17.3 Convexity

t and the supporting line t^* parallel to it on the same side of the origin O′ would define two stress vectors $\boldsymbol{\sigma}'$ and $\boldsymbol{\sigma}'^*$ for which the deformation rate $\boldsymbol{d}$ could be equal. On account of the inequality

$$\boldsymbol{\sigma}'^* \cdot \boldsymbol{d} \neq \boldsymbol{\sigma}' \cdot \boldsymbol{d} \tag{17.9}$$

the dissipation function would not be single-valued on the ray s in Figure 17.2. It follows that the yield locus F is convex.

If the yield locus is only piecewise smooth, it can be approximated with any desired accuracy by means of smooth curves as long as it does not contain cusps. Singularities of this type however violate either the definiteness of Φ or the condition that F is star-shaped with respect to O′. Thus, the yield locus is convex under the conditions stated at the beginning of this section.

It is clear that the result remains valid for the entire yield cylinder in the space $\sigma_{\mathrm{I}}, \ldots$. Incidentally, it can be shown [34] that the conditions to which the yield surface has been subjected here may be considerably relaxed and that, in particular, the result is not restricted to incompressible plastic materials.

In connection with Section 17.1 the foregoing discussion establishes the theory of the plastic potential as a consequence of the orthogonality condition in the sense of Section 14.3. It remains to be shown that it is not restricted to the spaces $d_{\mathrm{I}}, \ldots$ and $\sigma_{\mathrm{I}}, \ldots$ corresponding to principal axes.

In Section 10.2 we have seen that the theory of the plastic potential is equivalent to a principle of maximal dissipation rate, stating that the actual power of dissipation corresponding to a given deformation rate is never less than the fictitious power of dissipation calculated from the actual deformation rate and an arbitrary stress state at or below the yield limit. We thus know that

$$(\sigma_\mathrm{I}-\sigma_\mathrm{I}^*)d_\mathrm{I}+\ \ldots \geqslant 0, \tag{17.10}$$

where $d_\mathrm{I}, \ldots$ stands for the actual deformation rate, $\sigma_\mathrm{I}, \ldots$ for the corrresponding stress and $\sigma_\mathrm{I}^*, \ldots$ for an arbitrary stress state at or below the yield limit. The left-hand side of (17.10) is a scalar. If we use an arbitrary coordinate system, it takes the form $(\sigma_{ij}-\sigma_{ij}^*)d_{ij}$; consequently,

$$(\sigma_{ij}-\sigma_{ij}^*)d_{ij} \geqslant 0. \tag{17.11}$$

This is the expression of the principle of maximal power of dissipation in arbitrary coordinates, and it follows that the theory of the plastic potential is generally valid.

17.3. A generalization

We have restricted ourselves so far to the consideration of an element in a perfectly plastic body, and we have finally expressed the entire content of the theory of the plastic potential by means of the inequality (17.11), valid for arbitrary stresses σ_{ij} and σ_{ij}^* that are, respectively, at the yield limit and at or below this limit. Let us note that (17.11) remains valid (with the equality sign) for stresses σ_{ij} below the yield limit, since here $d_{ij} = 0$, and let us recall that stresses beyond the yield limit do not exist.

Proceeding to the consideration of a finite body V (cf. [35]) we note that its state of stress is defined by the stress field $\sigma_{ij}(x_k)$ within V. In a similar manner the field $d_{ij}(x_k)$ describes the global deformation rate. The fields $d_{ij}(x_k)$ and $\sigma_{ij}(x_k)$ may be represented by means of vectors $\boldsymbol{D}$ and $\boldsymbol{S}$ respectively in two function spaces (see, for instance [36]). If we define the scalar product of these vectors by the integral

$$\boldsymbol{S}\cdot\boldsymbol{D} = \int \sigma_{ij}d_{ij}\,\mathrm{d}V, \tag{17.12}$$

it represents the global power of dissipation $L^{(d)}$.

Let us assume now that the finite body is subjected to a slowly increasing load. As long as this load is sufficiently small, all local stresses lie below the local yield limit, and the entire body remains rigid. With increasing load the local yield limit is reached (but never exceeded) in certain parts of the body: they become plastified. In general the rigidity of the non-plastified portions of the body prevents plastic flow until the plastified zones have sufficiently spread out. When plastic flow finally sets in, we say that the global yield limit of the body is reached. The corresponding stress field is represented by a vector $\boldsymbol{S}$ in function space, and the end points of all such vectors define a hypersurface: the *global yield surface* in this space. If $\boldsymbol{S}^*$ represents an arbitrary stress state $\sigma_{ij}^*(x_k)$ at or below the global yield limit, σ_{ij}^* lies everywhere either at or below the local yield limit. It follows from (17.12) and (17.11) that

$$(\boldsymbol{S}-\boldsymbol{S}^*)\cdot\boldsymbol{D} = \int (\sigma_{ij}-\sigma_{ij}^*)d_{ij}\,\mathrm{d}V \geqslant 0. \tag{17.13}$$

The theory of the plastic potential thus remains valid for the body as a whole.

The result obtained here is mainly important in connection with a form of the theory which is due to Prager [37] and has proved highly fruitful for the solution of practical problems. It exploits (17.13) on the level of strength of materials, where the deformation can be described by a finite set of *generalized strains* q_k ($k = 1, 2, \ldots, n$) and the state of stress by the corresponding *generalized stresses* Q_k, the total dissipation rate being given by

$$L^{(d)} = Q_k\dot{q}_k\,. \tag{17.14}$$

According to Prager the theory of the plastic potential remains valid in the n-dimensional space of the stresses Q_k.

As an example let us discuss a problem already considered by Prager. Figure 17.4 shows a segment of unit length of a prismatic beam with rectangular cross section, subjected to tension and flexure. The states of stress and strain within this segment are actually complicated. Provided the unit of

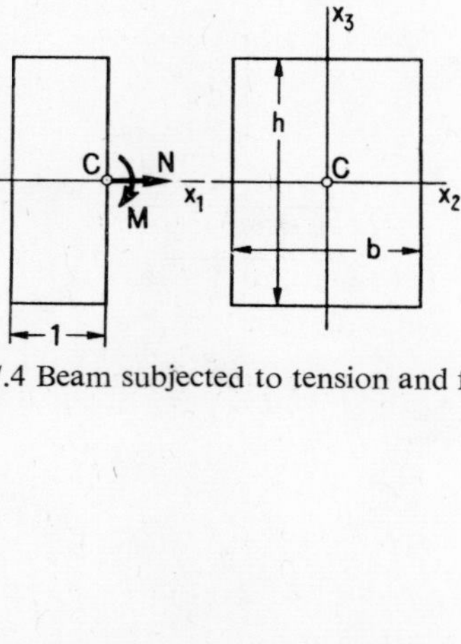

Fig. 17.4 Beam subjected to tension and flexure

length has been chosen sufficiently small, they are described by six stress and strain components, all of them functions of x_2 and x_3. It is well-known, however, that in a first approximation the state of stress is uniaxial with x_1 as axis and that the extension in this direction is a linear function of x_3. The deformation is described with sufficient accuracy (Fig. 17.5) by the extension ε of the center line and the angle $\varkappa$ between the two cross sections defin-

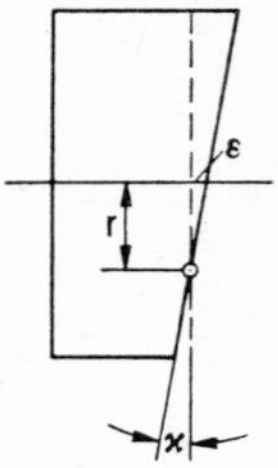

Flg. 17.5 Idealized deformation

ing the segment (the curvature of the center line), i.e. by two generalized strains $q_1 = \varepsilon$ and $q_2 = \varkappa$. The extension of a fibre with coordinates x_2, x_3 is $\varepsilon_{11} = \varepsilon + x_3\varkappa$. The total dissipation rate is thus

$$L^{(d)} = \int \sigma(\dot{\varepsilon} + x_3\dot{\varkappa})\,\mathrm{d}A = N\dot{\varepsilon} + M\dot{\varkappa}, \tag{17.15}$$

where $\sigma = \sigma_{11}$ is the only non-vanishing stress component and

$$N = \int \sigma\,\mathrm{d}A, \qquad M = \int x_3\sigma\,\mathrm{d}A \tag{17.16}$$

are the normal force and the bending moment respectively. It follows from (17.15) that the generalized stresses are $Q_1 = N$ and $Q_2 = M$.

Figure 17.6, where σ_0 denotes the yield stress in simple tension or compression, shows the stress distribution corresponding to the deformation of Figure 17.5. If r indicates the location of the neutral fibres, we obviously have

$$\varepsilon = r\varkappa. \tag{17.17}$$

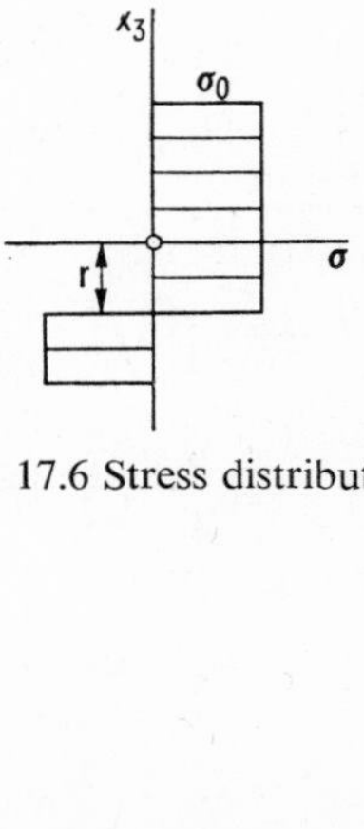

Fig. 17.6 Stress distribution

In terms of r we easily calculate

$$N = \pm 2br\sigma_0, \qquad M = \pm \frac{b}{4}(h^2 - 4r^2)\sigma_0 \quad \left(-\frac{h}{2} \leqslant r \leqslant \frac{h}{2}\right) \tag{17.18}$$

and

$$N = \pm bh\sigma_0, \qquad M = 0 \quad \left(r \leqslant -\frac{h}{2},\ r \geqslant \frac{h}{2}\right). \tag{17.19}$$

Eliminating r from the two equations (17.18) we obtain the yield condition

$$M = \pm \frac{bh^2}{4}\sigma_0 \left[1 - \left(\frac{N}{bh\sigma_0}\right)^2\right]. \tag{17.20}$$

Thus, the yield locus in the plane N, M consists of the two parabolic arcs of Figure 17.7, and it is easy to verify (P1) that the theory of the plastic potential holds in this plane.

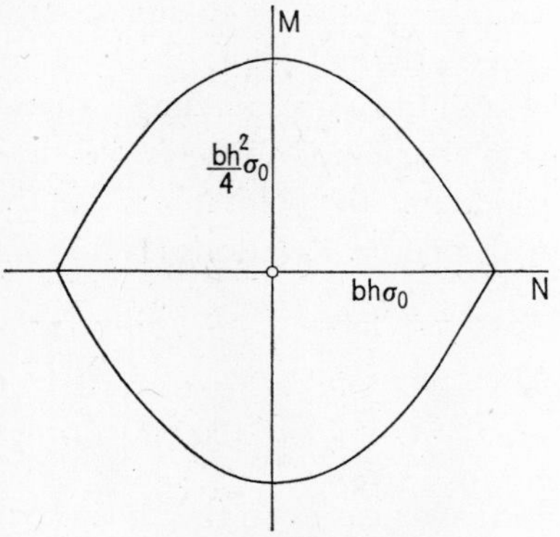

Fig. 17.7 Yield locus

The example considered here contains all elements that are typical in Prager's theory. We know that under normal conditions the state of stress in the beam is practically uniaxial. We may therefore replace the beam by a bundle of independent longitudinal fibres. We also know that the cross sections remain practically plane, and we therefore assume that the end sections of the segment are constrained so that the axial extension is proportional to x_3 (whereas the lateral contractions are free). In this manner we obtain an idealized model of the body. Subjecting this model to the theory of the plastic potential, we arrive at an approximation for the behavior of the actual beam.

It is clear that the deformations of the model are restricted compared to those of the actual segment. Let $\boldsymbol{E}^{(1)}$ denote all vectors in function space representing simple extension with $\varepsilon = 1$ of the model (with possibly different lateral contractions), and let $\boldsymbol{E}^{(2)}$ stand for all vectors representing simple flexure with $\varkappa = 1$. The deformation rate of the model has then the form

$$\boldsymbol{D} = \boldsymbol{E}^{(1)}\dot{\varepsilon} + \boldsymbol{E}^{(2)}\dot{\varkappa}, \tag{17.21}$$

and the dissipation rate is given by

$$L^{(d)} = \boldsymbol{S}\cdot\boldsymbol{D} = \boldsymbol{S}\cdot\boldsymbol{E}^{(1)}\dot{\varepsilon} + \boldsymbol{S}\cdot\boldsymbol{E}^{(2)}\dot{\varkappa}. \tag{17.22}$$

Comparing this to (17.15) we obtain

$$N = \boldsymbol{S}\cdot\boldsymbol{E}^{(1)}, \qquad M = \boldsymbol{S}\cdot\boldsymbol{E}^{(2)}, \tag{17.23}$$

where the fact that the lateral contractions have not been specified is irrelevant since they do not contribute to the dissipation rate.

If $\boldsymbol{D}$ is different from zero, the end point of the vector $\boldsymbol{S}$ lies on the global yield limit of the model, and the corresponding values of N and M define a point of the yield locus in the plane N, M. If $\boldsymbol{S}^*$ denotes an arbitrary stress state at or below the yield limit, the point N^*, M^* in the plane of the generalized stress has the same property. Relations (17.13) and (17.21) yield

$$(\boldsymbol{S} - \boldsymbol{S}^*)\cdot(\boldsymbol{E}^{(1)}\dot{\varepsilon} + \boldsymbol{E}^{(2)}\dot{\varkappa}) \geqslant 0 \tag{17.24}$$

or on account of (17.23)

$$(N - N^*)\dot{\varepsilon} + (M - M^*)\dot{\varkappa} \geqslant 0. \tag{17.25}$$

This explains in the particular case considered here why the theory of the plastic potential is satisfied in generalized stresses and strain rates.

The extension of the result to arbitrary generalized strains and the corresponding stresses is straightforward. First, the actual body must be replaced by a simplified model. This has to be done in such a way that the deformation, as far as it contributes to the work of dissipation, is described by a finite set q_k ($k = 1, 2, \ldots, n$) of generalized strains. In function space the corresponding vectors $\boldsymbol{E}^{(k)}$ define certain subspaces. The deformation rate of the model has the form

$$\boldsymbol{D} = \boldsymbol{E}^{(k)}\dot{q}_k, \tag{17.26}$$

and the dissipation rate is given by

$$L^{(d)} = \boldsymbol{S}\cdot\boldsymbol{D} = \boldsymbol{S}\cdot\boldsymbol{E}^{(k)}\dot{q}_k. \tag{17.27}$$

This value is independent of the particular choice of the vectors $\boldsymbol{E}^{(k)}$ within their corresponding subspaces, for the proper choice of the q_k as described above ensures that the scalar product $\boldsymbol{S}\cdot\boldsymbol{E}^{(k)}$ is the same for all vectors $\boldsymbol{E}^{(k)}$ corresponding to a given q_k. These scalar products define the generalized stresses

$$Q_k = \boldsymbol{S}\cdot\boldsymbol{E}^{(k)}, \tag{17.28}$$

and from (17.13), (17.26), and (17.28) we obtain

$$(\boldsymbol{S}-\boldsymbol{S}^*)\cdot\boldsymbol{E}^{(k)}\dot{q}_k = (Q_k - Q_k^*)\dot{q}_k \geqslant 0. \tag{17.29}$$

This proves the theory of the plastic potential in the space Q_k.

Problems

1. Verify the theory of the plastic potential for the problem of Figure 17.4.

2. Discuss the theory of the plastic potential for a circular shaft subjected to tension and torsion.

CHAPTER 18

VISCOELASTIC BODIES

An introduction to the treatment of viscoleastic bodies has been given in Chapter 11. There the concept of internal parameters has been developed with the aid of one-dimensional models, and it has been shown how the response of a linear model can be described by a differential or an integral equation. The constitutive relations of linear viscoelastic materials may be obtained as generalizations of these equations; to set them up it is not necessary to make further use of the model concept. A few examples have been discussed in Section 11.3, and it has been pointed out that in these cases the response is determined by the specific free energy and the specific dissipation function.

In the present chapter we will generalize the last result. In other words, we will accept the orthogonality principle formulated in Section 5.1, and we propose to show that in consequence the response of a viscoelastic body is governed by its free energy and its dissipation function.

To simplify the discussion, let us assume that the entropy production due to heat exchange is negligible compared to that caused by deformation. Besides, we restrict ourselves to the isothermal case, so that the temperature does not appear as a state variable.

18.1. Internal parameters

Viscoelastic materials are neither elastic bodies nor purely viscous fluids. Their response is characterized by the simultaneous presence of quasi-conservative and dissipative stresses, and as a rule they contain internal parameters. Like the strain components, these internal parameters appear in the form of symmetric second-order tensors, and the various types of viscoelastic bodies are distinguished by any number (including zero) of such tensors.

As a typical example, let us consider a viscoelastic body with a single set of internal parameters in the form of the components of a symmetric tensor α_{ij}. Since the temperature is assumed to be constant, the strains ε_{ij} and the internal parameters α_{ij} are the only independent state variables. The specific free energy is a function of the form $\psi(\varepsilon_{ij}, \alpha_{kl})$, and the specific dissipation function is given by $\varphi(\dot{\varepsilon}_{ij}, \dot{\alpha}_{kl})$ if we neglect a possible dependence on the state variables ε_{ij} and α_{kl}. According to (4.37)$_1$ the quasiconservative stress tensor is

$$\sigma_{ij}^{(q)} = \varrho \frac{\partial \psi}{\partial \varepsilon_{ij}}, \tag{18.1}$$

and on account of the orthogonality principle the dissipative stress is given by

$$\sigma_{ij}^{(d)} = \nu \varrho \frac{\partial \varphi}{\partial \dot{\varepsilon}_{ij}}. \tag{18.2}$$

The internal stresses β_{ij} corresponding to the α_{ij} have the quasiconservative and dissipative parts

$$\beta_{ij}^{(q)} = \varrho \frac{\partial \psi}{\partial \alpha_{ij}}, \qquad \beta_{ij}^{(d)} = \nu \varrho \frac{\partial \varphi}{\partial \dot{\alpha}_{ij}} \tag{18.3}$$

respectively, and the factor ν is determined by

$$\nu = \varphi \left(\frac{\partial \varphi}{\partial \dot{\varepsilon}_{kl}} \dot{\varepsilon}_{kl} + \frac{\partial \varphi}{\partial \dot{\alpha}_{kl}} \dot{\alpha}_{kl} \right)^{-1}. \tag{18.4}$$

Once the functions ψ and φ are specified, these relations define the quasiconservative stresses $\sigma_{ij}^{(q)}$ and $\beta_{ij}^{(q)}$ as functions of the independent state variables and the dissipative stresses $\sigma_{ij}^{(d)}$ and $\beta_{ij}^{(d)}$ as functions of their time rates.

If the internal parameters occur in the form of several symmetric tensors, we denote them by $\alpha_{ij}^{(m)}$ $(m = 1, 2, \ldots, n)$. The governing functions are now of the forms $\psi(\varepsilon_{ij}, \alpha_{kl}^{(m)})$ and $\varphi(\dot{\varepsilon}_{ij}, \dot{\alpha}_{kl}^{(m)})$. Equations (18.1) and (18.2) are still valid, whereas (18.3) has to be replaced by

$$\beta_{ij}^{(mq)} = \varrho \frac{\partial \psi}{\partial \alpha_{ij}^{(m)}}, \qquad \beta_{ij}^{(md)} = \nu \varrho \frac{\partial \varphi}{\partial \dot{\alpha}_{ij}^{(m)}}, \quad (m = 1, 2, \ldots, n) \tag{18.5}$$

and (18.4) by

$$\nu = \varphi \left(\frac{\partial \varphi}{\partial \dot{\varepsilon}_{kl}} \dot{\varepsilon}_{kl} + \frac{\partial \varphi}{\partial \dot{\alpha}_{kl}^{(m)}} \dot{\alpha}_{kl}^{(m)} \right)^{-1}, \tag{18.6}$$

where the summation convention includes the subscript m.

The internal parameters are auxiliary quantities. Their function has been discussed in Section 11.1. Let us assume that a material is given and that we are looking for an analytical description of its response. The easiest way of solving this problem is to subject a sample of the material to tensile or compressive stresses, e.g. to the standard test described in Section 11.1. Each spring-dashpot model exhibits a typical response as described, e.g. by Figures 11.5 or 11.6. As soon as we have found a model whose response is sufficiently close to the one of the given sample, the internal parameters are known, and we are thus in a position to set up approximate expressions for the specific free energy and the specific dissipation function. Equations (18.1) through (18.6) then supply the various stresses.

It is clear on the other hand that the internal parameters have no immediate physical significance and that they must be eliminated if we aim at a description of the material which is accessible to direct comparison with observation. Let us recall from Section 4.2 that the internal forces do not appear in the first fundamental law and hence are zero. It follows that the tensors $\beta_{ij}^{(mq)}$ and $\beta_{ij}^{(md)}$ are equal and opposite as shown in (4.42), and that therefore

$$\nu \frac{\partial \varphi}{\partial \dot{\alpha}_{ij}^{(m)}} = -\frac{\partial \psi}{\partial \alpha_{ij}^{(m)}}. \tag{18.7}$$

These relations may be used to eliminate the internal parameters from the constitutive equations. If use is also made of (4.36), the result is a set of differential equations connecting the tensors ε_{ij}, σ_{ij}, and their material derivatives. In this way we obtain a tensorial generalization of the differential equation (11.24).

The previous discussion is quite general and not restricted to linear or isotropic materials. In the particular case of a *linear* material we obtain $\nu = \frac{1}{2}$, and in an *isotropic* material the functions ψ and φ take on especially simple forms. As an example, we will discuss the *Maxwell material* considered in Section 11.3. Let us forget the model originally used, and let us define the body by the governing functions (11.55) and (11.56), i.e. by

$$\varrho\psi = \frac{\lambda}{2}\alpha_{ii}\alpha_{jj} + \mu\alpha_{ij}\alpha_{ij} + \frac{K}{2}(\varepsilon_{ii}-\alpha_{ii})(\varepsilon_{jj}-\alpha_{jj}), \tag{18.8}$$

and

$$\varrho\varphi = 2\mu'[(\dot{\varepsilon}_{ij}-\dot{\alpha}_{ij})(\dot{\varepsilon}_{ij}-\dot{\alpha}_{ij}) - \tfrac{1}{3}(\dot{\varepsilon}_{ii}-\dot{\alpha}_{ii})(\dot{\varepsilon}_{jj}-\dot{\alpha}_{jj})], \tag{18.9}$$

where the symmetric tensor α_{ij} defines a single set of internal parameters and where λ, μ, K, and μ' are constants. Since these expressions are quadratic, the material is linear; thus, $\nu = \frac{1}{2}$. The external stresses follow from (18.1) and (18.2). They are given by

$$\begin{aligned} \sigma_{ij}^{(q)} &= K(\varepsilon_{kk}-\alpha_{kk})\delta_{ij}, \\ \sigma_{ij}^{(d)} &= 2\mu'[\dot{\varepsilon}_{ij}-\dot{\alpha}_{ij}-\tfrac{1}{3}(\dot{\varepsilon}_{kk}-\dot{\alpha}_{kk})\delta_{ij}]. \end{aligned} \tag{18.10}$$

The internal stresses (18.3) are

$$\beta_{ij}^{(q)} = \lambda\alpha_{kk}\delta_{ij}+2\mu\alpha_{ij}-\sigma_{ij}^{(q)}, \qquad \beta_{ij}^{(d)} = -\sigma_{ij}^{(d)}. \tag{18.11}$$

These are the results already obtained in (11.58); they show that the material is isotropic.

In order to eliminate the internal parameters α_{ij}, let us use (18.7) in the form $\beta_{ij}^{(q)}+\beta_{ij}^{(d)} = 0$. On account of (18.11) we have

$$\sigma_{ij} = \sigma_{ij}^{(q)}+\sigma_{ij}^{(d)} = \lambda\alpha_{kk}\delta_{ij}+2\mu\alpha_{ij}. \tag{18.12}$$

This is Hooke's law (7.3) for the internal parameters α_{ij}. The inversion of (18.12) is

$$\alpha_{ij} = \frac{1}{2\mu}\left(\sigma_{ij}-\frac{\lambda}{3\lambda+2\mu}\sigma_{kk}\delta_{ij}\right), \tag{18.13}$$

and it follows immediately that

$$\alpha_{kk} = \frac{1}{3\lambda+2\mu}\sigma_{kk}. \tag{18.14}$$

Adding the two equations (18.10), we obtain

$$\sigma_{ij} = K(\varepsilon_{kk}-\alpha_{kk})\delta_{ij}+2\mu'[\dot{\varepsilon}_{ij}-\dot{\alpha}_{ij}-\tfrac{1}{3}(\dot{\varepsilon}_{kk}-\dot{\alpha}_{kk})\delta_{ij}]. \tag{18.15}$$

Substitution of (18.13) and (18.14) in (18.15) eventually yields

$$\begin{aligned} \frac{1}{2\mu'}\sigma_{ij}+\frac{K}{2\mu'(3\lambda+2\mu)}\sigma_{kk}\delta_{ij}+\frac{1}{2\mu}(\dot{\sigma}_{ij}-\tfrac{1}{3}\dot{\sigma}_{kk}\delta_{ij}) \\ = \frac{K}{2\mu'}\varepsilon_{kk}\delta_{ij}+\dot{\varepsilon}_{ij}-\tfrac{1}{3}\dot{\varepsilon}_{kk}\delta_{ij}. \end{aligned} \tag{18.16}$$

This is in fact a tensorial differential equation of the type (11.24).

It is not immediately obvious from (18.16) why the material considered here has been called a Maxwell body. However, if we resolve the various terms into their deviatoric and isotropic parts, (18.16) resolves into the two relations

$$\frac{1}{2\mu'}\sigma'_{ij}+\frac{1}{2\mu}\dot{\sigma}'_{ij}=\dot{\varepsilon}'_{ij}, \tag{18.17}$$

$$\sigma_{kk}=\frac{3K(3\lambda+2\mu)}{3K+3\lambda+2\mu}\varepsilon_{kk}. \tag{18.18}$$

The first, connecting the deviatoric parts of the strain and stress tensors, is clearly a generalization of (11.9); the second shows that the connection between the isotropic strain and stress is elastic.

Problems

1. Define the Maxwell body by means of (11.60) and (11.61) and obtain (18.16) by putting $\bar{\beta}_{ij}^{(q)}+\bar{\beta}_{ij}^{(d)}=0$.
2. Add the term

$$\frac{\lambda''}{2}(\varepsilon_{ii}-\alpha_{ii})(\varepsilon_{jj}-\alpha_{jj})+\mu''(\varepsilon_{ij}-\alpha_{ij})(\varepsilon_{ij}-\alpha_{ij})$$

to the right-hand side of (18.8) and find the corresponding generalizations of (18.17) and (18.18). Compare the result to the one of P1 in Section 11.1.

18.2. Hereditary integrals

In Section 11.2 the response of linear models has been described by integral equations. A similar description is possible for *linear* viscoelastic materials, even if they are anisotropic.

A convenient starting point is equation (11.35). Since it is to be expected that the value of an arbitrary stress component σ_{ij} at time t depends on the history of the entire strain tensor, the proper generalization of (11.35) is

$$\sigma_{ij}(t)=\int_0^t J_{ijkl}(t-t^*)\frac{\mathrm{d}\varepsilon_{kl}^*}{\mathrm{d}t^*}\,\mathrm{d}t^*. \tag{18.19}$$

In place of the single relaxation modulus J of the model we now have a fourth-order tensor J_{ijkl}, dependent on time, and the response of the material is known as soon as this *relaxation tensor* is given. Incidentally, it is obvious that (11.29) might be generalized in an analogous manner.

Since the stress and strain tensors are symmetric, it is possible to choose the tensor J_{ijkl} so that the symmetry relations

$$J_{jikl} = J_{ijlk} = J_{ijkl} \tag{18.20}$$

are satisfied. They are analogous to the symmetry conditions (7.61) and (7.62) of the anisotropic elastic solid. However, since the material considered here is more complex, there is no symmetry condition corresponding to (7.65).

According to (18.19) the contribution of the strain increment $d\varepsilon^*_{kl}$ at time $t^* < t$ to the stress $\sigma_{ij}(t)$ is

$$d\sigma_{ij}(t) = J_{ijkl}(t-t^*)\, d\varepsilon^*_{kl}. \tag{18.21}$$

If the material is *isotropic*, we conclude from (5.6) that the right-hand side has the special form

$$L(t-t^*)\, d\varepsilon^*_{kk}\delta_{ij} + 2M(t-t^*)\, d\varepsilon^*_{ij}, \tag{18.22}$$

where $L(t)$ and $M(t)$ are scalar relaxation moduli. In place of (18.19) we thus obtain

$$\sigma_{ij}(t) = \int_0^t \left[L(t-t^*) \frac{d\varepsilon^*_{kk}}{dt^*} \delta_{ij} + 2M(t-t^*) \frac{d\varepsilon^*_{ij}}{dt^*} \right] dt^*. \tag{18.23}$$

The deviatoric and the isotropic strains are given by

$$\sigma'_{ij}(t) = \int_0^t J'(t-t^*) \frac{d\varepsilon'^*_{ij}}{dt^*}\, dt^*, \qquad \sigma_{kk}(t) = \int_0^t J(t-t^*) \frac{d\varepsilon^*_{kk}}{dt^*}\, dt^*, \tag{18.24}$$

where

$$J' = 2M, \qquad J = 3L + 2M \tag{18.25}$$

are the two remaining relaxation moduli. It follows that the stress deviator is determined by the history of deviatoric strain and the isotropic stress by the history of dilatation.

As a simple example, let us consider the *Maxwell material* treated in Section 18.1. Comparison of $(18.24)_2$ and (18.18) shows that the relaxation modulus connecting the isotropic tensors is a constant

$$J = \frac{3K(3\lambda+2\mu)}{3K+3\lambda+2\mu}. \tag{18.26}$$

By means of Table 11.1 we obtain from (18.17) the relaxation modulus

$$J'(t) = 2\mu \exp\left(-\frac{\mu}{\mu'} t\right) \tag{18.27}$$

determining the connection between the deviatoric tensors.

The object of the last few chapters was to show how the constitutive equations of certain materials may be deduced from their free energies and their dissipation functions, provided one makes use of the orthogonality condition and accepts the orthogonality principle. In this way we obtained the theory of the plastic potential in Chapter 17 and a particular type of the Green–Rivlin liquid in Chapter 16. In Section 18.1 we assumed that the two governing functions are given in terms of external and internal parameters and their rates respectively; we have seen that the internal parameters may be eliminated by means of equations (18.7) and that the result of this process is a system of differential equations connecting the stresses with the strain components. In the case of a linear material, defined by quadratic functions $\varrho\psi$ and $\varrho\varphi$, these differential equations are linear, and they may be used as in Section 11.2 to construct the relaxation moduli (and the creep compliances) necessary for the representation of the response by hereditary integrals.

It is to be expected that the tensors characterizing the material in the differential equations and in the integrals have certain properties reflecting the fact that the material is determined by $\varrho\psi$ and $\varrho\varphi$ alone. One might also expect that the response of an arbitrary (linear or nonlinear) material might be deduced from two functionals representing $\varrho\psi$ and $\varrho\varphi$ in terms of the strain history. So far, however, these problems have not been investigated.

BIBLIOGRAPHY

[1] W. Flügge, *Tensor Analysis and Continuum Mechanics* (Springer, Berlin, 1972).
[2] D. C. Leigh, *Nonlinear Continuum Mechanics* (McGraw-Hill, New York, 1968) p. 73.
[3] E. and F. Cosserat, Sur la théorie de l'élasticité, *Ann. Toulouse* **10** (1896)
[4] H. Ziegler, *Vorlesungen über Mechanik* (Birkhäuser, Basel, 1970).
[5] W. Traupel, *Die Grundlagen der Thermodynamik* (Braun, Karlsruhe, 1971) p. 9.
[6] J. W. Gibbs, *Collected Works* vol. 1, p. 44. (Yale Univ. Press, New Haven, Conn., 1948).
[7] H. Ziegler, Systems with internal parameters obeying the orthogonality condition, *Z. Angew. Math Phys.* **23** (1972) 553.
[8] W. Noll, On the continuity of the solid and fluid states, *J. Rational Mech. Anal.* **4** (1955) 3.
[9] M. Reiner, A mathematical theory of dilatancy, *Amer. J. Math.* **67** (1945) 350.
[10] R. S. Rivlin, The hydrodynamics of non-Newtonian fluids, *Proc. Roy. Soc. A* **193** (1948) 260.
[11] K. Weissenberg, A continuum theory of rheological phenomena, *Nature* **159** (1947) 310.
[12] J. L. Ericksen, Overdetermination of the speed in rectilinear motion of non-Newtonian fluids, *Quart. Appl. Math.* **14** (1956) 318.
[13] A. E. Green and R. S. Rivlin, Steady flow of non-Newtonian fluids through tubes, *Quart. Appl. Math.* **14** (1956) 299.
[14] E. C. Bingham, *Fluidity and Plasticity* (McGraw-Hill, New York, 1922) p. 215.
[15] K. Hohenemser and W. Prager, Ueber die Ansätze der Mechanik isotroper Kontinua, *Z. Angew. Math. Mech.* **12** (1932) 216.
[16] R. v. Mises, Mechanik der plastischen Formänderungen von Kristallen, *Z. Angew. Math. Mech.* **8** (1928) 161.
[17] D. C. Drucker, A more fundamental approach to plastic stress-strain relations, *Proc. 1st U. S. Congr. Appl. Mech. Chicago, 1951* (New York, 1952) p. 487.
[18] G. I. Taylor, General theory of limit design, *Proc. 8th Int Congr. Appl. Mech. Istanbul, 1952* vol. 2, (Istanbul 1956) p. 65.
[19] R. Hill, A variational principle of maximum plastic work in classical plasticity, *Quart. J. Mech. Appl. Math.* **1** (1948) 18.
[20] W. T. Koiter, Stress-strain relations, uniqueness and variational theorems for elastic-plastic materials with a singular yield surface, *Quart. Appl. Math.* **11** (1953) 350.
[21] M. Sayir and H. Ziegler, Der Verträglichkeitssatz der Plastizitätstheorie und seine Anwendung auf räumlich unstetige Felder, *Z. Angew. Math. Phys.* **20** (1969) 78.
[22] W. Prager, The theory of plasticity: a survey of recent achievements, James Clayton Lecture, *Proc. Inst. Mech. Eng.* **169** (1955) 41.
[23] H. Ziegler, A modification of Prager's hardening rule, *Quart. Appl. Math.* **17** (1959) 55.
[24] W. Flügge, *Viscoelasticity*, 2nd ed., (Springer, Berlin, 1975).
[25] F. K. G. Odqvist, *Mathematical Theory of Creep and Creep Rupture* (Clarendon, Oxford, 1966).

[26] D. B. Macvean, Die Elementararbeit in einem Kontinuum und die Zuordnung von Spannungs- und Verzerrungstensoren, *Z. Angew. Math. Phys.* **19** (1968) 157.

[27] R. Hill, New horizons in the mechanics of solids, *J. Mech. Phys. Solids* **5** (1956) 66.

[28] H. Ziegler, Thermodynamik und rheologische Probleme, *Ing. Arch.* **25** (1957) 58.

[29] J. Kestin and J. R. Rice, Paradoxes in the application of thermodynamics to strained solids, in E. B. Stuart et al., editors (Mono Book Corp., Baltimore, 1970).

[30] L. Onsager, Reciprocal relations in irreversible processes, *Phys. Rev.* **37** (II) (1931) 405, **38** (II) (1931) 2265.

[31] H. Ziegler, J. Nänni and Ch. Wehrli, Zur Konvexität der Dissipationsflächen, *Z. Angew. Math. Phys.* **25** (1974) 76.

[32] M. A. Biot, Variational principles in irreversible thermodynamics with application to viscoelasticity, *Phys. Rev.* **97** (1955) 1463.

[33] S. R. De Groot, *Thermodynamics of Irreversible Processes*, (North-Holland, Amsterdam, 1952) p. 196.

[34] H. Ziegler, J. Nänni and Ch. Wehrli, Zur Konvexität der Fliessfläche, *Z. Angew. Math. Phys.* **24** (1973) 140.

[35] H. Ziegler, On the theory of the plastic potential, *Quart. Appl. Math.* **19** (1961) 39.

[36] J. L. Synge, *The Hypercircle in Mathematical Physics* (Cambridge Univ. Press, Cambridge, 1957).

[37] W. Prager, *An Introduction to Plasticity* (Addison Wesley, Reading, Mass., 1959) p. 13.

[38] H. Ziegler, A possible generalization of Onsager's theory, in H. Parkus and L. I. Sedov, ed., *IUTAM Symp. on Irreversible Aspects of Continuum Mechanics, Vienna 1965*, (Springer, Berlin, 1968).

SUBJECT INDEX